学术引领系列

国家科学思想库

中国学科发展战略

板块俯冲带

国家自然科学基金委员会
中 国 科 学 院

科 学 出 版 社
北 京

审图号：GS 京（2022）0482 号

内 容 简 介

板块俯冲带是地球区别于太阳系其他行星的重要标志，是地球圈层相互作用的关键纽带。它深刻影响了地球圈层之间物质和能量循环，对地球的宜居性演化起了重要作用。

本书阐述了板块俯冲带学科的科学意义与战略价值，总结了学科的发展规律与研究特点，论述了发展现状与发展态势，阐明了发展思路与发展方向，提出了面临的关键科学问题，并就未来学科发展给出了资助机制与政策建议。

本书适合战略和管理专家、相关领域的高等院校师生、研究机构的研究人员阅读，有助于读者洞悉板块俯冲带学科发展规律、把握汇聚板块边缘前沿领域和重点方向。同时，本书可供科技管理部门制定政策时参考，也可作为社会公众了解板块俯冲带学科发展现状及趋势的读本。

图书在版编目（CIP）数据

板块俯冲带 / 国家自然科学基金委员会，中国科学院编. —北京：科学出版社，2022.11

（中国学科发展战略）

ISBN 978-7-03-073459-4

Ⅰ. ①板…　Ⅱ. ①国…　②中…　Ⅲ. ①板块－俯冲带　Ⅳ. ①P544

中国版本图书馆 CIP 数据核字（2022）第 191143 号

丛书策划：侯俊琳　牛　玲
责任编辑：杨婵娟　吴春花 / 责任校对：韩　杨
责任印制：李　彤 / 封面设计：黄华斌　陈　敬

科学出版社 出版
北京东黄城根北街 16 号
邮政编码：100717
http://www.sciencep.com
北京九州迅驰传媒文化有限公司印刷
科学出版社发行　各地新华书店经销
*
2022 年 11 月第　一　版　开本：720×1000　1/16
2024 年 9 月第三次印刷　印张：21
字数：423 000

定价：168.00 元

（如有印装质量问题，我社负责调换）

中国学科发展战略

联合领导小组

组　长: 高鸿钧　李静海

副组长: 包信和　韩　宇

成　员: 张　涛　裴　钢　朱日祥　郭　雷　杨　卫
王笃金　石　兵　王长锐　姚玉鹏　董国轩
杨俊林　徐岩英　于　晟　王岐东　刘　克
刘作仪　孙瑞娟　陈拥军

联合工作组

组　长: 石　兵　姚玉鹏

成　员: 范英杰　龚　旭　孙　粒　刘益宏　王佳佳
李鹏飞　钱莹洁　薛　淮　冯　霞　马新勇

中国学科发展战略·板块俯冲带

专 家 组

组 长：郑永飞

成 员（以姓氏拼音为序）：

陈 骏　丁 林　高 锐　金振民　李曙光
吴福元　肖文交　徐义刚　许志琴　杨经绥
杨树锋　张国伟　赵国春　朱日祥

总体撰写组

组 长：郑永飞

成 员（以姓氏拼音为序）：

陈伊翔　李三忠　李忠海　宋述光　万 博
王 强　王 勤　张贵宾　张宏福　朱弟成

分章节撰写组

成　员（以姓氏拼音为序）：

陈华勇　陈立辉　陈　凌　陈　龙　陈仁旭
陈　意　陈伊翔　戴立群　胡修棉　冷　伟
李继磊　李三忠　李忠海　刘福来　刘　良
刘晓春　倪怀玮　宋述光　孙道远　万　博
王　强　王　勤　魏春景　吴元保　肖益林
熊小林　许文良　姚华建　张贵宾　张海江
张宏福　张建新　张立飞　章军锋　赵　亮
赵子福　郑建平　郑永飞　朱弟成

秘书长：陈伊翔

秘书组（以姓氏拼音为序）：

陈伊翔　万　博　张贵宾

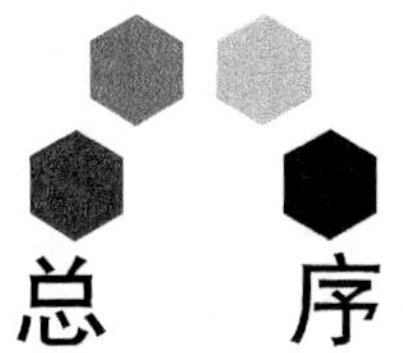

总　序

白春礼　杨　卫

17 世纪的科学革命使科学从普适的自然哲学走向分科深入，如今已发展成为一幅由众多彼此独立又相互关联的学科汇就的壮丽画卷。在人类不断深化对自然认识的过程中，学科不仅仅是现代社会中科学知识的组成单元，同时也逐渐成为人类认知活动的组织分工，决定了知识生产的社会形态特征，推动和促进了科学技术和各种学术形态的蓬勃发展。从历史上看，学科的发展体现了知识生产及其传播、传承的过程，学科之间的相互交叉、融合与分化成为科学发展的重要特征。只有了解各学科演变的基本规律，完善学科布局，促进学科协调发展，才能推进科学的整体发展，形成促进前沿科学突破的科研布局和创新环境。

我国引入近代科学后几经曲折，及至上世纪初开始逐步同西方科学接轨，建立了以学科教育与学科科研互为支撑的学科体系。新中国建立后，逐步形成完整的学科体系，为国家科学技术进步和经济社会发展提供了大量优秀人才，部分学科已进入世界前列，有的学科取得了令世界瞩目的突出成就。当前，我国正处在从科学大国向科学强国转变的关键时期，经济发展新常态下要求科学技术为国家经济增长提供更强劲的动力，创新成为引领我国经济发展的新引擎。与此同时，改革开放 30 多年来，特别是 21 世纪以来，我国迅猛发展的科学事业蓄积了巨大的内能，不仅重大创新成果源源不断产生，而且一些学科正在孕育新的生长点，有可能引领世界学科发展的新方向。因此，开展学科发展战略研究是提高我国自主创新能力、实现我国科学由“跟跑者”向“并行者”和“领跑者”转变的一项基础工程，对于更好把握世界科技创新发展趋势，发挥科技创新在全面创新中的引领作用，具有重要的现实意义。

学科发展战略研究的核心是结合科学技术和经济社会的发展需求，在分析科学前沿发展趋势的基础上，寻找新的学科生长点和方向。在这个过程中，战略科学家的前瞻引领作用十分重要。科学史上这样的例子比比皆是。在1900年8月巴黎国际数学家代表大会上，德国数学家戴维·希尔伯特发表了题为“数学问题”的著名讲演，他根据过去特别是19世纪数学研究的成果和发展趋势，提出了23个最重要的数学问题，即“希尔伯特问题”。这些“问题”后来成为许多数学家力图攻克的难关，对现代数学的研究和发展产生了深刻的影响。1959年12月，美国物理学家、诺贝尔奖得主理查德·费曼在加利福尼亚理工学院举行的美国物理学会年会上发表了题为“物质底层大有空间——一张进入物理新领域的请柬”的经典讲话，对后来出现的纳米技术作出了天才的预见。

学科生长点并不完全等同于科学前沿，其产生和形成不仅取决于科学前沿的成果，还决定于社会生产和科学发展的需要。1841年，佩利戈特用钾还原四氯化铀，成功地获得了金属铀，可在很长一段时间并未能发展成为学科生长点。直到1939年，哈恩和斯特拉斯曼发现了铀的核裂变现象后，人们认识到它有可能成为巨大的能源，这才形成了以铀为主要对象的核燃料科学的学科生长点。而基本粒子物理学作为一门理论性很强的学科，它的新生长点之所以能不断形成，不仅在于它有揭示物质的深层结构秘密的作用，而且在于其成果有助于认识宇宙的起源和演化。上述事实说明，科学在从理论到应用又从应用到理论的转化过程中，会有新的学科生长点不断地产生和形成。

不同学科交叉集成，特别是理论研究与实验科学相结合，往往也是新的学科生长点的重要来源。新的实验方法和实验手段的发明，大科学装置的建立，如离子加速器、中子反应堆、核磁共振仪等技术方法，都促进了相对独立的新学科的形成。自20世纪80年代以来，具有费曼1959年所预见的性能、微观表征和操纵技术的仪器——扫描隧道显微镜和原子力显微镜终于相继问世，为纳米结构的测量和操纵提供了“眼睛”和“手指”，使得人类能更进一步认识纳米世界，极大地推动了纳米技术的发展。

作为国家科学思想库，中国科学院（以下简称中科院）学部的基本职责和优势是为国家科学选择和优化布局重大科学技术发展方向提供科学依据、发挥学术引领作用，国家自然科学基金委员会（以下简称基金委）则承担着协调学科发展、夯实学科基础、促进学科交叉、加强学科建设的重大责任。继基金委和中科院于2012年成功地联合发布“未来10年中国学科发展战略研究”报告之后，双方签署了共同开展学科发展战略研究的长期合作协议，通过联合开展学科发展战略研究的长效机制，共建共享国家科学思想库的研究咨询能力，切实担当起服务国家科学领域决策咨询的核心作用。

基金委和中科院共同组织的学科发展战略研究既分析相关学科领域的发展趋势与应用前景，又提出与学科发展相关的人才队伍布局、环境条件建设、资助机制创新等方面的政策建议，还针对某一类学科发展所面临的共性政策问题，开展专题学科战略与政策研究。自2012年开始，平均每年部署10项左右学科发展战略研究项目，其中既有传统学科中的新生长点或交叉学科，如物理学中的软凝聚态物理、化学中的能源化学、生物学中生命组学等，也有面向具有重大应用背景的新兴战略研究领域，如再生医学，冰冻圈科学，高功率、高光束质量半导体激光发展战略研究等，还有以具体学科为例开展的关于依托重大科学设施与平台发展的学科政策研究。

学科发展战略研究工作沿袭了由中科院院士牵头的方式，并凝聚相关领域专家学者共同开展研究。他们秉承“知行合一”的理念，将深刻的洞察力和严谨的工作作风结合起来，潜心研究，求真唯实，“知之真切笃实处即是行，行之明觉精察处即是知”。他们精益求精，“止于至善”，“皆当至于至善之地而不迁”，力求尽善尽美，以获取最大的集体智慧。他们在中国基础研究从与发达国家“总量并行”到“贡献并行”再到“源头并行”的升级发展过程中，脚踏实地，拾级而上，纵观全局，极目迥望。他们站在巨人肩上，立于科学前沿，为中国乃至世界的学科发展指出可能的生长点和新方向。

各学科发展战略研究组从学科的科学意义与战略价值、发展规律和研究特点、发展现状与发展态势、未来 5～10 年学科发展的关键科学问题、发展思路、发展目标和重要研究方向、学科发展的有效资助机制与政策建议等方面进行分析阐述。既强调学科生长点的科学意义，也考虑其重要的社会价值；既着眼于学科生长点的前沿性，也兼顾其可能利用的资源和条件；既立足于国内的现状，又注重基础研究的国际化趋势；既肯定已取得的成绩，又不回避发展中面临的困难和问题。主要研究成果以“国家自然科学基金委员会—中国科学院学科发展战略”丛书的形式，纳入“国家科学思想库—学术引领系列”陆续出版。

基金委和中科院在学科发展战略研究方面的合作是一项长期的任务。在报告付梓之际，我们衷心地感谢为学科发展战略研究付出心血的院士、专家，还要感谢在咨询、审读和支撑方面做出贡献的同志，也要感谢科学出版社在编辑出版工作中付出的辛苦劳动，更要感谢基金委和中科院学科发展战略研究联合工作组各位成员的辛勤工作。我们诚挚希望更多的院士、专家能够加入到学科发展战略研究的行列中来，搭建我国科技规划和科技政策咨询平台，为推动促进我国学科均衡、协调、可持续发展发挥更大的积极作用。

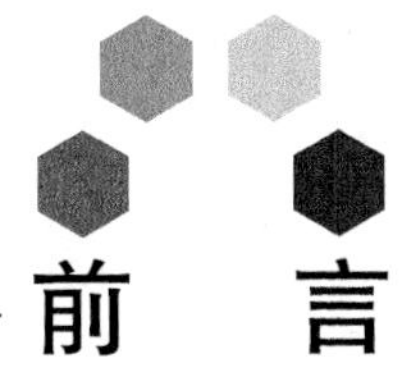

前 言

板块俯冲带是典型的前沿交叉学科，研究领域涉及地质学、地球化学、地球物理学以及地球动力学和地球系统科学等各个方面。诞生于20世纪60年代的板块构造理论改变了人类对地球整体运行行为的认识，是20世纪最重大的科学理论之一。板块俯冲带过程是地球各圈层物质交换和能量循环的关键纽带，在地球的演化过程中起了关键作用。板块俯冲带深刻影响了地球深部和表生圈层的演化，是重大的基础科学前沿领域。同时，它也是火山喷发、地震活动和资源能源聚集的主要场所，与人类的生存和发展密切相关。因此，对板块俯冲带的系统深入研究，不仅有助于理解地球内部运作机制及宜居性演化历史，还对经济社会发展具有重要意义。

鉴于板块俯冲带在重大基础前沿和国家战略需求方面的重要性，我们于2018年启动了国家自然科学基金委员会和中国科学院联合支持的板块俯冲带学科发展战略调研项目，中国科学院院士郑永飞教授为负责人。该项目于2018～2019年完成联合调研，2020年提交正式报告。为了有效完成调研任务，我们成立了专家组、撰写组、秘书组。专家组由10余位从事固体地球科学研究的中国科学院院士组成，撰写组以板块俯冲带领域30余位杰出的中青年科研骨干为主体，秘书组以中国科学院壳幔物质与环境重点实验室为依托。为了保证调研质量，项目执行以来举行了6次专题研讨会，邀请国内多个单位50余位相关领域的专家做学术报告、参与项目调研与研讨。同时，召开了多次研讨会专门讨论板块俯冲带学科发展战略报告的撰写工作，系统探讨了该领域的发展历史、发展趋势和前沿研究方向，凝练了板块俯冲带领域的重大前沿科学问题。在此基础上，组织专家总结调研成果并撰写综述文章，2020年在《中国科学：地球科学》上出版了“板块俯冲带与汇聚边缘”的专辑，在此基础上撰写组和秘书组一起撰写了《中国学科发展战略·板块俯冲带》。

本书共分五章，分别从板块俯冲带学科的科学意义与战略价值、发展规律与研究特点、发展现状与发展态势、发展思路与发展方向、资助机制与政策建议五个方面进行了详细阐述。通过成果集成和专题研讨，本报告阐述了板块俯冲带研究在地球系统科学和国家战略需求方面的重大意义，分析了国内外研究历史和发展规律，特别是国内学者的贡献以及与国际先进水平的差距，总结了目前的发展现状和态势，着重提出了未来的发展思路和发展方向，进而提出了学科发展战略资助机制与政策建议。在发展思路与发展方向方面，本报告围绕俯冲带的结构、过程、产物和动力学机制，遴选出板块俯冲带研究4个方面的18个关键科学问题。对这些科学问题的深入探究，将极大地深化板块俯冲带研究，推动对地球内部运行机制的认识，促进地球系统科学的发展。

本书认为，板块俯冲带是典型的前沿交叉学科，需要加强俯冲带的学科布局，进一步明确板块俯冲带的战略定位。需要推动汇聚板块边缘学科交叉，加快复合型交叉人才培养和队伍建设，建立持续稳定的资助机制，支持重大平台建设，加强国际学术交流合作，力争在未来5～10年使该学科领域达到国际先进水平。

本书每一章由报告总体撰写组专人负责协调，熟悉该章各小节的专家具体撰写，由秘书组汇总、初审，报告总体撰写组所有成员进一步通读、修改，并经项目负责人最终审校成稿。参与各章撰写的专家包括（按姓氏拼音排序）：陈华勇、陈立辉、陈凌、陈龙、陈仁旭、陈意、陈伊翔、戴立群、胡修棉、冷伟、李继磊、李三忠、李忠海、刘福来、刘良、刘晓春、倪怀玮、宋述光、孙道远、万博、王强、王勤、魏春景、吴元保、肖益林、熊小林、许文良、姚华建、张贵宾、张海江、张宏福、张建新、张立飞、章军锋、赵亮、赵子福、郑建平、郑永飞、朱弟成。南京大学王孝磊教授在报告撰写方面提出了诸多建议，图件承蒙付璐露女士清绘，在此一并表示衷心感谢。本书经秘书长陈伊翔教授统稿初审，总体撰写组组长郑永飞院士最终审校定稿。

郑永飞　陈伊翔

2022年10月18日

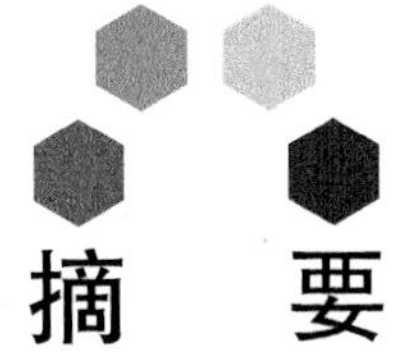

摘　要

1. 引言

板块构造理论的核心是俯冲带，板块俯冲是将地球浅表物质输送到地球内部的最主要方式，因此是地球圈层物质和能量交换的关键机制。随着俯冲过程中温度和压力条件的变化，俯冲地壳不仅经历变质脱水和部分熔融，而且发生一系列矿物相变。更为重要的是，当板片或其衍生物质（如变质脱水释放的富水溶液或者地壳熔融形成的含水熔体）在不同深度与地幔楔岩石接触时，物理条件和化学成分的巨大差异会导致二者之间发生复杂的化学反应，形成具有特殊成分和性质的新矿物相或组合，并改造地幔楔的化学成分，导致地幔的化学不均一性。板块俯冲带与地球壳幔系统演化、矿产资源聚集、大气和海洋的长期化学演化等重大地球科学问题紧密相连，同时也是火山喷发和地震活动的主要场所，与人类的生存和发展息息相关。

2. 俯冲带过程与结构

在俯冲带浅部，俯冲板片发生变质脱水乃至部分熔融。一方面，经历高压－超高压变质的地壳岩石会沿着俯冲隧道折返。另一方面，释放出的流体向上运移交代地幔楔，所形成的地幔交代岩比正常橄榄岩易于熔融，结果引起弧岩浆作用，这些弧岩浆喷出地表完成整个物质循环。肇始于板块俯冲的壳幔挥发分循环又对地表环境演化和宜居性产生了深刻的影响。相反，俯冲到地幔深部、超过折返极限的地壳岩石则不再以超高压变质岩的形式折返，而是发生进一步的矿物相变，将地壳组分迁移进入地幔过渡带乃至下地幔。对板块俯冲带产物的认识和理解取决于我们对俯冲带结构和过程的认识和理解。板块俯冲带的结构包括俯冲板片的几何形态、内部结构和地质特征，以及俯冲板片及其周围的温度压力变化等。俯冲带过程可以体现为浅部过程和深部过程，包括构造作用、变质作用、交代作用、岩浆作用、成矿作用等，深刻影响地球内部的物质和能量循环及全球构造格局。

俯冲带变质作用将引起岩石发生一系列物理和化学变化，这些变化不仅是导致板块进一步俯冲的主要驱动力，而且控制着俯冲到地球深部物质的组成，对俯冲带化学地球动力学过程产生重要影响。传统板块构造理论认为，大洋板块俯冲带是壳幔相互作用最为活跃的地方，在这里洋壳（由镁铁质火成岩和海底沉积物组成）经历变质脱水和部分熔融后进入地幔。鉴于大陆地壳物质在组成上与地幔物质存在巨大差异，大陆地壳的深俯冲和折返必然会对上覆大陆岩石圈和大陆板块汇聚边缘的结构、组成、变形和演化进程造成显著影响。洋壳和陆壳在物质组成上的差异，必然造成二者与地幔楔相互作用性质上的差异。

俯冲带流体活动和元素迁移是俯冲带研究的热点和前沿。由于大洋地壳相对富含流体，大洋板块俯冲普遍导致大规模的地幔交代和岛弧岩浆活动。此外，经历过脱水和化学变化的残留板片物质继续下沉并对地幔的地球化学组成造成影响，最终会形成板内玄武质岩浆活动的源区物质来源。由于俯冲镁铁质洋壳的密度高于周围的地幔岩石，大洋俯冲带深部的洋壳极少能折返到地表。因此，研究与洋壳俯冲和洋壳物质再循环有关的流体活动和元素分异主要借助有限折返的大洋地壳（高压蓝片岩和榴辉岩）和浅层地幔（橄榄岩包体和蛇绿岩）岩石，以及间接利用弧岩浆岩和高温高压实验来研究。但是，目前对深俯冲大洋板块内部流体活动和元素分异的具体物理化学过程和机制还缺乏足够的直接证据，对大洋板片组分进入地幔楔的方式、反应过程和机制还缺乏充分认识。地幔楔中弧岩浆产生的深度通常在80～160 km的弧下深度，对应于进入柯石英稳定域乃至金刚石稳定域的超高压变质作用。只有对在弧下深度形成的超高压变质岩和相应的地幔楔岩石进行直接观察，才能全面认识和理解俯冲带深部流体活动和壳幔相互作用。

由于陆壳具有相对古老、干和冷的特征，一般认为其在俯冲过程中释放流体极为有限。与大洋俯冲带相比，大陆俯冲带之上的岩石圈以缺乏同俯冲大陆弧岩浆作用为特征。先前将这个特征归因于深俯冲大陆地壳相对缺水，因此大陆俯冲带相对缺乏流体活动。但是对俯冲带地壳脱水行为的研究发现，大陆和大洋地壳在大于30km的深度都是以含水矿物作为水的主要储库，名义上无水矿物只是水的次要储库。大洋和大陆俯冲隧道的温度在同等俯冲条件下（主要是板块汇聚速率和俯冲角度）相似，弧下深度超高压变质温度也相似。因此，大陆和大洋俯冲带地壳岩石在弧下深度发生了类似程度的脱水作用，但是上覆地幔楔温度的升高才是引起弧岩浆作用的决定因素。地质上，大陆岩石圈地幔楔橄榄岩在俯冲隧道界面受到俯冲陆壳及其碎块产生的流体交代，形成了富化、富集的造山带岩石圈地幔交代岩。根据这个观察，结合俯冲带温压结构和地壳脱水行为研究，可见大陆俯冲带之上缺乏同俯冲弧岩浆作用并不是由俯冲大陆地壳缺水造成的，而是上覆大陆岩石圈太厚导致地幔楔温度太低，因此其中的交代岩在大陆俯冲阶段难以发生部分熔融进而引发大陆弧岩浆作用。在碰撞后阶段，这些交代岩发生部分熔融，在碰撞造山带形成了各种镁铁质－超镁铁质火成岩。

地幔楔作为俯冲系统中连接俯冲盘和仰冲盘的关键构造单元，在地球圈层之间的物质循环和能量交换等方面起着重要作用。造山带橄榄岩直接记录了俯冲带多种性质的流体（硅酸盐熔体、碳酸盐熔体、含硅酸盐组分的富水溶液乃至超临界流体）交代作用，以及复杂的壳幔物质循环过程等。通过对造山带橄榄岩的系统研究，不仅可以从微观尺度上制约复杂的化学交代过程和变质变形历史，而且可以与宏观构造的时空演化相联系，结果对认识和理解俯冲带壳幔相互作用至关重要。对大洋弧玄武岩和大陆弧安山岩的研究已经证实，俯冲洋壳脱水熔融交代上覆地幔楔中的橄榄岩，地幔楔中的交代岩发生部分熔融引起弧岩浆作用。与洋壳俯冲带相比，陆壳俯冲带同样发生了显著的变质脱水作用。进一步，俯冲进入地幔深度的地壳岩石会发生不同程度的脱水熔融，所产生的含水熔体会交代不同深度和不同性质的地幔。因此，这些板片流体与其上覆地幔楔之间的化学反应，会记录在不同类型的变质岩和岩浆岩中。汇聚板块边缘的造山带橄榄岩可能保存了这些过程的直接记录，而碰撞后岩浆作用产生的岩浆岩则间接记载了这些过程。

对现代洋壳俯冲带温压结构与地震和弧火山作用之间的关系研究发现，大洋俯冲带具有“冷”和“暖”之分，由此造成在俯冲板片地震和弧火山岩分布上出现显著差别。在“冷”的大洋俯冲带，俯冲地壳岩石在小于 60km 的弧前深度不会发生显著脱水，只是发生蓝片岩相－榴辉岩相变质作用，而在大于 80km 的弧下深度才发生大规模脱水引起弧岩浆作用。相反，在“暖”的大洋俯冲带，俯冲地壳岩石在弧前深度就发生大规模脱水，结果在弧下深度脱水较少，没有产生大规模弧岩浆作用。无论如何，深俯冲洋壳释放的流体交代上覆地幔楔橄榄岩，而后其中的交代岩发生部分熔融形成大洋弧玄武岩或大陆弧安山岩。经历脱水熔融的板片将继续俯冲进入深部地幔，从而对地幔地球化学组成产生显著影响。俯冲进入弧下深度的大洋地壳衍生出富水溶液和含水熔体，它们与地幔楔橄榄岩反应形成大洋弧玄武岩或者大陆弧安山岩的地幔源区。经历脱水熔融的大洋地壳进一步俯冲到大于 200km 的后弧深度发生部分熔融，所产生的熔体与软流圈地幔楔橄榄岩反应形成板内玄武岩的地幔源区。因此，俯冲地壳随深度增加，从在岩石圈深度发生变质脱水到在软流圈深度发生部分熔融，所产生的流体进入地幔楔是实现俯冲带地球化学传输的核心机制。

3. 俯冲带物质再循环

大陆俯冲带具有低的地热梯度，在大陆地壳俯冲 / 折返过程中也存在显著的流体活动。在这类“冷”的大陆俯冲带，俯冲地壳岩石在弧前深度没有发生显著的变质脱水，只是发生蓝片岩相－榴辉岩相变质作用；到弧下深度才发生显著的变质脱水形成超高压变质岩，但是由于上覆大陆岩石圈地幔较冷而未能引发同俯冲弧岩浆作用。不过，在深俯冲大陆地壳折返过程中，超高压变质岩会发生降压脱水乃至部分熔融，所产生的流体不仅交代上覆地幔楔橄榄岩，而且形成内部流体引起角闪岩相退变质和各种成分的脉体形成。因此，大陆深俯冲不但能够引起超高压变质作用，而且会在大陆俯冲隧道内发生强烈的壳幔相互作用。折返到地表的高压-超高压变质岩是大洋/大陆板块俯冲物质再循环的第一种表现形式，它们受到俯冲带温度压力变化的影响最大，但是受到地幔楔物质影响的程度不一。因此，研究大陆俯冲隧道内的流体活动和壳幔相互作用，正在成为备受国际地球科学界关注的前沿领域。

大洋俯冲带之上镁铁质弧岩浆岩是大洋/大陆板块俯冲物质再循环的第二种表现形式。这些岩石样品记录了俯冲带从深部地幔到地表的过程，为认识地球深部物质循环提供了理想的天然样品。自从板块构造理论建立以来，国际上对现代大洋俯冲带壳幔相互作用的研究相对较多，对大洋弧和大陆弧岩浆岩成因中的地壳物质再循环获得了相对成熟的认识。与此相比，国际上对古大洋俯冲带壳幔相互作用的研究相对薄弱。由于在这类俯冲带出露有不同成分的火成岩，不仅包括古洋壳俯冲过程中产生的大洋弧玄武岩和大陆弧安山岩，而且包括汇聚板块边缘的俯冲后洋岛型和岛弧型镁铁质火成岩。这些岩浆岩为我们研究地质历史时期地球深部物质再循环提供了良好的研究对象。

碰撞造山带同折返或碰撞后岩浆岩是大洋/大陆板块俯冲物质再循环的第三种表现形式。在超高压变质岩折返阶段，深俯冲大陆地壳可以发生大规模部分熔融，形成花岗质岩体。这些花岗质岩浆岩具有岛弧型微量元素和富集的同位素组成特征，可以作为俯冲带深部壳幔相互作用的长英质熔体来源。一些同折返基性岩可能是俯冲板片衍生熔体交代地幔楔之后部分熔融的产物。这些岩石中还保存有俯冲地壳的特征矿物（如残留锆石核）以及元素和同位素记录。在古汇聚板块边缘产出的基性岩中，有的具有大离子亲石元素（large ion lithophile elements，LILE）和轻稀土元素（light rare earth elements，LREE）富集、高场强元素（high field strength elements，HFSE）和重稀土元素（heavy rare earth elements，HREE）亏损等类似岛弧玄武岩的微量元素配分特征，有的则具有轻稀土元素富集、铅负异常和高场强元素正异常等类似洋岛玄武岩的微量元素配分特征。这个差别指示了两者在地幔源区性质上的差异，可能与俯冲陆壳、俯冲洋壳及上覆沉积物在不同深度脱水熔融产生熔体交代地幔楔有关，但是其地球动力学机制尚有待进一步澄清。

4. 俯冲带科学体系

在20世纪下半叶，化学地球动力学致力于研究板块俯冲对地幔地球化学成分的影响，力图区分不同类型地壳和不同成因沉积物脱水熔融对岛弧和洋岛玄武岩源区成分的贡献。进入21世纪以来，人们将注意力转到俯冲带结构、过程和产物上来，发现俯冲带地热梯度有冷与暖之分，动力来源有重力与浮力之分，动力体制有挤压与拉张之分，时空演化有进行时与过去时之分，板片再循环机制有断离与拆沉之分，地壳再循环形式有流体与固体之分。因此，板块俯冲带研究不仅需要确定俯冲带地壳物质再循环的机制和形式，而且需要确定俯冲带动力来源和地热梯度及其随时间的变化。为了识别不同类型壳源流体对地幔楔的交代作用、寻求板片－地幔界面交代反应的岩石学和地球化学证据、理解汇聚板块边缘岩石圈俯冲和拆沉对地幔不均一性的影响，我们必须将俯冲带构造作用、变质作用、交代作用和岩浆作用作为一个地球科学系统来考虑。

自岛弧玄武岩地球化学成分与俯冲大洋地壳变质脱水和部分熔融联系起来以后，一般认为俯冲地壳再循环的形式是流体，包括富水溶液、含水熔体和超临界流体。虽然洋岛玄武岩地球化学成分中也含有地壳信息，但是学界对进入洋岛玄武岩地幔源区的地壳组分是以流体还是固体的形式未加区分。另外，俯冲进入弧下深度的上地壳会受到上覆地幔楔的刮削作用而与下伏岩石圈发生拆离，所拆离的地壳物质可能作为“冷柱”底辟进入地幔楔内部（而不是沿俯冲隧道折返到弧前深度），在受到地幔楔加热后发生部分熔融并与橄榄岩反应，由此成为镁铁质弧火山岩的岩浆源区。这意味着，俯冲地壳再循环形式是固体而不是流体。由于流体只溶解有少量硅酸盐物质和大量不相容元素（如大离子亲石元素和轻稀土元素），它与地幔楔橄榄岩反应后形成的蛇纹石化和绿泥石化橄榄岩以及辉石岩和角闪石岩依然具有超镁铁质成分，因此其部分熔融产物可以具有玄武岩成分。对于固体而言，其熔融产物包含大量硅酸盐物质

（其中也富集不相容元素），它与地幔楔橄榄岩之间反应的产物可能不再具有超镁铁质成分。因此，究竟俯冲地壳是化学分异产生流体上升还是物理分离产生固体上升，还有待进一步研究。

在汇聚大陆边缘大洋板块牵引大陆板块俯冲，由于大陆岩石圈相对浮力较大，当其俯冲到岩石圈/软流圈边界的弧下深度发生超高压变质时俯冲阻力达到最大，这时在俯冲板片的洋陆过渡带可能发生断离，结果大陆板片在浮力作用下发生折返，而大洋板片在重力牵引下继续俯冲进入软流圈地幔。但是对板片断离是否发生和何时发生这两个问题，目前仍然缺乏明确答案。一方面，对于正在进行的板块俯冲，由于板片断离后软流圈会沿着断离空隙上涌，不仅会发生降压熔融形成玄武质熔体，而且会加热正在折返的大陆地壳形成超高温变质岩，结果会出现超高温变质作用叠加在超高压变质岩之上，并且这两种极端变质事件的时间间隔较小。另一方面，汇聚板块边缘岩石圈会发生碰撞加厚，加厚的造山带根部岩石圈由于重力不稳定而发生拆沉作用，结果是将部分岩石圈地幔（有时可能包括上覆的镁铁质下地壳）带入软流圈地幔。这个过程一般发生在古俯冲带，那里一旦造山带发生去根作用，汇聚板块边缘的岩石圈就会发生减薄，接下来就会出现大陆张裂作用（rifting）。软流圈上涌一方面会发生降压熔融形成玄武质熔体，另一方面会加热减薄的岩石圈地幔使上覆大陆地壳深部岩石发生高温－超高温麻粒岩相变质作用，结果会引起超高温变质作用叠加在超高压变质岩之上，不过这两种极端变质事件的时间间隔较大。如何区分断离和拆沉这两种板片再循环机制，已经成为汇聚板块边缘研究的前沿。

板块俯冲角度与地热梯度之间的关系随时间而变化，一般分为两个阶段。早期阶段是低角度俯冲，汇聚板块边缘热梯度较低；晚期阶段变成高角度俯冲，汇聚板块边缘地热梯度升高。俯冲板片回卷是俯冲角度由低变高的基本原因。前人在大洋俯冲带发现双变质作用，在靠近海沟一侧发现蓝片岩相－榴辉岩相组合带，形成于低的地热梯度（＜10℃/km）；在靠近岛弧一侧发现角闪岩相－麻粒岩相组合带，形成于高的地热梯度（＞30℃/km）。对于正在进行的板块俯冲，俯冲板片回卷引起弧后位置岩石圈拉张减薄，导致弧后盆地打开，结果软流圈地幔发生上涌，不仅发生降压熔融形成玄武质熔体，而且加热深部地壳岩石形成高温－超高温变质岩。对碰撞造山带变质岩而言，虽然也出现双变质作用，但是两者在时间和空间上常常呈现叠加，指示俯冲带在地热梯度上发生了变化，早期形成于低地热梯度的变质岩受到晚期高地热梯度变质作用的叠加。对于正在汇聚的板块边缘，俯冲板片回卷能够导致地幔楔底部受到软流圈加热发生部分熔融，但是这个加热过程是否就是俯冲带弧岩浆作用的动力学机制，还有待进一步确定。对于不再汇聚的板块边缘，加厚的造山带岩石圈在根部拆沉之后会发生大陆张裂作用，导致软流圈上涌加热减薄的岩石圈地幔，引起上覆大陆地壳发生高温－超高温变质作用，结果高地热梯度变质作用叠加到低地热梯度变质岩之上，这个过程是否就是汇聚板块边缘叠加造山作用的动力学机制，还有待进一步研究。

俯冲板片与上覆板块之间的耦合关系随时间而变化，一般也分为两个阶段。早期阶段处于耦合状态，汇聚板块边缘在动力上处于挤压体制；晚期阶段处于解耦状态，汇聚板块边缘在动力上处于拉张体制。在挤压体制下，板块界面具有低的地热梯度，引起蓝片岩相－榴辉岩相变质作用，不会发生弧岩浆作用。在拉张体制下，板块界面具有高的地热梯度，地幔楔底部受到软流圈加热引起弧岩浆作用，弧后岩石圈减薄导致上覆地壳出现角闪岩相－麻粒岩相变质作用。如果碰撞加厚的造山带根部岩石圈发生拆沉作用，减薄的岩石圈就会发生大陆张裂作用，结果软流圈上涌加热减薄的岩石圈地幔使上覆大陆地壳发生高温－超高温变质作用，引起角闪岩相－麻粒岩相变质作用叠加到榴辉岩相高压－超高压变质带之上，伴有混合岩化和花岗质岩浆作用。在挤压体制向拉张体制转换的过程中，俯冲隧道中的岩石在浮力驱动下发生逆冲折返，不仅蓝片岩相－榴辉岩相变质岩受到高压角闪岩相－麻粒岩相变质叠加，而且折返的地壳岩石会发生降压熔融引起同折返岩浆作用。因此，俯冲带动力体制的变化导致了俯冲带地热梯度的变化，结果引起了不同类型的变质作用和岩浆作用。如何认识俯冲带动力体制随时间的变化，已经成为俯冲带动力学研究的关键科学问题。

在20世纪60年代板块构造理论建立之初，一般将地幔对流作为板块运动的驱动力。一方面，随着地球动力学研究的深入，人们逐渐认识到，俯冲的大洋岩石圈在进入弧下深度后玄武质地壳发生榴辉岩化，由于榴辉岩的密度比橄榄岩大，其与低温板块的高密度特征相结合，使得重力（负浮力）成为板块俯冲的主要驱动力。另一方面，在海底扩张之处软流圈地幔在浮力驱动下上涌并发生降压熔融，形成洋中脊玄武岩，洋中脊推力也是引起板块俯冲的动力之一。在大陆张裂带，汇聚板块边缘加厚的造山带岩石圈在减薄后软流圈也会在浮力驱动下上涌。如果岩石圈减薄速率大于软流圈上涌速率，不仅上涌的软流圈能够发生部分熔融，而且减薄的岩石圈可以发生垮塌作用，导致双峰式岩浆作用。如果岩石圈减薄速率小于或者等于软流圈上涌速率，则上涌的软流圈难以发生部分熔融（相当于超慢速扩张的洋中脊），并且减薄的岩石圈也不会发生垮塌作用；但是减薄后的岩石圈会受到上涌软流圈的加热，结果在大陆张裂带出现角闪岩相-麻粒岩相高温-超高温变质岩、花岗岩和混合岩等变质核杂岩组合。因此，在汇聚板块边缘既有重力主导的岩石圈俯冲，也有浮力主导的软流圈上涌。两者在地球历史上各自发挥了多大的作用，已经成为板块构造启动机制研究的前沿和热点。

俯冲带与造山带之间在结构和成分上具有继承与发展的关系，板块汇聚导致造山带形成，板内张裂导致造山带破坏。在板块汇聚过程中，如果是一个大洋板块俯冲到一个大陆板块之下，新的地壳物质增生到大陆边缘，结果就产生了增生造山带，典型实例就是东太平洋俯冲带。如果是一个大陆板块俯冲到另一个大陆板块之下，只是两个板块之间的地壳发生碰撞加厚，没有新的地壳物质增生到大陆边缘，结果就产生了碰撞造山带，典型实例就是阿尔卑斯-喜马拉雅造山带、大别-苏鲁造山带。无论是增生造山还是碰撞造山，造山带形成与板块俯冲过程同步，造山旋回属于俯冲进行时。在两个板块不再汇聚以后，原来的板块边缘转化为板块内部并处于不活动状态，直到加厚的造山带根部在重力作用下发生拆沉，或者受到软流圈地幔的对流侵蚀，然后减薄的造山带岩石圈地幔发生大陆张裂作用，导致造山带破坏。前人对俯冲带的研究基本上注重的是正在进行的俯冲带，忽视了俯冲带的结构和过程会随时间而变化。例如，早期在挤压体制下的低角度冷俯冲在晚期会变成拉张体制下的高角度暖俯冲，古俯冲带在岩石圈减薄后会发生大陆张裂作用，其中成功的张裂带就成为大陆裂解和海底扩张带，而夭折的张裂带则成为陆内再活化带、热造山带或超热造山带等。因此，造山带破坏机制不同于造山带形成机制，两者分别对应于俯冲过去时和俯冲进行时。威尔逊旋回由海底扩张、大洋俯冲到大陆碰撞组成，但是如何再从大陆碰撞带演化到海底扩张带，已经成为板块构造研究的前沿。汇聚板块边缘在俯冲进行时与俯冲过去时之间如何发生时间和空间上的演化，已经成为俯冲带研究的关键科学问题。

在板块构造的识别上，一般将其鉴定性标志确定为岩石圈边界发生了地壳俯冲作用，有的甚至强调只有形成蓝片岩相－榴辉岩相变质岩的冷俯冲带才能作为板块构造的标志。但是，汇聚板块边缘既有俯冲作用也有张裂作用。一般来说，俯冲板块边缘在低地热梯度下表现出刚性行为，表壳岩石能够俯冲到大陆岩石圈地幔深度；在中等地热梯度下则表现出韧性行为，表壳岩石只能俯冲到大陆下地壳深度。根据俯冲带地热梯度的高低，可以区分出两种体制的板块构造：一是现代板块构造，以冷俯冲为特征，典型产物是阿尔卑斯型蓝片岩相－榴辉岩相变质系列；二是古代板块构造，以暖俯冲为特征，典型产物是巴罗型角闪岩相－麻粒岩相变质系列。在太古宙广泛出现的是暖俯冲，在古元古代开始出现局域性冷俯冲，到显生宙才出现全球性冷俯冲。此外，软流圈上涌将热能从地幔传输到地壳，不仅引起高温低压变质作用，而且使大陆下地壳发生部分熔融产生长英质岩浆。这些高温低压过程是在先前俯冲带的基础上发展起来的，由此所产生的大陆张裂带构成古汇聚板块边缘的活动带，其性质受控于软流圈地幔上涌的效率。显生宙地幔温度较低，俯冲陆壳在汇聚边缘发生碰撞加厚；太古宙地幔温度较高，俯冲洋壳在汇聚边缘也可以发生碰撞加厚。

5. 结语

俯冲带科学已经成为板块构造理论的核心组成部分。通过我们的系统调研和概括总结，可以概括出4个关键科学问题18个科学挑战：①俯冲带结构，包括几何结构、温压结构和地质结构；②俯冲带过程，包括变质脱水和部分熔融、双向交代作用、地幔楔部分熔融、流体活动与元素迁移；③俯冲带产物，包括洋壳俯冲带变质岩、陆壳俯冲带变质岩、镁铁质岩浆岩、长英质岩浆岩、热液矿床；④俯冲带动力学，包括地壳再循环形式、板片再循环机制、地热梯度演化、俯冲带构造体制、俯冲带动力来源、俯冲带时空演化。

尽管国际上在板块俯冲带领域针对地球浅部与深部过程之间关系的研究方面取得了一系列重要成果，但是由于不同学科之间的研究工作缺乏密切的协同配合，结果对于汇聚板块边缘物质的物理化学性质、俯冲带壳幔相互作用的机制和过程、缝合带壳源和幔源岩浆活动的物质来源和启动机制以及深部地幔过程对浅部地壳物质的影响等许多关键科学问题尚未得到根本解决。将来的研究需要聚焦汇聚板块边缘岩石圈物质循环和再造这两个核心科学问题，进一步查明俯冲带构造作用、变质作用、交代作用、岩浆作用等过程的各自特征和相互联系，包括挥发性组分在地球深部的迁移过程及其资源和环境效应，着力考察研究相对薄弱的古俯冲带，认识和理解俯冲带向造山带转换的机制和过程，在地球系统科学的框架内阐明汇聚板块边缘岩石圈与软流圈之间在能量和物质上的交换机制与效应。

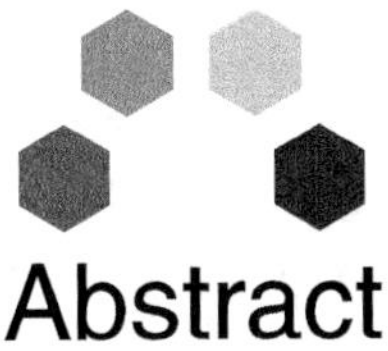

Abstract

1. Introduction

Subduction zones are a key to plate tectonics. Subduction is the most important way to transport crustal materials to the mantle; thus it is the most efficient mechanism for mass and energy exchange between the Earth's spheres. With increasing temperature and pressure during subduction, the subducting crust may experience not only metamorphic dehydration and partial melting but also a series of mineral phase transitions. More importantly, when the slab or slab-derived material (such as aqueous solutions released by metamorphic dehydration or hydrous melts produced by crust partial melting) becomes in contact to the mantle wedge at different depths, a series of chemical reactions take place due to significant differences in physicochemical conditions and geochemical composition between the crustal and mantle materials. These reactions can lead to the formation of new mineral phases or mineral associations with special compositions and properties, which may modify the composition of the mantle wedge to cause the geochemical heterogeneity of the mantle. Subduction zones are closely related to key questions in geoscience, such as the evolution of the crust and mantle systems, the formation of mineral resources, and the long-term chemical evolution of the atmosphere and ocean. At the same time, they are also the major tectonic setting for volcanic eruption and seismic activity, which are important for the survival and development of human beings.

2. Processes and structures of subduction zones

In the shallow part of subduction zones, the subducting crust experiences metamorphic dehydration or even partial melting. On one hand, the crustal rocks are subjected to high-pressure (HP) to ultrahigh-pressure (UHP) metamorphism at lithospheric depths and then exhume along subduction channels. On the other hand, the released fluid migrates upwards to metasomatize the mantle wedge, forming the hydrous metasomatites in the early stage of oceanic subduction. Such mantle metasomatic rocks are fertile and enriched and thus more susceptible to partial melting than the normal mantle peridotite. This induces mafic arc magmatism in the mature stage of oceanic subduction, leading to eruption of arc lavas to the surface and thus completeness of crustal recycling. The volatile cycling in the crust-mantle system starts from subduction and has a profound influence on the environmental evolution and habitability of the Earth's surface. The crustal rocks subducting into the deep mantle and exceeding the depth limit of exhumation no longer return in the form of UHP metamorphic rocks back to the surface, but undergo further mineralogical transformation to transfer the crustal components into the mantle transition zone or even the lower mantle. Understanding the products of subduction zone processes depends on our knowledge of the structure and process of subduction zones. The structure of subduction zones includes the geometry, thermal and geological ones, which have bearing on the temperature and pressure of the subducting slab and its overlying mantle wedge. The subduction zone processes may take place at different depths, varying from shallow at lithospheric depths to deep at asthenospheric depths. They may occur in the form of deformation, metamorphism, metasomatism, magmatism and mineralization. In either case, they have a profound impact on the mass and energy cycle inside and the global tectonic pattern on Earth.

Metamorphism in subduction zones leads to a series of physicochemical changes at different depths. Such changes at eclogite facies are major driving force for further subduction. They also exert an important influence on the composition of the subducting crust at mantle depths, which has great bearing on the chemical geodynamics of subduction zones. The traditional theory of plate tectonics assumed that only oceanic subduction zones are the most active place for the crust-mantle interaction, where the oceanic crust (composed of mafic igneous rocks and seafloor sediments) is carried into the mantle after metamorphic dehydration and partial melting. However, recognition of continental deep subduction to mantle depths has brought new attention to crustal recycling at convergent plate boundaries. Similar to effects above oceanic subduction zones, the subduction of continental crust has inevitably affected the structure, composition, deformation and evolution of the overlying continental lithosphere. The difference in composition between the oceanic crust and continental crust also results in the difference in the feature of their interaction with the mantle wedge.

The fluid action and element mobility in subduction zones are the hotspot and frontier of subduction zone research. It is generally assumed that the subducting oceanic crust is relatively rich in fluids, releasing significant amounts of fluids at subarc depths for chemical metasomatism of the mantle wedge and subsequent mafic arc magmatism. The residual slab material after the dehydration continues sinking to affect the composition of the deep mantle, eventually forming the source of intraplate basalts. Because the density of the subducting mafic oceanic crust is higher than that of the surrounding mantle rock, the oceanic crust in oceanic subduction zones rarely returns to the surface. Thus, the study of fluid action and element mobility during the subduction of oceanic crust is primarily realized by investigating not only metamorphic rocks from the rarely exhumed oceanic crust (HP blueschist and eclogite) and shallow mantle rocks (peridotite xenolith and ophiolite), but also arc magmatic rocks from active plate margins in addition to high-temperature and HP experiments under subduction zone conditions. Nevertheless, there is insufficient evidence for the specific physicochemical processes and mechanisms of fluid action and element mobility at postarc depths. It also remains unclear on the style, reaction and mechanism for incorporation of crustal components into the mantle wedge. Mafic arc magmas are generally generated in the mantle wedge at subarc depths of 80-160 km, corresponding to the UHP metamorphism in the coesite to diamond stability field. The direct observation of UHP metamorphic rocks and their surrounding mantle lithologies is still an important way to understand the chemical geodynamics of subduction zones at subarc to postarc depths.

Due to relatively old, dry and cold characteristics, the continental crust is generally considered to release very limited amounts of fluids during subduction. Compared to oceanic subduction zones, the continental lithosphere above continental subduction zones is characterized by the lack of syn-subduction continental arc magmatism. This feature is ascribed to the relative lack of fluids in the deeply subducting continental crust, and therefore the relative lack of fluid action in continental subduction zones. However, the study of crustal dehydration during subduction indicates that the water in both continental and oceanic crust is mainly stored in hydrous minerals at depths of >30 km, and nominally anhydrous minerals are only a minor water reservoir. The temperatures of oceanic and continental subduction channels are similar under the same subduction conditions (mainly plate convergence rate and subduction angle), and the UHP metamorphic temperatures are also similar under subarc depths. Therefore, the crustal rocks in continental and oceanic subduction zones undergo similar degrees of dehydration for UHP metamorphism at subarc depths, leading to similar metasomatism of the mantle wedge. Nevertheless, the mantle wedge will be significantly heated in the mature stage of oceanic subduction zones, which is the decisive factor of arc magmatism. In continental subduction zones, the mantle wedge is also metasomatized by fluids from the subducting crust and its fragments to form fertile, enriched domains, but its temperature cannot be elevated enough during continental collision to cause partial melting for arc magmatism. According to the thermal structure of subduction zones and the dehydration behavior of crustal rocks, it is reasonable to infer that the lack of syn-subduction arc magmatism above continental subduction zones is not caused by the lack of water in the subducting continental crust, but rather by the lack of high heat flow into the mantle wedge by the overthickened continental lithosphere. In the post-collision stage, mafic-ultramafic metasomatites in orogenic lithosphere can acquire high heat flow from the underlying asthenospheric mantle for partial melting, giving rise to various mafic igneous rocks in collision orogens.

The mantle wedge, as a key lithotectonic unit connecting the subducting slab and the obducting plate in the subduction system, plays an important role in the material recycling and energy exchange. Orogenic peridotite directly records chemical metasomatism of the mantle wedge by subduction zone fluids such as aqueous solution, hydrous silicate melt and carbonatitic melt. The systematic study of orogenic peridotite can provide constraints not only on the chemical metasomatism, metamorphic and deformation history on the microscopic scale, but also on the spatiotemporal evolution of macroscopic tectonics. The results are of great importance to understanding the crust-mantle interaction in subduction zones. The study of oceanic arc basalts and continental arc andesites has confirmed that peridotite in the mantle wedge is metasomatized by fluids derived from the subducting oceanic crust, and partial melting of the metasomatitic domains in the mantle wedge results in mafic arc magmatism. Compared to oceanic subduction zones, rocks in subducting continental crust also underwent significant metamorphic dehydration at subarc depths. Furthermore, the crustal rocks sinking to the subarc depths may undergo different degrees of dehydration melting, giving rise to hydrous melts to metasomatize the mantle wedge. Therefore, the chemical reactions between these slab-derived fluids and their overlying mantle wedges are recorded in different types of metamorphic and magmatic rocks. The orogenic peridotite at convergent plate boundaries holds a direct record of these processes, whereas magmatic rocks produced by post-collisional magmatism provide an indirect record of such processes.

The study of the relationship between thermal structure, earthquake and arc volcanism in modern oceanic subduction zones indicates their category into cold and warm in thermal state. On one hand, in cold oceanic subduction zones, the subducting crustal rocks do not significantly dehydrate at forearc depths of <60 km, but only experience blueschist to eclogite facies metamorphism; the large-scale dehydration occurs at subarc depths of >80 km and then forms the mantle source of arc magmatism. On the other hand, in warm oceanic subduction zones, the subducting crustal rocks experience large-scale dehydration at forearc depths but low degrees of dehydration at subarc depths, resulting in the lack of large-scale arc magmatism. In either case, fluids are released from the deeply subducting oceanic crust to metasomatize the mantle wedge peridotite, and then the metasomatites undergo partial melting to form oceanic arc basalts or continental arc andesites. After dehydration and melting, the subducting slab will continue to descend into the deep mantle, which will have a significant influence on the composition of the mantle. Aqueous solutions and hydrous melts derived from the subducting oceanic crust at subarc depths react with the mantle wedge peridotite to form the mantle source of oceanic arc basalts or continental arc andesites. The residual oceanic crust further subducts to postarc depths of >200 km for further partial melting, and the resulting melts are less hydrous but more enriched with HFSE and their reaction with the asthenosphere mantle wedge peridotite generates the mantle source of intraplate basalts. With an increase in the depth of crustal subduction from metamorphic dehydration to partial melting, metasomatic reaction of the resulted fluids with the mantle wedge is the key mechanism for the geochemical transport in both oceanic and continental subduction zones.

3. Recycling of subducted crustal material

Continental subduction zones are characterized by low geothermal gradients, and there is significant fluid action in subduction channels. In the cold subduction zones, the subducting crustal rocks do not undergo significant metamorphic dehydration at forearc depths, but only undergo HP blueschist to eclogite facies metamorphism. They experience significant metamorphic dehydration during UHP metamorphism at subarc depths, but no syn-subduction arc magmatism can take place because of the cold mantle wedge. Nevertheless, much more profound dehydration and melting may occur in UHP metamorphic rocks as soon as they undergo decompressional exhumation. The released fluids would not only metasomatize the overlying mantle wedge peridotite, but also form the internal fluids for amphibolite facies retrograde metamorphism and veining of various materials. Therefore, the continental deep subduction not only causes the UHP metamorphism, but also induces the crust-mantle interaction in continental subduction channels. In fact, HP to UHP metamorphic rocks are the first manifestation of the recycling of the subducted crustal material to the surface. They are primarily affected by the changes of temperature and pressure in subduction zones, liberating the fluids for metasomatism of the mantle wedge. Therefore, the study of fluid action and crust-mantle interaction in continental subduction channels is becoming a forefrontier of convergent plate boundaries.

Mafic arc magmatic rocks above oceanic subduction zones are the second manifestation of the recycling of subducted oceanic material. These rocks record the subduction zone processes from the subarc mantle wedge to the crust, providing a natural laboratory to investigate the recycling of crustal materials into the mantle. Since the establishment of the plate tectonics theory, there have been many studies of the crust-mantle interaction in modern oceanic subduction zones. As a result, we have obtained relatively mature understanding of the crustal recycling in the origin of oceanic arc basalts and continental arc andesites. However, it still remains unresolved whether there was the same property of crust-mantle interactions in ancient oceanic subduction zones. Mafic igneous rocks with different compositions are common above such subduction zones, including not only arc-like basalts and andesites generated during subduction of the paleo-oceanic slab, but also oceanic island basalts (OIB)-like and island arc basalts (IAB)-like mafic igneous rocks at convergent plate boundaries that formed after the paleo-oceanic subduction. These magmatic rocks provide another nature laboratory to study the recycling of crustal materials into deep mantle in the geological history.

The third manifestation of the recycling of subducted materials in both oceanic and continental subduction zones is syn-exhumation and post-collisional magmatic rocks. In the exhumation stage of UHP metamorphic rocks, the deeply subducted continental crust may undergo significant partial melting, giving rise to felsic rocks on different scales. These felsic rocks generally show the characteristics of arc-like trace element and isotopically enriched compositions, suggesting that their parental melts can serve as a kind of metasomatic agents for the crust-mantle interaction in deep subduction zones. Some syn-exhumation mafic rocks may be the melting product of mafic-ultramafic metasomatites in the mantle wedge. These rocks still preserve the diagnostic minerals (such as residual zircon cores) as well as the element and isotope records from the deeply subducted crust. For mafic rocks at ancient convergent plate boundaries, some of them show arc-like trace element distribution patterns, i.e., enrichment in large ion lithophile elements (LILE) and light rare earth elements (LREE) but depletion in high field strength elements (HFSE) and heavy rare earth elements (HREE), whereas the others show OIB-like trace element patterns, i.e., LREE enrichment, negative Pb anomaly and positive HFSE anomalies. This difference indicates variations in the composition of mantle sources, which is related to metasomatism of the mantle wedge by hydrous melts derived from the dehydration and melting of subducting continental crust and oceanic crust, respectively, at different depths. However, its geodynamic processes still remain to be clarified with respect to different physicochemical mechanisms.

4. Subduction zone science system

In the second half of the 20th century, chemical geodynamics was devoted to studying the effect of plate subduction on the geochemical composition of the mantle, trying to distinguish contributions from dehydration and melting of different types of crustal rocks (including sediments) to the composition of island arc and ocean island basalt source regions. Since the 21st century, attention has turned to the structure, process and product of subduction zones. It is found that subduction zones have a series of difference in thermal state between cold and warm, in dynamic source between gravity and buoyancy, in stress realm between compression and extension, in time between past and present, in recyling mechanism between breakoff and delamination, and in the style of crustal recycling between fluid and solid. Therefore, the study of subduction zones requires determination of not only the mechanism and style of crustal recycling, but also the source of the driving force and the thermal state of subduction zones as well as their changes with time. In order to identify the different types of crust-derived fluids that metasomatize the mantle wedge, it is important to find the petrological and geochemical evidence for the metasomatic reaction at the slab-mantle interface, understand the impact of lithosphere subduction and delamination at convergent plate boundaries on the mantle heterogeneity. In doing so, we have to integrate the tectonism, metamorphism, metasomatism and magmatism of subduction zones into a whole in the future study.

Since the geochemical composition of island arc basalts is associated with the metamorphic dehydration and partial melting of subducting oceanic crust at subarc depths, it is generally considered that the recycled material from the subducting crust is a kind of liquid phases in the form of aqueous solution, hydrous melt and supercritical fluid. Although the crustal signature is clearly identified in the geochemical composition of oceanic island basalts, it is not distinguished yet whether the crustal material entering the mantle source of such basalts is in the form of fluid or solid. The supracrustal rocks subducted to the subarc depths can be offscrapped by the overlying mantle wedge, which may migrate as a "cold plume" to diapir into the interior of the mantle wedge rather than directly return to the forearc depths along the subduction channel. These diapiric materials can undergo partial melting and then react with the mantle wedge peridotite, becoming the magma source of mafic arc volcanics. This means that the recycled material is in the form of solids rather than fluids. Because the fluids dissolve only a small amount of silicates and a large amount of incompatible elements (such as LILE and LREE), their reaction with peridotite forms not only serpentinized and chloritized peridotites but also pyroxenite and amphibolite in the mantle wedge. These metasomatites are still of ultramafic composition, and their partial melting produces basaltic melts. For the solids, their melting product contains a large amount of crustal silicates that are significantly enriched in not only silica and aluminum but also incompatible elements. Its reaction product with the mantle wedge peridotite may no longer have ultramafic composition. Therefore, it remains to be further constrained whether the subducted crust is chemically differentiated to produce the fluids or physically separated to produce the solids that migrate and react with the mantle wedge.

In continental subduction zones, continental subduction is dragged by subduction of the oceanic slab. Because continental lithosphere is of the relatively buoyant property, the resistance becomes the largest when it is subducted to the lithosphere/asthenosphere boundary for UHP metamorphism. At this time, the ocean-continent transition part of the subducting slab may break off, leading to exhumation of the continental slab at the action of buoyancy and continuing subduction of the oceanic slab under dragging of gravity into the deeper asthenospheric mantle. However, it still remains unresolved whether the slab breakoff does occur and when this process takes place in continental subduction zones. For the ongoing subduction, upwelling of the asthenospheric mantle along the gap after slab breakoff will not only enable its decompressional melting to form the basaltic melts, but also facilitate extensional exhumation of the continental crust to form the ultrahigh temperature (UHT) metamorphic rocks. The result will be superimposition of UHT metamorphism on the UHP metamorphic rocks, and both extreme metamorphic events show a small time interval. In addition, the lithosphere at the convergent plate boundaries will be thickened due to the continental collision. The lithosphere at the root of the thickened orogen will be delaminated due to gravitational instability, bringing part of the lithospheric mantle (sometimes including the overlying mafic lower crust) into the asthenospheric mantle. This process typically occurs in the fossile subduction zones, where once the orogens loss their roots, the mantle lithosphere at the convergent plate boundaries would be thinned to cause continental rifting. The asthenospheric upwelling, on one hand, leads to the decompressional melting to form the basaltic melts. On the other hand, it will heat the thinned lithospheric mantle and enable the deep-sited rocks of the overlying continental lower crust experience HT to UHT granulite facies metamorphism. This process causes the superimposition of UHT metamorphism on the UHP metamorphic rocks, but both extreme metamorphic events have relatively large time intervals. How to distinguish the two kinds of slab recycling mechanism, i.e., slab breakoff and delamination, has become the forefront field in the study of convergent plate boundaris.

The temporal relationship between subduction angle and geothermal gradient is generally categorized into two stages at converging plate boundaries. The early stage is the low angle subduction at low geothermal gradients, and the late stage is the high angle subduction at elevated geothermal gradients. The rollback of subducting slab is the basic reason for the change of subduction angles from low to high. Paired metamorphic belts are common in oceanic subduction zones, where blueschist to eclogite facies assemblages occur close to the trench and form at low geothermal gradients of <10 ℃/km, whereas sillimanite-present amphibolite to granulite facies assemblages occur beneath the continental arc and form at high geothermal gradients of >30 ℃/km. For the ongoing subduction, the rollback of subducting slab results in lithospheric thinning at backarc sites, which in turn leads to the opening of backarc rift basins. As a result, the asthenosphere upwelling occurs, which can not only partially melt to form the basaltic melts, but also heat the deep crustal rocks to form HT to UHT metamorphic rocks. For the metamorphic rocks in collisional orogens, although there are also bimodal metamorphism, they are often superimposed in time and space, indicating that the continental subduction zones have changed their geothermal gradients in due course. The metamorphic rocks formed in the early stage of cold metamorphism at lower geothermal gradients are often superimposed by the late stage of warm metamorphism at higher geothermal gradients. For the converging plate boundaries, the rollback of subducting slab can lead to partial melting of the mantle wedge bottom under asthenospheric heating, but whether this heating process is the geodynamic mechanism for mafic arc magmatism in the subduction zones remains to be further determined. For plate boundaries that are no longer converging, the thickened orogenic lithosphere will experience continental rifting after the delamination of orogenic roots, resulting in heating of the thinned lithosphere mantle by the upwelling asthenosphere. This process leads to HT to UHT metamorphism of the overlying continental crust due to the superimposition of the high geothermal gradient rocks on the low geothermal gradient rocks. However, it remains to be further studied whether this process is the geodynamic mechanism for the composite orogeny at convergent plate boundaries.

The coupling relationship between the subducting slab and the mantle wedge varies with time, which is generally divided into two stages at converging plate boundaries. The early stage is the slab-wedge coupling in the realm of compression. In the late stage, the slab-wedge decoupling leads to an extensional realm. In the compressional system, the plate interface has low geothermal gradients, leading to blueschist-eclogite facies metamorphism but no arc magmatism. In the extensional system, the plate interface has high geothermal gradients and the bottom of the mantle wedge is heated by the asthenosphere. This results in partial melting of the metasimatic domains in the mantle wedge for arc magmatism, and the backarc crust undergoes amphibolite-granulite facies metamorphism at high geothermal gradients. If delamination takes places for the collision-thickened orogenic lithosphere roots, continental rifting occurs along the thinned lithosphere due to upwelling of the asthenospheric mantle. This heats the thinned lithospheric mantle, enabling HT to UHT metamorphism of the overlying continental crust and thus superimposition of the amphibolite to granulite facies metamorphism on the eclogite facies HP to UHP metamorphic rocks with contemporaneous migmatization and granitic magmatism. During the transformation from the compressional to the extensional realm, the rocks in the subduction channel undergo exhumation at the action of crustal buoyancy. As a consequnce, not only the blueschist-eclogite metamorphic rocks are subjected to the superimposition of HP amphibolite to granulite facies metamorphism, but also the exhumed crustal rocks undergo decompressional melting for syn-exhumation magmatism. Therefore, the change of tectonic realms leads to a change in the thermal state of subduction zones, resulting in different types of metamorphism and magmatism at convergent plate boundaries. How to understand the tectonic regime changes with time in the lifetime of subduction zones has become a key problem in the study of subduction geodynamics.

When plate tectonics was firstly established in the 1960s, mantle convection was generally considered as the driving force for plate movement. With the deepening of geodynamic studies, it is gradually realized that the basaltic crust in the subducting oceanic lithosphere undergoes eclogitization at subarc depths. Because the density of eclogite is higher than that of peridotite, in combination with the high density feature of subducting low-T slabs, gravity (negative buoyancy) is acknowledged to be the major driving force for plate subduction. Where the seafloor spreads, upwelling of the asthenospheric mantle is driven by its thermal buoyancy for successful rifting, resulting in decompressional melting to form mid-oceanic ridge basalts. Ridge push is also one of the driving forces for plate subduction. In continental rift zones, the asthenosphere upwelling occurs due to its thermal buoyancy at the sites where the collision-thickened orogenic lithosphere is thinned significantly. If the rate of lithospheric thinning is greater than the rate of asthenosphere upwelling, not only the upwelling asthenosphere can partially melt, but also the thinned lithosphere can collapse to cause bimodal magmatism. If the rate of lithospheric thinning is less than or equal to the rate of asthenosphere upwelling, then it is difficult for the upwelling asthenosphere to partially melt (equivalent to the ultraslow spreading mid-oceanic ridge), and the thinned lithosphere will not collapse. As a result, metamorphic core complexes are composed of amphibolite to granulite facies metamorphic rocks together with granites and migmatites form in continental rift zones. At converging plate boundaries, thus, there are both gravity-dominated lithospheric subduction and buoyancy-dominated asthenospheric upwelling. Although the two forces have played variable roles in the history of convergent plate boundaries, our understanding of them is still not sufficient. Therefore, this issue has become the forefront and hotspot in the study of the initiation mechanism of plate tectonics.

Subduction zones and orogenic belts have inherited and developed relationships in both structure and composition. Plate convergence leads to the formation of orogenic belts, and intraplate rifting leads to the destruction of orogenic belts. During the processes of plate convergence, if an oceanic plate subducts beneath a continental plate, the result is the formation of an accretionary orogen, and a typical example is the Andean orogen above the eastern Pacific subduction zone. If one continental plate subducts beneath another continental plate, the result is the formation of a collisional orogen, and typical examples are the Alpine-Himalayan orogenic belt and the Dabie-Sulu orogenic belt. Although the formation of both accretionary and collisional orogens is contemporaneous with the process of plate subduction, the orogenic cycle belongs to the late stage of subduction zone process. In the case of oceanic subduction beneath the continental plate, the orogenic cycle is divided into two stages of coupling and decoupling between the subduction slab and mantle wedge, with continental arc magmatism for accretionary orogeny in the late stage. In the case of continental collision, it is divided into two stages of lithospheric subduction and crustal exhumation, with the exhumation of subducted crustal slices for collisional orogeny in the late stage. When the two plates are no longer converging, the original plate margin is transformed to the continental interior, which remains inactive until the thickened orogenic roots are foundering for continental rifting. This leads to destruction of the orogenic belts after thinning of the orogenic lithospheric mantle. Previous studies of subduction zones were mainly focused on the ongoing subduction zones, overlooking changes in the structure and process of subduction zones with time. For example, the low-angle cold subduction for compression is prominent in the early stage, but it may become high-angle warm subduction for extension in the late stage. At this time, continental rifting will occur after the lithospheric thinning in fossil subduction zones. Successful continental rift zones become continental rupturing for seafloor spreading, whereas aborted continental rift zones also undergo intracontinental reactivation to form hot to suprahot orogens. Therefore, the formation and destruction of orogenic belts not only have different mechanisms but also take place at different times, leading to the distinction between the fossil and ongoing subductions. The Wilson cycle is composed of seafloor spreading, oceanic subduction and continental collision, but it remains to be resolved how a continental collision zone is evolved into a seafloor spreading zone. The evolution in time and space between the fossil and ongoing convergent plate boundaries has become a key issue in the study of subduction zones.

In the identification of plate tectonics in the Precambrian, it is generally thought that the diagnostic indicator is the crustal subduction along the lithospheric boundaries, and some researchers even emphasize that only the cold subduction zones forming blueschist to eclogite facies metamorphic rocks can be regarded as the indicator of plate tectonics. However, the convergent plate boundaries experience both subduction and rifting. Generally speaking, subducting plate margins exhibit the rigid behavior at low geothermal gradients, and supracrustal rocks can subduct to subcontinental lithospheric mantle depths. At moderate geothermal gradients, supracrustal rocks can only subduct to continental lower crust depths. This difference results in the distinction between two styles of plate tectonics in the geological history. The modern plate tectonics is characterized by cold subduction and its typical product is Alpine-type blueschist to eclogite facies metamorphic facies series. The ancient plate tectonics is characterized by warm subduction and its typical product is the Barrovian-type amphibolilte to granulite facies metamorphic facies series. The warm subduction would widely occur in the Archean, the local cold subduction would appear in the Paleoproterozoic, and the global cold subduction has appeared only since the Phanerozoic Eon. In addition, the asthenospheric upwelling transfers high heat flow from the mantle to the crust, leading to not only Buchan-type metamorphism at high geothermal gradients but also partial melting of the continental lower crust for felsic magmatism. These high-T/low-P processes are developed on the basis of previous subduction zones. The resulting continental rift zones form the active zones at converged plate boundaries, and their properties are controlled by the efficiency of the upwelling asthenosphere. The Archean mantle has the high temperatures and the subducting crust undergoes significant high-degree dehydration melting at subarc depths. In contrast, the Phanerozoic mantle has the low temperature and the subducting crust only undergoes low-degree dehydration melting at subarc depths.

5. Concluding remarks

Subduction zone science has been a central part of the plate tectonics theory. Through our systematic survey, comprehensive summary and conceptual generalization, we have outlined four key questions consisting of 18 scientific challenges: ①the structures of subduction zones, including geometric, thermal, and geological structures; ② the processes of subduction zones, including metamorphic dehydration and partial melting, bidirectional metasomatism, partial melting of the mantle wedge, and fluid action and element transport; ③ the products of subduction zones, including the metamorphic rocks of oceanic and continental protoliths, mafic igneous rocks, and felsic magmatic rocks; ④the geodynamics of subduction zones, including the forms of crustal recycling, the mechanisms of slab recycling, the evolution of thermal state, dynamic regime, dynamic source, and the evolution in both space and time.

Although the worldwide studies have made a series of important achievements on the relationship between Earth's superficial and deep processes in the field of subduction zones, there are still many problems remaining to be solved due to the lack of integrated studies between different subjects and approaches. These problems include: the chemical and physical properties of convergent margin materials, the mechanism and process of crust-mantle interaction, the source and production of crust-derived and mantle-derived magmatism in suture zones, the influence of deep mantle processes on the shallow crustal material, and so on. Future studies need to focus on the two key issues: ① recycling and reworking of the lithospheric material at convergent plate boundaries, ②the characteristic features of subduction zone tectonism, metamorphism, metasomatism and magmatism and their relationships, including the migration process of volatile components in the deep mantle and its effects on mineral resources and surface environments. Future studies should also focus on the less-known fossil subduction zones, not only understanding the mechanism and process for the transformation from subduction zones to orogenic belts but also documenting the mechanism for and effect on the exchange of energy and material between lithosphere and asthenosphere within the framework of Earth system science.

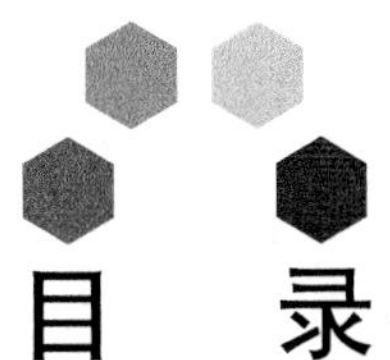

目　录

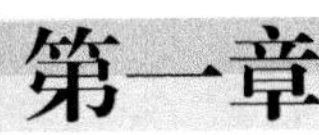

第一章 科学意义与战略价值

板块构造理论是人类理解行星地球整体运行行为的一座思想高峰，其重要性比肩相对论和量子力学之于物理学、DNA 之于生命科学的意义，是 20 世纪最重大的科学理论之一（Zheng，2018）。在历史沿革上，板块构造理论由大陆漂移、海底扩张和岩石圈俯冲发展而来。板块构造理论以地球整个岩石圈的活动方式为依据，建立了全球尺度的构造运动模式（Le Pichon et al.，1973；Cox and Hart，1986；Frisch et al.，2011；Livermore，2018）。板块运动是地球内部的一级运行机制，也是地球充满活力的重要象征。板块运动既是地球内动力地质作用的结果，又反过来影响地球从表层到内部的演化进程。在 20 世纪 60 年代板块构造理论建立之时，人们强调的是三个要素（Zheng，2021a）：板块整体具有刚性行为、离散板块边缘海底扩张产生新的洋壳、汇聚板块边缘岩石圈俯冲使洋壳消亡。最近半个世纪的研究发现，刚性板块边缘在高温下表现出韧性特点、离散板块边缘夭折的大陆张裂（rift）并不产生新的洋壳、汇聚板块边缘岩石圈俯冲产生不同性质的活动带（Zheng，2018）。板块边界以活动带为标志，其中变形、变质、地震和岩浆作用呈带状分布。越来越多的研究发现，板块边界活动带既是岩石圈重力驱动的俯冲带，也是软流圈地幔浮力驱动的岩石圈张裂带（Zheng，2021a）。

海底扩张、板块俯冲和大陆漂移这三大要素“三位一体”构成了板块构造（图 1-1）。海底扩张产生新的大洋岩石圈，而板块俯冲使已有的大洋岩石圈消亡，在这个大洋岩石圈产生 - 消亡的链条上出现大陆漂移。尽管现代板块边界活动带具有全球网络性质（Hawkesworth and Brown，2018），但这个并不是认识和区分板块构造与非板块构造的重要标志。活动带具有全球网络性质属于板块构造，只具有局部网络性质的活动带也属于板块构造。板块构造是通过刚性板块的全球性运动来运行的，这些运行包括在板块边界发生的俯冲作用和张裂作用，并且通过不同类型的造山作用形成了活动带（Cawood et al.，2018；Hawkesworth and Brown，2018；Zheng，2021b）。因此，并不是俯冲带和张裂带形成全球网络才能作为板块构造的判别标志。

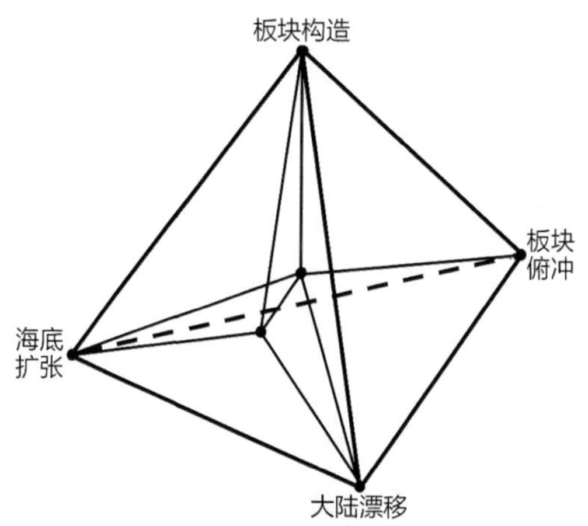

图 1-1　板块构造三大要素"三位一体"示意图

板块俯冲属于自上而下的物质运动，海底扩张属于自下而上的物质运动，大陆漂移是两者之间的水平运动。在地球刚性外层，板块俯冲与海底扩张这两大类运动所涉及的岩石圈物质消亡和生长处于质量守恒状态

板块构造是地球区别于太阳系内其他类地行星的独特标志。板块构造在地球的出现，促进了地球表层与深部圈层的物质循环和能量传输。特别是碳、氮、硫和水等关键挥发分在地球深部的再循环，使地球逐渐演化为宜居星球，促进了生命的诞生和演化。板块构造理论深刻揭示了地球内部的运作机理，是 20 世纪地球科学的巨大成就。尽管板块构造理论起源于海洋地质学，但在过去 50 年中通过对大陆地质的研究，板块构造理论得到了很大的发展（Kearey et al.，2009；Moores et al.，2013；郑永飞等，2015；Zheng，2018，2021a）。因此，板块构造理论的成功极大地促进了地球科学各方面的发展，从而成为地球科学，特别是固体地球科学领域的指导性学术思想。

第一节　板块俯冲带研究是地球科学的核心和前沿

板块构造理论的核心是俯冲带，俯冲带过程是地球圈层演化的核心机制。板块俯冲是连接地球表生圈层和深部圈层的关键纽带，对整个地球演化起到了关键作用。板块俯冲带与地球壳幔系统演化、矿产资源聚集、大气和海洋的长期化学演化等重大地球科学问题都紧密相连，同时也是火山喷发和地震活动的主要场所，与人类的生存和发展密切相关。因此，对板块俯冲带的系统深入研究，不仅有助于理解地球内部运作机制及其表层响应，还对国民经济和社会发展具有重要意义。板块俯冲带不仅是地球科学基础研究的核心和前沿，而且攸关国家经济社会发展全局。板块构造研究极大推动了相关学科的进步，是自然科学的关键交叉学科，为国家安全和战略发展提供了关键支撑。

一、板块构造理论是 20 世纪地球科学的巨大成就

板块构造理论认为，地球的外壳是由地壳与岩石圈地幔组成的，属弱构造变形的刚性块体。地球的岩石圈被许多大型断裂或构造带分割成若干块体，即岩石圈板块。刚性的岩石圈板块位于软流圈之上，并以大规模水平运动为主。地幔上涌产生的岩浆从洋中脊喷发，向两侧增生，形成对称的磁异常条带，构成不断增长的大洋地壳，与下伏的岩石圈地幔一道构成大洋岩石圈板块。大洋岩石圈向两侧运动到达板块边界，就沿俯冲带俯冲消减，部分大洋岩石圈物质进入地幔。之后，幔源岩浆再次沿洋中脊对流上涌，形成新的大洋岩石圈，再次在俯冲带俯冲消亡，从而构成了大洋岩石圈的产生－消亡演化旋回。

板块构造由三个关键要素组成（Le Pichon et al.，1973；Cox and Hart，1986；Frisch et al.，2011；Livermore，2018）：①地球上刚性板块的存在；②海底扩张使刚性板块相互离散，产生新的洋壳；③岩石圈俯冲使刚性板块发生汇聚，导致洋壳的消亡。在古板块构造的识别上，很多学者将其鉴定性标志确定为岩石圈边缘发生了俯冲作用。一方面，只有形成蓝片岩相－榴辉岩相变质岩的冷俯冲带才能作为现代板块构造的标志（Stern，2005）；另一方面，板块边界既有俯冲作用也有张裂作用。在板块构造之前是停滞层盖构造（stagnant lid tectonics），缺乏活动层盖构造（mobile lid tectonics）。从太古宙起地球上已经出现丰富的活动层盖构造，但是板块边缘的热性质和连通度在地球历史上的不同时期存在显著差别，板块边缘性质随动力体制和地热梯度的变化而变化（Zheng，2021a）。

自从 20 世纪 60 年代板块构造理论建立以来，它不仅促使地球科学家思维方式发生重大转变，而且给地球科学带来了革命性的进展和突破。板块构造理论的建立旨在认识地球内部运行机制。半个多世纪以来，该理论

在地球科学的许多关键基础科学问题上取得了举世瞩目的巨大成就（Le Pichon et al.，1973；Cox and Hart，1986；Oreskes，2003；Frisch et al.，2011；Livermore，2018；Zheng，2021a）。

首先，板块构造理论不仅完美地解释了大洋板块从产生到消亡的全过程，而且成功描绘了现今全球构造格局。七大板块的划分进一步明确了地球现今的洋－陆格局。以钙碱性安山质火山岩为代表的岛弧岩浆活动与地震多出现在板块边缘或古板块缝合带，如环太平洋火山弧和地震带。同时，板块构造理论也为地球历史上多次出现的超大陆（如古元古代时期的哥伦比亚超大陆、新元古代时期的罗迪尼亚超大陆、晚古生代时期的潘吉亚超大陆）提供了一种合理的动力解释。

其次，由大洋板块俯冲消亡建立起来的板块构造理论不仅阐释了大洋岩石圈从产生到消亡的威尔逊旋回，而且发现密度相对较小的大陆岩石圈板块同样能够俯冲到地幔深度，从而形成以出现金刚石和柯石英为代表的超高压变质带，如中国东部的大别－苏鲁造山带、西部的柴北缘造山带。由大洋玄武岩研究揭示的地幔化学不均一性被认为是 20 世纪 80 年代以来地幔地球化学领域的重要发现之一，但其形成原因一直存在争议。板块构造理论为其提供了很好的解释。板块俯冲必然会将大洋或大陆岩石圈物质，特别是地壳物质带入地幔，这是造成地幔化学不均一性的主要原因。

再次，20 世纪 90 年代以来，矿床学研究的一个突破性进展是发现许多大型矿床的形成与大陆边缘弧岩浆活动有关，如南美西海岸的铜矿带、中国特提斯新生代铜矿带。这些矿床多出现在大洋板块俯冲－碰撞形成的造山带内。因此，板块构造理论给自然资源勘查带来了新的突破。

最后，板块构造理论成功地将地球各圈层的物质循环与能量交换联系起来，如地核－下地幔－地幔过渡带－软流圈－岩石圈－土壤圈－生物圈－水圈－大气圈，从而促进了地球系统科学的诞生。同时，板块构造促进了地表与地球深部的物质循环和能量交换，特别是导致了碳、氮、硫和水等关键挥发分在地球内部与地表之间的循环，促使生命的诞生和演化，最终使地球逐渐演化为人类宜居的唯一星球。地质历史上全球尺度的气候变化也可能与板块活动有联系，如东亚季风的形成与板块俯冲－碰撞导致的青藏高原的隆升有密切关系。

二、俯冲带是板块构造的核心

在大陆漂移学说和海底扩张学说的基础上，20 世纪 60 年代建立了板块构造学说，并很快得到大部分地质学家的重视和认同。板块构造理论认为，整个地球表层可以划分为大小不等的刚性板块，即包括大洋和大陆地壳及软流圈以上的上地幔的岩石圈板块，板块之间发生持续不断的大规模水平运动。在地质历史中，不断有大洋板块的消亡和新的板块形成，构成了大陆张裂－大洋产生－大洋俯冲闭合－大陆碰撞造山的威尔逊旋回（Wilson，1966，1968；Dewey and Spall，1975）。板块运动也造成了大陆地壳的持续生长，以及超大陆的聚合和裂解。岩石圈板块之间的边界有三种类型（Condie，1997）：其一为汇聚型板块边界，即岩石圈板块消亡的位置；其二为离散型板块边界，即洋中脊，为岩石圈板块新生的位置；其三为使洋中脊错位的转换型板块边界或转换断层（图 1-2）。

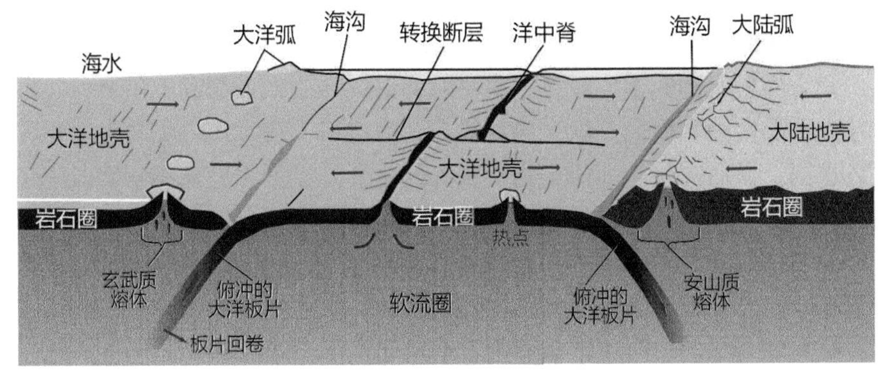

图 1-2　三种类型的岩石圈板块边界

海沟－汇聚型板块边界、洋中脊－离散型板块边界、转换断层－转换型板块边界

作为离散型板块边界的洋中脊是被动的，洋中脊的形成是外力拉张撕裂造成的，软流圈地幔降压上升并熔融形成新的大洋地壳。板块运动的主要动力来源是在负浮力作用下冷且密度大的大洋板块在俯冲带位置上重力下沉所引发的拖曳力（Forsyth and Uyeda，1975；Conrad and Lithgow-Bertelloni，2002）。因此，俯冲带岩石圈板块的下沉在板块运动中起主导作用。

板块俯冲带的结构包括几何形态、温压结构和地质结构（Zheng and Chen，2016）。俯冲带过程包括变质作用、交代作用、岩浆作用、成矿作用等。由岩石圈俯冲和软流圈上涌构成的自上而下和自下而上两类运动体系，是板块边缘物质和能量传输的关键过程，使得地球成为与其他行星不同的、具有生命力的星球。因此，从板块运动的动力来源，到地球内部的物质和能量的交换，俯冲带在板块构造系统中起主导作用。理解俯冲带的结构和过程是板块构造及全球构造理论的关键。

俯冲带过程深刻影响大陆地壳的形成演化。大陆地壳的平均成分为安山质，与现代大洋弧的岩浆岩平均成分相近，因此大陆地壳的生长与俯冲带岩浆作用密切相关。从地球早期到现代，随着地球内部温压结构的变化，俯冲带结构和岩浆成分也发生相应的变化。俯冲带结构和过程及其岩浆作用的研究对于揭示大陆地壳的形成和生长过程具有重要意义。

俯冲带在威尔逊旋回各阶段都发挥着重要作用（Wilson et al.，2019；Zheng et al.，2019a）。威尔逊旋回可以划分为大洋扩张期和闭合期（图 1-3）。扩张期代表大洋岩石圈的形成和发育，这一过程与相邻地区的大洋板块俯冲作用具有密切关系，持续的俯冲作用能够造成弧后扩张，并进一步发育成新的大洋，如欧洲古瑞亚克洋（Rheic Ocean）的形成是巨神洋（Iapetus Ocean）俯冲过程中由弧后盆地逐渐发育而成的（Stampfli and Borel，2002）。闭合期代表大洋岩石圈的消失和大陆岩石圈的形成，俯冲带在古大洋消亡、大洋俯冲向大陆俯冲转换、壳幔物质和能量相互作用、新大洋形成过程中发挥了关键作用。周而复始的大洋扩张和俯冲闭合造成了超大陆聚集和离散；陆内造山带在拉张构造作用下的再活化是大陆张裂作用的表现形式，其中成功的张裂就是大陆裂解，而夭折的张裂就是陆内再造（Zheng and Chen，2017，2021）。

超大陆裂解导致海底扩张，大洋板块俯冲引起弧岩浆作用，大陆碰撞导致超大陆聚合，缝合带加厚岩石圈减薄引起主动张裂。超大陆裂解是成功的显性张裂，大陆张裂但未裂解是夭折的显性张裂。

俯冲带多期次流体和岩浆活动为各类矿床的形成提供了优越的条件。地球表面 70% 以上的金属矿产与汇聚板块边缘的俯冲－碰撞带有关，如智利超大型斑岩型铜矿床、中国东部中生代多金属硫化物成矿带、古亚洲洋成矿带、特提斯成矿带等。另外，俯冲带流体活动和碳、氢、氮、硫等挥发性元素循环对地球的表层环境有重大影响。随着板块构造和俯冲带结构的不断变化，地球环境也发生系统变迁，逐渐从早期非宜居地球演化到目前的宜居地球。

三、俯冲带过程是地球圈层演化的核心机制

俯冲带过程是解决很多重大地球科学问题的关键，是地球圈层演化的核心机制。俯冲带包括洋－洋俯冲带、洋－陆俯冲带和陆－陆俯冲带三类（Cox and Hart，1986；Frisch et al.，2011），每类都涵盖大量重大地球科学问题。可以说，俯冲过程是地球圈层演化的加速器，是地球深－浅部系统的纽带和桥梁，是物质与能量激烈交换的场所。无论哪类俯冲带，都涉及地球深部碳循环、水循环

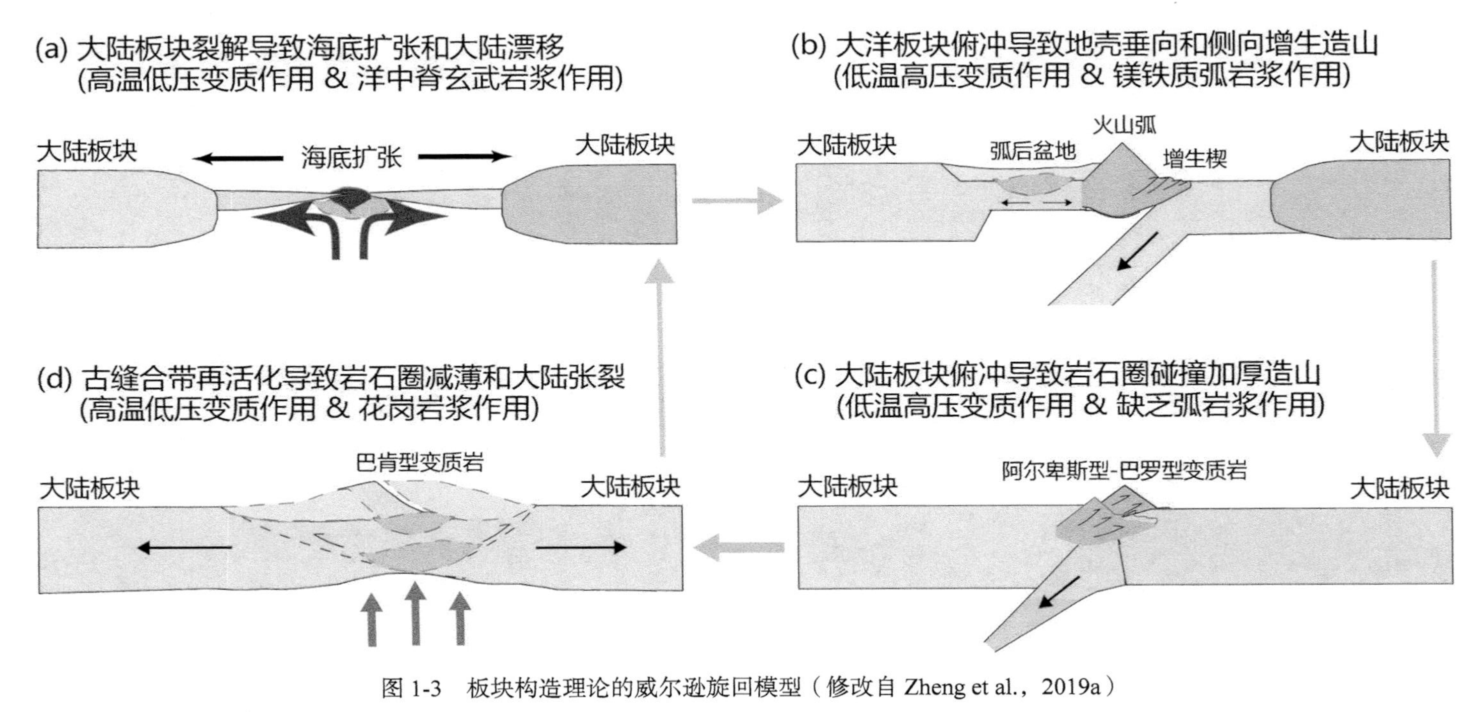

图 1-3　板块构造理论的威尔逊旋回模型（修改自 Zheng et al.，2019a）

等挥发分循环，以及地幔不均一性、地球壳幔系统演化、大陆增生、大陆风化与表生过程等重大问题。这些关键问题在20世纪后半叶就得到了高度重视。

20世纪后半叶以来，俯冲带研究产出了很多重要成果，如俯冲工厂、俯冲隧道与物质循环、俯冲侵蚀与拼贴、俯冲脱水和挥发分循环、俯冲拆沉与造山作用等（Zheng and Chen，2016；Zheng and Zhao，2017）。进入21世纪以来，俯冲系统的整体运行受到越来越多的重视。俯冲带过程的研究涉及地球各个圈层，需要从地球系统科学的角度分析俯冲带的系统构成（系统论认识）、力学控制因素（流变学结构和行为）、信息传递方式（如俯冲系统精细的地震波传播方式）、能量耗散结构（热、元素、流体分配结构）、多物理量协同机制（如浅部和深部流－固耦合机制）、灾变事件过程（如连续俯冲过程中的间断性地震、地震海啸、滑坡等灾变行为，地震、火山等灾害的随机性分布和不确定性）等。这些领域的研究更好地揭示了俯冲带的复杂性、非线性、整体性行为。

俯冲系统涉及地球在极端环境下的物理过程、化学过程和生命过程。这不仅涉及各种相变过程，还涉及各种脱水、脱碳、脱气主通道与微循环机制及其宏观效应。构建各圈层之间物质微观循环机理与宏观构造体制的关联，有助于增强对俯冲过程中各种重要变量的准确预测能力。俯冲带过程的三维物理模拟和耦合变质反应、不同组分熔融或相变模式、元素亏损－富集－循环模式的俯冲带岩浆动力学、变质动力学、成矿成灾动力学的综合研究，也可服务公益性防灾减灾及国家急需的战略性资源能源勘探。

俯冲带是地球系统特异于其他行星的重要构造单元，俯冲作用作为地球圈层演化的核心机制，涉及以下几个重要方面。①俯冲板片多维流变学演变及其对俯冲动力学的影响。俯冲带结构的复杂多样源于俯冲过程的多样性，俯冲板块的形态不仅与板块间相对汇聚速率、仰冲板块的运动方向和绝对速度、俯冲板块的年龄以及是否存在外来地体（包括洋底高原、无震海岭、海山、微陆块、岛弧等）有关，还受到地幔楔构造特性的影响。②俯冲带多圈层相互作用。地幔柱－俯冲带、板片窗－地幔楔、洋中脊－俯冲带、水圈－岩石圈、地壳－岩石圈地幔、岩石圈－软流圈两两之间的相互作用，导致俯冲带系统的地热梯度、密度、速度、挥发分和元素分配等变化，产生了复杂多样的资源和灾害效应。③俯冲带多尺度物质循环。俯冲带物质循环涉及多个圈层、多种形式，不仅影响海沟系统的生命元素和营养盐循环，还影响地球深部碳循环、水循环等过程，特别是板块起源、板块俯冲过程中元素循环、岩石循环、流体和熔体过程、壳幔相互作用等。④俯冲带多层次耦合过程。俯冲系统不同深度层次的微板块行为、成因机制和演化历程，研究从早期到现代不同时代俯冲体制机制的差异和转变过程，有助于真实揭示早期地球俯冲启动的机制，约束超大陆背景下的全球俯冲体系及其在超大陆形成与裂解过程中的效应，揭示地幔深部大规模低剪切波速异常体与俯冲过程的关联。⑤地球系统科学框架下的整体性研究有助于构建完整的俯冲系统理论，推动固体地球科学发展。

四、俯冲带研究是多学科交叉典型领域

俯冲带研究是多学科交叉的典型领域。俯冲带复杂的结构与物质组成、丰富的地质过程与产物、多个复杂动力系统相互作用等使得俯冲带研究需要多学科交叉融合。地球科学的发展态势也要求俯冲带研究范式发生变革，需要多学科交叉来探究俯冲带过程在地球系统科学中的关键作用。

（1）俯冲带复杂的结构、物质组成使得俯冲带研究需要多学科交叉融合

俯冲带具有复杂的几何、温压、地质结构和物质组成。理解俯冲带的过去、现在和未来，关键科学问题之一是查明俯冲带的结构特征。一个完整的俯冲带包括俯冲板片（下盘）和上覆板块（上盘）及其相互作用产物出现的区域（Zheng and Chen，2016），在大洋俯冲带通常表现为前陆褶皱冲断带、海沟外缘隆起、海沟、增生楔、弧前区和弧前盆地、火山弧、弧后区和弧后盆地等（Stern，2002；Zheng，2021b），这些要素在许多俯冲带都有表现（图1-4）。然而，很多现代大洋俯冲带的组成部分都在海平面以下几米到上万米，不能直接现场观测。查明俯冲带复杂的结构和物质组成，必须依靠地球物理探测、大洋深钻或深潜技术等手段才能识别或取得样品。此外，还必须结合地质学、地球化学和地球动力学等多种研究方法，多学科融合，才能揭示俯冲带的几何结构、温压结构、地质结构和物质组成。

大洋板块俯冲到大陆板块之下，在80～160 km的弧下深度析出流体交代地幔楔，由此形成的交代岩部分熔融形成镁铁质火山弧。①弧前增生楔形成，洋壳发生变质作用形成低级变质岩；②俯冲板片释放流体交代地幔楔，形成地幔交代岩；③地幔交代岩部分熔融形成弧岩浆，它们迁移到地壳表面形成弧火山岩，弧岩浆结晶分异导致斑岩型矿床形成。

根据板块构造理论，大洋板块在洋中脊形成以后，随着洋中脊的扩张向两侧运动；当到达海沟时，大洋板块俯冲进入地幔深部（Turcotte and Schubert，2002）。俯冲板块在俯冲过程中与周围地幔不断发生热交换（图1-5），该热演化过程主要由热传导和平流物质交换两种作用构成。俯冲带温压结构的演化是控制俯冲过程中物理化学性质变化的关键因素之一，直接影响矿物

脱水、岩石部分熔融、岛弧火山喷发、俯冲带地震等地质过程。同时，在板块俯冲过程中，由于其相对低温产生的负浮力是板块运动和地幔对流的主要驱动力。因此，俯冲带温压结构是板块构造理论的重要研究方向。

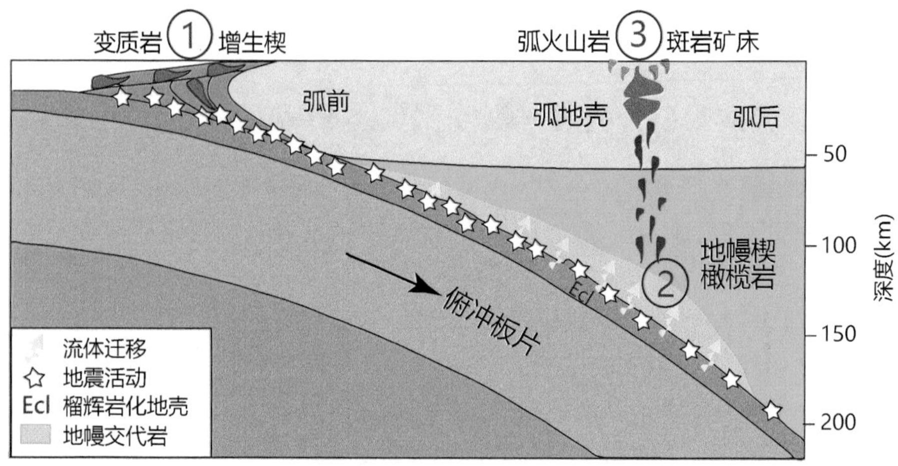

图 1-4　俯冲带结构、过程和产物示意图

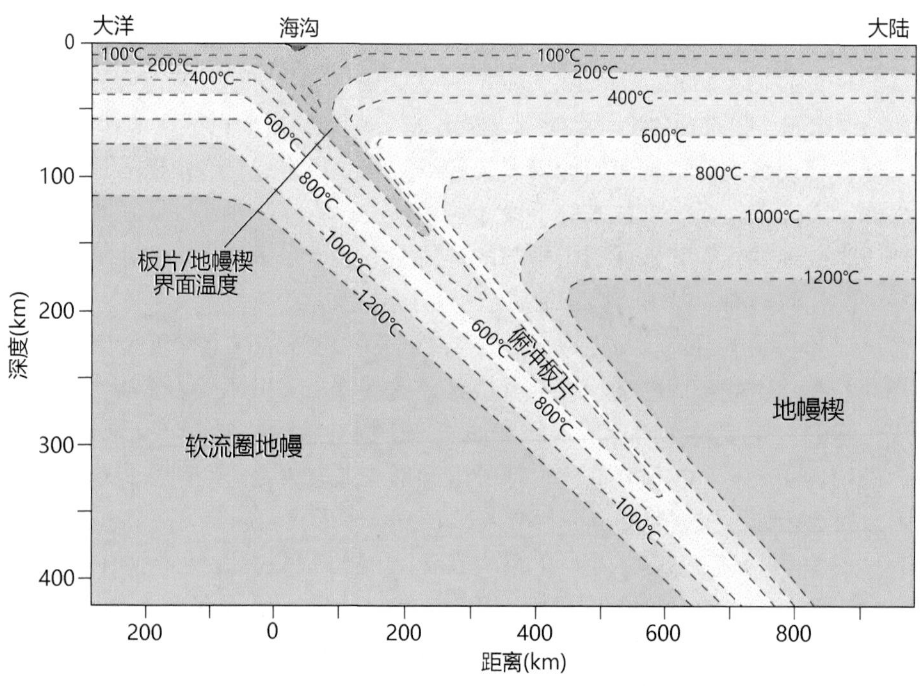

图 1-5　俯冲带温压结构示意图及主要控制因素（修改自 Frisch et al.，2011）

当两个汇聚板块之间处于耦合状态时，板块界面温度总是最低

（2）俯冲带丰富的地质过程和产物要求俯冲带研究必须多学科交叉融合

俯冲带有丰富的地质过程和产物（Zheng and Zhao，2017；Zheng，2021b）。俯冲带过程是理解俯冲板片和地幔楔之间物质循环和能量交换的关键所在，在很大程度上决定了地球壳幔系统的长期演化进程。俯冲带过程包括俯冲带流体活动与元素迁移、俯冲带双向交代作用、地幔楔部分熔融等多个过程。俯冲带的产物也非常复杂，包括各种类型的变质岩、岩浆岩、火山、地震、矿床和油气资源等。查明俯冲带产物的矿物、岩石或矿石的成因和形成机制，是揭示俯冲过程、全球物质循环和指导找矿（或油气）勘探的不可或缺的理论基础。俯冲带丰富的地质过程和产物要求俯冲带研究必须利用地质学、地球化学、地球物理和地球动力学多学科交叉融合，才有可能查明俯冲带地质过程和产物的形成机制。

（3）俯冲带多个复杂动力系统驱动俯冲带研究需要多学科交叉融合

俯冲带包括如下多个复杂的动力系统。①流体系统，指俯冲板片在俯冲变质过程中释放出多种流体，向上迁移并交代上覆地幔楔。此外，C-H-N-S 等关键挥发性元素也可以通过俯冲带再循环进入地球深部。这些流体系统运行期间都经历了复杂的物理、化学和流体动力学过程。②地震系统，由于地震绝大部分发生于板块边界，因此大洋岩石圈板块的俯冲以及大陆岩石圈板块的碰撞造山、侧向滑移是诱发强烈地震的主要地质过程。③俯冲隧道系统，俯冲隧道中除了俯冲刮削、俯冲侵蚀、沉积物底辟、高压－超高压变质岩折返等过程外，还有俯冲带上盘、下盘以及俯冲带物质间各种复杂的物理、化学交换过程。④岩浆系统，指岩浆从产生、源区分离到形成岩浆房或岩浆储库以及最后的侵位和喷发，其间经历了复杂的物理、化学和岩浆动力学过程。⑤成矿系统，指成矿物质从源区分离、演化到最后的富集、沉淀成矿所经历的物理、化学和成矿动力学过程。查明上述复杂动力系统运行机制，要求俯冲带研究必须利用地质学、地球化学、地球物理和地球动力学多学科交叉融合，才有可能明确俯冲带复杂动力系统的运行机制。

（4）地球科学的发展态势要求俯冲带研究范式发生变革

地球是太阳系类地行星中唯一具有水圈、生命活动和板块构造的行星。地球早期从停滞层盖构造转变到活动层盖构造可能导致了地幔物质向地壳物质的转化，而太古宙汇聚板块边缘加厚地壳的部分熔融使其成分出现由基性向酸性的转化，新形成的酸性地壳出露海面发生化学风化导致地球自由氧的增加，而火山喷发、地幔岩石蛇纹石化有可能导致氨基酸、有机大分子的形成和生命的诞生。因此，板块构造可能是导致地球向宜居性转化的重要驱动力。

地球科学在20世纪的突破在于板块构造理论，特别是大洋板块构造，但是对大陆地质还有许多基本问题亟待解决。进入21世纪以来，人们对板块构造的启动时间和运作方式有了更多的认识，发现以暖俯冲为主导的板块构造活动可能从始太古代（40亿～38亿年前）就已经开始，分别在古元古代中期和新元古代晚期出现局域性冷俯冲，到显生宙才出现全球性冷俯冲，由此划分出古代和现代板块构造（Zheng and Zhao，2020）。俯冲带是地球圈层在物质和能量上发生交换的关键场所，汇聚板块边缘大陆岩石圈的形成和演化是大陆板块构造的核心。如何将地球的各个圈层作为统一的整体来认识深部与浅部圈层间的联动与互馈机制、地球系统的历史演化过程及其对地球宜居性的控制作用，是地球科学研究中极具挑战性的前沿，亟须建立适应板块构造发展的地球科学理论知识体系和方法技术体系。俯冲带属性研究是建立板块构造研究新范式最有可能的领域。

俯冲带是板块构造理论的核心，俯冲带过程是地球圈层相互作用的核心机制和物质循环最重要的场所（Stern，2002；Zheng and Chen，2016）。大量物质通过俯冲带进入地幔甚至核幔边界，改变地球深部的物理化学性质和组成。同时，板块俯冲形成弧岩浆，形成大陆地壳和矿产、油气资源，同时也诱发了地震，并导致大规模火山爆发，释放了碳、氢、氮和硫等关键挥发分，影响大气环境。板块俯冲导致了地球宜居环境的形成，也给人居环境带来了影响。由板块俯冲引发的深部物质循环过程是地球内部的一级运行机制，主宰了地球从内到外、从古到今的演化进程。

开展板块俯冲带这一前沿领域的研究，必须要进行多学科的交叉研究，通过跨学科（地质学、地球化学、地球物理学、地理学、大气科学和海洋科学）、跨学部（如地球科学、化学、物理学、数学、生物学等）和多种方法（构造变形与盆地分析、同位素定年和示踪、地球物理深部探测、数值模拟与高温高压实验模拟、大数据分析等）的综合研究，有可能从基本原理和逻辑思维上取得创新，促进地球科学研究范式变革，推动地球科学向前发展。

第二节　板块俯冲带研究极大地推动了地球科学的进步

俯冲带研究极大地推动了地质科学、地球物理学与地球化学等学科的发展，同时也推动了微区原位高精度地球化学分析测试、高温高压实验、地球物理探测等技术的进步。这些技术反过来应用在其他学科中，推动了如比较行星学等其他交叉学科的发展。

一、促进了地质科学的发展

俯冲带研究极大地推动了地质学和地球化学领域的发展。在地质学研究方面，很多关键科学问题，如造山带的形成和演化、大陆的形成与再造、地幔不均一性、地球挥发分循环与表生环境变化等都和俯冲过程密切相关。对俯冲带结构、俯冲过程和产物的深入认识，极大地推动了对上述问题的深刻理解。

俯冲带研究加深了对地幔不均一性成因的认识。20 世纪 60 年代开始，地球化学家和地球物理学家就认识到了地幔的不均一性。这种不均一性主要体现在地幔分层，如上地幔和下地幔物质组成的差异，以及玄武岩同位素地球化学成分，如 He、Sr、Nd、Pb 和 Hf 等同位素揭示的原始地幔、亏损地幔、富集地幔等源区多样性。这种多样性常被归因于地幔源区中的再循环陆壳与洋壳物质组成的差异（Zindler and Hart，1986；Hofmann，1997；Stracke et al.，2005）。近年来，随着大洋钻探计划的持续开展和分析技术的变革，新的同位素地球化学手段，如 Os、W、Si、Mg、Fe 和 Zn 同位素等，进一步揭示了地幔不均一性存在多级尺度的规模和多种成因，如再循环地壳、岩石圈地幔和俯冲沉积物等组分影响（Kogiso et al.，2004a；Rampone and Hofmann，2012；Lambart et al.，2019）。地幔不均一性可能由俯冲到深部地幔的板片引起，但仍然不清楚大尺度地幔不均一性的形成过程和不同地幔端元成分的成因等重要问题（White，2010；Rampone and Hofmann，2012；Stixrude and Lithgow-Bertelloni，2012）。

俯冲带研究促进了对陆壳生长、再造过程和机制的深刻理解。陆壳生长不仅包括活动陆缘弧岛弧杂岩的侧向增生，还可以通过幔源岩浆底侵引起初始陆壳的垂向增生，伴随着陆壳内部化学分异。通常认为，陆壳生长主要是大洋俯冲过程的结果，而在大陆俯冲带主要是新生或者古老陆壳的再造，通过同碰撞或碰撞后镁铁质岩浆作用引起的大陆地壳生长是相对次要的（郑永飞等，2015；Zheng and Chen，2016）。但也有研究认为，由于俯冲带幔源物质添加到陆壳中的量与大陆物质进入地幔的量相当，俯冲带总体上不会发生地壳净增长，而碰撞带物质更易于保存，更有利于大陆地壳的生长（Stern and Scholl，2010；Niu et al.，2013；郑永飞等，2015）。对俯冲带物质迁移过程的深入认识，能更好地理解陆壳的物质来源，从而促进对陆壳生长、再造过程和机制的理解。

俯冲带岩石学和地球化学的研究，需要对矿物进行精细的微区原位高空间分辨率的元素和同位素分析，以及对微量样品的高精度同位素分析。为了适应这个需求，研究人员开发了先进的微区原位元素和同位素分析技术，极大地推动了诸多地球化学领域的不断进步。以激光剥蚀－等离子体质谱仪（LA-ICPMS）、激光剥蚀－多通道等离子体质谱仪（LA-MC-ICPMS）、离子探针 U-Pb 定年和稳定同位素分析、热电离质谱仪（TIMS）高精度同位素分析等为代表的前沿分析测试技术，大大拓展了地球化学的研究领域和应用范围。这些重大分析测试技术的突破，推动了如元素地球化学、同位素地球化学、化学地球动力学等领域的新发展和新跨越。可见，俯冲带的研究深入到地球化学领域的方方面面，为地球化学研究带来了新思想和新视角，是推动地球化学发展的“源头活水”。

尽管目前学界对地球上板块构造的启动时间和过程仍有很大争议，但不可否认板块构造在地球的演化过程中发挥了重大作用。俯冲作用作为板块构造的核心过程，对地球各个圈层演化都起着关键作用。俯冲带研究是前沿交叉学科，地球化学是实现多学科交叉的桥梁，发挥了独特作用。它不仅能提供俯冲带目前的物质组成特征，还能通过年代学和地球化学组成变化规律查明俯冲带的演化历史。同时，针对与俯冲带相关的重大地球科学问题，如地球内部挥发分含量及其对地幔对流的影响、地幔化学不均一性、俯冲带过程对地球大气和海洋成分演化的影响等，地球化学作为关键的研究手段之一，对这些影响提供了重要制约。例如，在洋岛玄武岩中发现硫同位素非质量分馏信号（Cabral et al.，2013），对地幔对流方式和效率、俯冲时间和壳幔相互作用等问题都提供了新的关键证据。总之，俯冲带的研究加深了对地球内部运行机制的认识，推动了地球化学的发展；同样，地球化学的发展也推动了俯冲带研究的深入。

二、推动了地球物理学的发展

俯冲带是全球构造最活跃的区域，主导了地球表层和深部的物质循环，并发育世界上最强最多的地震、最具破坏性的海啸、最猛烈的火山喷发，以及体量巨大的滑坡等自然灾害。因此，俯冲带不仅是认识全球构造和深部过程的窗口，也一直是人类识灾、防灾、减灾研究的重点地区。地球物理探测是认识和研究俯冲带结构的主要手段之一。俯冲带复杂的地质特征和构造过程，包括强烈的构造变形、地震、火山、流体活动，造成其结构从浅到深表现出不同尺度的复杂性和多样性。这就给地球物理探测带来了挑战，既要求对地震空间定位、浅部地壳关键构造带和俯冲界面等结构具有精细分辨能力（百米至千米级），又要求能够有效制约俯冲板片与上覆板块、地幔楔及周围地幔数百至上千千米范围的宏观结构和性质变化（十至百千米分辨率），还要求获取整个地幔深度俯冲板片形态的演变及其与数千至上万千米尺度对流地幔相互作用的结构证据（数百至上千千米分辨率）。

俯冲带多尺度、多分辨率的结构探测要求的提出，极大地促进了地球物理观测技术，如宽频全海深多波束技术、三维重磁场分离技术、三维便携式电缆地震反射技术、可控源大地电磁探测技术、多尺度层析成像技术等的进步。以探测地球内部结构最基本、最有效的地震学研究为例。为了实现对俯冲带地震活动的连续实时监测和高精度定位，对俯冲板片形态的高分辨率成像以及流－熔体运移改变壳幔结构和性质等的有效探测，要求获取宽频带、全方位的地震信号，并要求实施高密度、大孔径的地震观测。从 20 世纪 70 年代以来，地震仪器的快速发展，使得地震信号记录更丰富、更全面并具有更高精度。随着地震仪技术的发展，硬件成本降低，特别是便携式地震仪技术的成熟，极大地推动了全球范围的大规模密集流动地震台阵和固定台网观测，台站间距由早期的数百至上千千米小了一个量级，在一些俯冲带关键区域甚至达 10～20 km，而基于短周期节点式地震仪的密集流动观测台间距已缩小至千米甚至百米量级。针对俯冲带特定研究目标，海底地震仪观测、基于海底通信光缆的地震监测技术（Nishikawa et al.，2019）也逐步分别向漂移式海底地震仪、海底光纤网监测技术发展，为海域和洋陆过渡带地震活动和深部结构研究提供了更灵活、更经济、更智能的技术支撑。此外，为了获得全方位、高分辨的俯冲带地震监测和分析能力，已促成建立了全球地震台阵 / 台网、全球定位系统（global positioning system，GPS）观测、海啸测量、雷达卫星影像以及重力测量等多学科、多手段联合观测台网（Lay，2015）。

密集地震观测的开展和海量高精度数据的获得，进一步促进了地震资料分析和成像方法的进步，如先后发展了有限频层析成像、全波形反演、背景噪声干涉成像、融合勘探地震学和天然地震学的界面结构偏移成像方法等。此外，在基于密集地震观测开展多角度、全方位俯冲带结构研究趋势下，多资料（面波 + 体波）、多参数（波速 + 波速比 / 各向异性 / 衰减）联合地震反演技术以及多学科观测联合约束（地震 + 重力 / 大地电磁测深 / 热流 / 地形 / 形变 / 地质－地化）的思路和方法也应运而生。特别是随着越来越密集的观测积累，地球科学进入了大数据时代，应用人工智能、移动互联网等新技术开展俯冲带和其他地学领域的研究已成为当前国际学科前沿。

俯冲带孕育了全球绝大多数破坏性浅源地震（震源深度数千米至30～40 km）和中深源地震（震源深度70～700 km）。俯冲带地震时空分布和震源特征信息、发震机制和控制因素，是俯冲带研究的重要内容。近年来，基于密集地震台阵和大地测量观测的俯冲带地震震源相关研究取得了一系列新认识。例如，沿俯冲板块界面发震层的地震破裂行为表现出随深度和横向位置显著变化的特征，进而导致触发地震的大小、诱发海啸的能力等方面存在明显区域差异（Lay，2015）。再如，对俯冲带的研究发现了慢地震现象，即介于正常地震快速破裂和缓慢蠕变之间的断层活动（Peng and Gomberg，2010）。它与正常地震不仅沿俯冲板块界面存在深度互补关系，而且在水平方向也存在互补关系，体现了沿俯冲板块界面断层力学行为的空间差异性。这种差异性不仅与俯冲板块界面本身特征（如洋底地形或沉积厚度）有关，而且受到上覆板块结构和刚性的影响，并与俯冲带温压结构、流体活动及其控制的介质流变性质紧密联系（Gao and Wang，2017；Nishikawa et al.，2019；Halpaap et al.，2019）。

因此，对俯冲带正常地震和慢地震及其发生机制的细致研究，对深入认识俯冲带浅部（<60 km）结构和物理化学性质具有重要意义；而厘清慢地震与正常地震的相互关系，也有望在揭示俯冲带地震活动规律、预报大震及相关地质灾害（如海啸）等方面取得突破。中深源地震是发生在俯冲带的特有地质现象，其空间分布可有效约束俯冲板片的深部形态，而发震机制则是研究俯冲带动力状态、物质组成、含水量、矿物相变等问题的基本依据。在板块构造理论建立之初，通过中深源地震震源机制研究获得的俯冲带深部动力特征，就为俯冲板片自身重力是地表板块运动重要驱动力的论断提供了地震学证据。但中深源地震的具体力学成因问题一直存在争议（Guest et al.，2003），一些研究提出了包括脱水脆化、塑性剪切失稳、相变致裂等观点，也有研究者认为复杂的中深源大地震可能是由多种机制共同作用而诱发产生的。尽管尚未形成共识，但对中深源地震发震机制的研究已成为理解俯冲过程中板片物理化学性质变化和物质运移的重要途径。

三、催生了内生与表生交叉研究的新领域

板块俯冲带极大地改变了地球内部和表面的运行方式，对地球表层环境产生重要影响。板块俯冲带是地球散热的重要方式，促进了地幔的冷却，在

维持地球表层温度稳定性方面扮演着至关重要的角色。板块构造中的俯冲作用、脱挥发分过程和岩浆作用是连通地球内部和表面的重要通道，促进了水和多种生命必需元素的再循环过程；板块构造带来的表生作用，在漫长的地球演化历史中不断调节大气的成分（CO_2、O_2、CH_4 等），导致地球逐渐演化为宜居行星。

大陆地壳的形成和演化是从表层识别地球的独特标志之一，是地球内部动力学的响应，与俯冲作用息息相关。表生圈层与深部圈层相互作用深刻影响了地球的大气和海洋组成，同时也为陆生生物提供了栖息地，与生命演化息息相关。例如，在新太古代—古元古代地球构造体制的转型期，地球的表生环境也发生了巨大的变化，如大氧化事件和第一次雪球地球事件。地球大气从无氧状态迅速演化到现今大气氧含量的 10%～20%，并促进了真核生物等的大量出现（Lyons et al.，2014）。大氧化事件的主导原因还不清楚，但地球内部的动力学机制必然起了非常大的作用，如大陆地壳的成分变化、太古宙俯冲方式改变、火山喷发模式差异等机制可能造成了大气的氧化。

伴随着板块俯冲的进行，紧邻俯冲带位置的海沟、海沟斜坡、弧前和弧后盆地记录了深部物质与表层物质的动态演化过程，共同构成了俯冲体系的沉积储库（Dickinson，1995；Zheng，2021b）。进入海沟的俯冲沉积物主要包括两种来源：一种来源是俯冲大洋板块从海沟下潜时被上盘刮削下来的深海洋盆物质的碎片，包括洋中脊玄武岩、远洋风成软泥、硅质岩，以及海山型玄武岩、灰岩等，在海沟处形成增生楔和混杂岩；另一种来源为海沟形成后通过浊流、滑塌、等深流等作用充填海沟的沉积物，包括深海黏土、陆源碎屑、火山灰、生物物质、海洋自生物质和宇宙物质等。大洋钻探对全球海沟附近大洋沉积的定量计算表明，现今海沟系统中的各沉积物组分包括：陆源碎屑（黏土和砂岩）每年有 10.8 亿 t，占总沉积物通量的 75.5%，碳酸钙、蛋白石输入通量分别为每年 2.2 亿 t（占 15.4%）、1.3 亿 t（占 9.1%）（Rea and Ruff，1996）。海沟盆地陆源碎屑物质通常来自上覆板块，可由上覆板块腹地、岩浆弧或俯冲增生楔提供，多为中粗粒陆源碎屑沉积，指示大陆来源的物质，明显区别于洋壳的物质组成。俯冲带沉积物记录了当时岩浆喷发、剥蚀、地震、滑坡、搬运、剥露过程，是当时表生海洋－大气环境综合的产物。对这些沉积盆地的研究，极大地促进了内生作用与表生过程及其相互作用的研究。

除沉积作用外，活动的海沟也发生沉积物俯冲消减作用，形成俯冲型造山带。造山带隆升引起构造－气候耦合的风化过程，通过碳－硅循环过程不仅反馈消耗大气圈组分，进而对全球气候变化有重要的影响；而且上覆板块的物质，通过前端侵蚀或底部侵蚀的作用还会进入俯冲带深处（Noda，2016），伴随关键元素的深循环。统计表明，在现代大洋俯冲带内，沉积物俯冲作用表现有所区别：约 8200 km 的俯冲带内约 70% 的沉积物被俯冲至增生楔之下，在东太平洋发育较大规模的俯冲增生楔；约 16 300 km 的俯冲带内约 80% 的沉积物发生俯冲，发育中小规模的俯冲增生楔；而约 19 000 km 的俯冲带内沉积物完全发生俯冲，在西太平洋俯冲增生楔不发育（von Huene and Scholl，1991）。这些俯冲的物质通过俯冲带被带入地幔，深刻影响和改造地幔物质组成，导致地幔不均一性，同时也决定了火山弧物质的组成（Plank and Langmuir，1993；Zheng，2019）。另外，沉积物质俯冲的多寡还影响着地震的震级，并决定地震是否诱发海啸等（Geersen，2019）。

俯冲带内物质流和能量流的交换也反馈到水圈及大气圈，构成闭环的俯冲再循环（李三忠等，2020）。俯冲板片随着温度、压力升高，释放水和其他挥发物，部分流体和气体返回水圈和大气圈，部分流体和沉积物则加入火山弧下的岩浆源区，并作为火成岩组分返回大陆或海沟系统。俯冲再循环过程中释出的水和挥发物，深刻影响着弧火山活动。另外，俯冲板片释放的元素和流体，不仅控制了海沟系统中热液、冷泉系统的分布和产物，而且涉及矿床的形成过程。俯冲带上方往往形成富含金、银的热液矿床。在海沟系统的最浅部，水和碳通量影响到水合物的形成和破坏，并为海底深部生物圈提供营养。这一再循环过程也改造了俯冲板块和上覆板块，涉及地表和地球内部的所有圈层（Zheng，2019）。

总之，作为俯冲工厂的浅部构成，海沟系统再循环释放的流体进入海洋，影响到海洋中化学成分的收支平衡，滋养着一些极端环境下的生物群落。释放的 CO_2 等气体和微量元素进入大气圈，对大气圈的成分也产生重要影响。源于下插板片的水等流体，导致上覆地幔楔乃至增生楔部分熔融，年轻的俯冲板片本身也可以部分熔融，之后经冷却或喷出，形成埃达克岩等，由此产生的弧岩浆活动形成了新的陆壳。经俯冲工厂加工改造后的大洋板片下潜至地幔深处，甚至抵达核幔边界，并且部分通过深部循环加入上涌的地幔底辟中；下潜的板片连同残存的挥发物及生热元素，会对地幔的流变性质、氧逸度和对流造成深刻的影响，进而影响火山喷发产物及海沟系统沉积建造类型。

四、加速了高温高压实验科学的新变革

俯冲带研究也显著促进了高温高压实验地球科学的发展。俯冲带广泛发育的岩浆作用、变质、变形、成矿、成灾等地质过程，地震波波速各向异性和温压结构等地球物理现象及其各种物理化学过程，为高温高压实验提供了丰富的模拟对象和应用场景。实验学者瞄准俯冲带物质组成、结构和构造、过程和效应的诸多科学问题开展了深入研究，这些研究包括深俯冲洋壳和陆壳的矿物组成及俯冲最大深度、俯冲相关的壳幔相互作用和挥发分循环、俯冲带元素配分和迁移、俯冲带波速结构、地震活动特征及成因等。

为了解决上述关键科学问题，实验地球科学领域不断涌现出新的技术和方法。早期的高温高压实验一般在常压或近地表的低压下进行，且不含碳、氢、氮、硫等挥发性组分。然而俯冲带是一个富含挥发分的地质环境，且很多过程发生在地球深部的高温高压条件下。俯冲带的流体活动和部分熔融十分发育，高温高压下硅酸盐－水体系的熔体和流体（特别是超临界流体）具有很强的物理和化学活性。这给实验研究带来了巨大挑战。实验学者迎难而上，开发了以离心式超重力活塞圆筒压机、摇摆式多面砧压机、金刚石阱结合冷冻样品台 LA-ICPMS 分析、基于多面砧压机平台的 DDIA 流变仪、2000t 以上级别的大腔体压机为代表的多种变革性实验平台和技术，在俯冲带物质组成、部分熔融、流体行为、元素配分、流变和波速结构等方面的研究中取得了重要突破。

将高温高压实验技术与多种现代分析测试技术，特别是与同步辐射相关分析技术相结合，实现在高温高压条件下对实验体系进行原位观察和测量，更已成为俯冲带相关实验研究的重要发展趋势。实验学者开发了多面砧压机和金刚石压腔中高压矿物（特别是致密含水富镁硅酸盐）同步辐射 X 射线衍射谱原位测量等技术，这为认识俯冲板片矿物相变奠定了基础。人们还开发了多面砧压机或金刚石压腔与同步辐射 X 射线吸收谱相结合的技术，实现了在高温高压条件下原位测定物质的密度和流变强度、元素价态和配位状况。同时，开发了在活塞圆筒压机、多面砧压机和金刚石压腔中，矿物、熔体和流体电导率、热导性质和波速的多种原位测量实验方法。这些实验技术的成功开发为深化对俯冲带的成分、结构、构造和过程的认识奠定了基础。

俯冲带研究的旺盛需求还推动实验地球科学向更广、更深方向发展，形成了实验岩石学、实验地球化学、实验矿床学、矿物物理学、实验流变学等多门目标有别但又紧密相关的分支学科。实验岩石学和实验地球化学的温

度及压力条件相对较低，但研究体系一般相对接近真实地质环境，往往涉及多种矿物组分和挥发分。相对侧重相平衡和相转变及主量元素特征研究的实验岩石学，实验地球化学更侧重微量元素和同位素研究。实验矿床学作为实验岩石学的分支，以认识岩浆矿床与岩浆热液矿床的成矿机制为主要研究目标。矿物物理学一般以单矿物的物理和化学性质（密度、相变、波速、电导、热导）为主要研究对象，实验温度和压力可以达到地核的条件。实验流变学研究矿物和岩石在差异动力条件下的流变强度和变形行为，目前能达到的温度和压力条件不高。总之，实验地球科学的发展极大地促进了地球深部物质研究的新发展，为解释俯冲带地质学和地球物理观测结果提供了重要基础，为深入认识俯冲带物质组成、结构和构造、地质过程和动力学成因提供了必要依据，使俯冲带研究变得更加系统和严谨。

五、夯实了比较行星学等新兴学科的理论基础

比较行星学脱胎于地球科学，形成于地球科学、天文学、空间科学等学科的交叉，其实质是以地球为基准，对比研究各行星的物质组成、表面特征、物理场、内部构造和演化历史的学科。基于此学科的特点，人类对地球本身的理解深度直接决定了对其他行星理解的高度，而对其他行星的探索不仅能够验证人类已有的知识体系，还能反过来促进人类对地球本体的理解，这也是比较行星学学科的重要意义。当前的比较行星学发展以地球为参照，俯冲带的研究成果对比较行星学有较大的推动作用。

在固体地球科学研究中，一个前沿问题是板块构造何时且如何启动（Condie and Kröner，2008；Hawkesworth and Brown，2018；Stern and Gerya，2018）。关于板块构造启动的时间，争论的时间范围可以从冥古宙（40 亿年前）到新元古代（5.41 亿年前）（Korenaga，2013；Zheng，2021a），主要争论焦点是俯冲构造与扩张构造的启动时间和规模，其中对太古宙某个时期启动的讨论较多。对板块构造启动的机制也存在多种模型，如层盖破裂、冷却自发、地幔柱触发、星体碰撞等（Tackley，1998；Gerya et al.，2015；O'Neill et al.，2019；Tang C A et al.，2020），主要争论焦点是停滞层盖向活动层盖转变的构造原因及其识别标志。Zheng 和 Zhao（2020）基于板块边缘流变学性质与俯冲带地热梯度之间的对应关系，将地球历史划分为前板块构造时期（≥40 亿年）、古代板块构造（40 亿～5.41 亿年）和现代板块构造（5.41 亿年至今）三大阶段（图 1-6）。查明地球早期浅部地质记录最关键的特征，识别俯冲带的关键特征和效应，进而反演地球早期海洋和大气组成，并将其与

现有类地行星的表面观察结果进行对比，不仅能够回答俯冲带何时在地球上出现的问题，而且有助于验证已有行星演化模型（Moore and Webb，2013；Bédard，2018）。两者的相互比较、相互验证有利于人类对行星演化的整体认识，也有助于理解板块构造的启动机制。

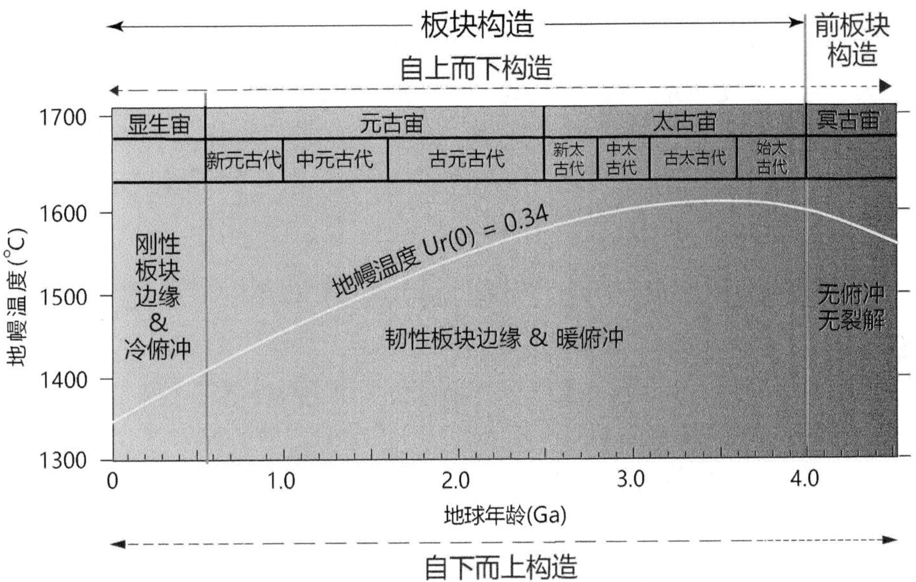

图 1-6　板块构造体制随地球热状态演化变化示意图（修改自 Cawood，2020）

用于热模型的周围地幔 Urey 比值 [Ur(0)] 为 0.34（取自 Herzberg et al.，2010）。板块构造不仅包括岩石圈俯冲这类自上而下构造，而且包括软流圈及其衍生物质上涌这类自下而上构造。在地质历史上板块俯冲深度主要受控于地幔温度（Zheng and Zhao，2020）：显生宙时期地幔温度较低，岩石圈边缘处于刚性状态，板块冷俯冲可以到达地幔深部。太古宙时期地幔温度较高，岩石圈边缘处于韧性状态，板块暖俯冲只能到达地幔浅部

从当前的空间探测结果来看，地球是太阳系行星中唯一具有活跃板块构造的行星，而其他行星（如火星）大多存在过板块构造活动的痕迹或表现（Yin，2012; Kattenhorn and Prockter，2014）。地球的独特性在板块构造，板块构造的独特性在俯冲带。因此，对俯冲带的研究为比较行星学提供了更广阔的思路和探索空间。固体地球科学对地球的研究注重从地球浅部的记录来认识地球内部的演化。板块构造框架下板块俯冲带是地球物质能量交换最强烈、火山 - 地震活动最活跃、矿产资源最丰富的地区，也是地球科学研究内容最集中的地区，因此有学者将板块构造称为俯冲构造（Anderson，2001; Stern，2007），甚至将地球孕育生命的本质归因于地球板块构造的活动（Korenaga，2012）。

板块构造时期的地质记录和特征，是判断其他行星是否存在过板块构造最关键的依据。太阳系其他类地行星（如火星）早期是否存在板块构造？如果其他行星存在过板块构造，为什么板块构造停滞了？厘清这些问题有助于理解地球历史上层盖构造范式的变化。从现有的观测可知，其他行星存在生命的可能性很低，而具有板块构造的地球发育大量不同形式的生命。板块构造对宜居星球的形成和演化到底起到了什么作用？针对这些问题，学界建立了地球化学成分分析测试、地球物理探测和地球动力学数值模拟等前沿技术，积累了对俯冲带研究的丰富经验，提供了关于地球内部运行机制的新认识，这些都为行星科学等新兴学科的孕育和发展提供了绝佳的便利。

第三节　板块俯冲带是国家自然科学的关键交叉学科

俯冲带研究由于其重要意义，在国家自然科学总体学科发展布局中具有重要的地位。它是典型的多学科交叉的前沿学科，其研究领域分布在地质学、地球化学、地球物理学等多个方向。在与众传统学科的交叉中，不断衍生新的前沿发展方向，如比较行星学、地球宜居性的形成和演化、地球深部过程与表层响应、地球深部挥发分循环、极端环境下的生物圈等，推动了传统学科的发展。

一、在国家总体学科发展布局中的历史定位

板块俯冲带研究需要多学科交叉。板块俯冲带构造复杂，矿产资源集聚、灾害多发，因此它历来是固体地球科学包括大陆动力学、矿床学、地震学的重点研究地区。

中国是造山带众多的国家，中国大陆就是小陆块（克拉通）被各类造山带拼接起来的，著名造山带有华北中部古元古代碰撞型造山带、江南新元古代增生型造山带、中亚古生代增生型造山带、秦岭－祁连－昆仑古生代－中生代增生－碰撞型造山带、大别－苏鲁中生代碰撞型造山带、喜马拉雅新生代碰撞型造山带。甚至一些陆内构造过程，如华北克拉通破坏、华南下地壳再造等，都与周边板块的俯冲作用存在密切联系。

板块俯冲带研究是地球科学的重要部分，在整个地质学、地球化学、地球物理学学科中一直处于引领地位。俯冲带与关键金属矿床关系密切，一直

是中国矿床学研究和勘探的重要目标区域。俯冲带的地震和火山活动不仅对人类社会发展有重要影响，而且提供了探索深部地质过程的关键信息，是地球物理学研究的重要内容。因此，与俯冲带研究相关的基金申请占比一直很高，据估计占整个固体地球科学的40%左右。1991～2021年，中国科学院增选的地学部院士中，与板块俯冲带研究有关的院士占新当选院士总数的29%。这足以显示板块俯冲带学科的重要性和引领性。

二、在国家总体学科发展布局中的当代定位

板块构造理论为20世纪自然科学的重大进展之一，而俯冲带在板块构造中起主导作用，是地球科学领域的重要研究方向和热点。从板块构造理论诞生以来，全球科学家在地质学、地球化学、地球物理学等不同领域对俯冲带进行综合研究。在中国，青藏高原、中央造山带（祁连－秦岭－桐柏－红安－大别－苏鲁）、中亚造山带、中国东部大陆边缘等与板块俯冲有关的岩石学、地球化学、成矿作用等都是受到重点资助的研究区域。俯冲带科学研究已经逐渐扩展到板块构造起源和早期地球演化、俯冲带与矿产资源、俯冲带碳循环、大陆生长和地球动力学、地球不同圈层相互作用和深部过程、俯冲带与环境变迁、宜居地球形成和演变等不同方面。

进入21世纪以来，国家自然科学基金委员会和科学技术部先后部署了重大研究计划、国家重点基础研究发展计划（973计划）和深地项目，重点进行了以下与俯冲带相关领域的深入研究和探索：①华北克拉通破坏过程和机制；②大陆俯冲带壳幔相互作用；③中亚造山带构造演化及其成矿作用；④西太平洋俯冲带动力学系统及中国东部岩浆作用和成矿作用；⑤喜马拉雅造山带俯冲碰撞动力学机制、青藏高原隆升过程和资源环境效应；⑥特提斯构造域的形成演化、资源和能源效应、古环境效应与生物演化、多陆块裂解和聚合动力学。国家自然科学基金委员会在“十三五”期间，重点支持了矿产资源和化石能源形成机理研究、地球环境演化与生命过程研究、地球深部过程与动力学、人类活动和地球环境相互作用机理与调控研究、典型地区圈层相互作用与资源环境效应研究、全球环境变化与地球圈层相互作用研究、重大灾害形成机理及其减灾对策研究。这些研究都与板块俯冲带和地球动力学密切相关。因此，俯冲带动力学研究与其他学科的交叉将主导今后地球科学的发展，有助于解决与资源供应、人类生存环境等相关的重大地球科学问题。

根据中国科学院地学部的院士名录，自1991年以来当选的、与俯冲带科学研究相关的地学部院士中，其研究领域包括俯冲带变质作用、岩浆作用、地球化学、成矿作用、几何结构和大地构造等研究领域。自1994年以来，与俯冲带研究相关的杰出青年科学基金获得者专业包括岩石学、地球化学、矿床学、地球物理、构造地质学等，约占地学部杰出青年科学基金获得者总数的1/6。可见，板块俯冲带学科从20世纪末以来一直处于固体地球科学的核心位置。

三、在国家总体学科发展布局中的未来定位

俯冲带问题既是当代地球科学的全球性研究热点和前沿，也是中国国家自然科学基金委员会未来“三深一系统”战略中深地、深海、地球系统研究的重要突破口。俯冲带研究不仅是揭示地球物质深部循环过程的重要研究途径，更是探索地球系统演化规律的“金钥匙”。古老俯冲过程中伴生的成藏成矿作用对保障国家能源安全、解决国家战略性矿产需求有着至关重要的意义。现今俯冲带研究还涉及地震、海啸、海底滑坡等重大地质灾害的预防、预测、预报和预警，关系到人民生命和财产安全，为公众所关注。

从20世纪80年代以来，中国在陆－陆俯冲带研究领域取得了世界性成就，但对洋－陆俯冲带和洋－洋俯冲带的研究依然相对薄弱。当前，中国急需走向深海大洋，其中西太平洋是中国的战略必争海域，也是全球俯冲带的集中海域。在深海探测技术上，2010年以来中国实现了“蛟龙”探海，“科学号”和“东方红3号”综合科考船世界领先，而且中国大洋钻探船“梦想号”正在快速建设当中，深海获取新鲜样品能力将一跃进入国际先进行列。近年来，实验平台的建设使中国科学家对样品的化学分析能力大幅提升，世界领先的超算设施也为俯冲过程的计算模拟提供了强大的平台保障。同时，在人才储备方面，中国培养了一支高素质地质、地球化学和地球物理专业人才队伍。这些都为深入研究各种类型的俯冲带提供了良好条件。

岩石圈俯冲是地球演化的加速器，是驱动板块运动的发动机，深部碳循环和深部水循环是俯冲工厂的催化剂。在深海科学发展的新背景下，以地球系统科学理念统筹多学科优势，发挥陆－陆俯冲带领域的研究人员优势和研究积累，开拓洋－陆俯冲带和洋－洋俯冲带研究新领域，发展俯冲系统、海沟系统理论体系，在国际深海科学、地球系统科学新领域抢占先机，形成诸多重大前沿发展方向，建立中国科学家提出的地球系统科学理论，尤为迫切。

新的发展趋势为俯冲带研究带来了新的挑战，同时也提供了解决学科重

大问题的新机遇。在技术红利的推动下，理论上的新突破往往能够应运而生。为此，围绕国家发展战略和科研走向，未来的俯冲带研究要拓宽思路、集成方法，注重原始创新研究。

1）多学科交叉开展原始创新研究。发展陆－陆俯冲带、洋－陆俯冲带深部高精度层析成像技术、电磁探测技术、剪切波分裂技术等，建立电、震、热、磁等结构特征，以约束俯冲带深部精细三维结构、流体分布结构，为俯冲过程的数值模拟提供更具体约束。随着计算与模拟能力的提升，针对俯冲带的高分辨率大规模数值模拟研究已逐步展开。

2）俯冲带多样性和共同性研究并举。布局开展典型俯冲带对比研究，选择马里亚纳海沟洋－洋俯冲系统、巴布亚－新几内亚大火成岩省－岛弧俯冲系统、琉球海沟洋－陆俯冲系统、大别－苏鲁陆－陆俯冲系统、东亚环形俯冲系统等开展深入调查和对比研究。同时，有必要平行开展对应海沟的生命科学研究，以认识深渊生物多样性演替规律，揭示其代谢互作机制、生命特征、环境适应机制及生态效应，丰富人类对深部生物圈的理解和对深渊生命科学的认知，提升中国在深渊生命过程研究领域的总体水平和国际地位。

3）坚持以地球系统科学为理念开展集成创新。深入元素循环机制与圈层相互作用、生命元素循环、热液和冷泉的环境效应、极端生命过程研究，开展生命科学与地球科学的跨学科交叉，实现俯冲带研究的系统化和整体化，更好地满足学科发展和国计民生需要。这要求俯冲带研究更加多样化、交叉化、集成化，在大数据支撑下技术复合化。通过大数据技术支撑，未来俯冲带研究能够将系统的科学知识与海量的地质数据转化为生产资料，推动相关产业革新，催发相关技术出现。未来俯冲带研究不仅要解决地质学的基础问题，也要解决地球系统科学的关键问题。它能够逐步揭示板块水平运动所反映的地球内部演化，也为地球系统科学带来了新的研究方向。

4）实施典型深海沟系统国际大科学计划。推动海沟对比国际计划，利用各国学科优势和地域特色，开展综合俯冲系统的立体综合调查，利用新一代载人深潜器实施海沟分段式系统调查，凝练深潜的科学目标，利用钻探平台实施海沟分段式系统采样，采用多种地球物理探测技术实现海沟分段式系统三维结构、水分布、地热梯度等物理属性调查。

四、对国家中长期科技规划和科技政策的支撑作用

未来10～20年是全球经济、政治、军事和创新格局深刻变革期，是中国从大国迈向强国的战略机遇期。党的十九大报告中提出了诸多具体领域

的强国目标，如科技强国、海洋强国、人才强国等。在新的历史条件下，2021～2035年国家中长期科技发展规划和“十四五”规划明确，科技创新是高质量发展的强大动能，中国科技创新事业未来10～20年将进入一个快速发展的关键阶段，国家中长期科技发展要聚焦科技前沿、国家重大战略需求和国民经济主战场。俯冲带研究不仅面向国际科技前沿，同时也契合国家发展的战略需求，能够为社会经济发展提供重要科学支撑。

（1）加强俯冲带研究是推动地球科学加速发展的关键

基于大洋岩石圈研究发展起来的板块构造理论是当今固体地球科学最重要的理论。由于大陆岩石圈和大洋岩石圈在成分、厚度和力学强度上存在明显的差别，现有的板块构造理论主要是以大洋地质为基础，还不能完全解释大陆构造问题，需要发展大陆板块构造理论。将地球的各个圈层作为统一的整体来认识深部与浅部圈层间的联动与互馈机制、地球系统的历史演化过程及其对地球宜居性的控制作用，是俯冲带科学研究中极具挑战性的前沿。

地球是太阳系类地行星中唯一具有水圈、生命活动、长英质大陆地壳和板块运动的行星。板块构造是导致地球从非宜居向宜居转化的重要驱动力。板块构造理论深刻揭示了地球内部的运作机理并影响地球表层的宜居性，而俯冲带过程是地球圈层演化的核心机制。板块俯冲带与地球壳幔系统演化和物质循环、大气和海洋的长期化学演化以及地球宜居性等重大地球科学问题都紧密相连。对板块俯冲带的系统深入研究，对深入理解地球内部运作机制及其表层响应具有重要科学意义。俯冲带的研究在推动板块构造理论发展和地球系统科学理论的完善中发挥了不可替代的作用。因此，紧紧围绕俯冲带开展研究，将是推动中国地球科学研究未来加速赶超并引领国际前沿的关键切入点。

（2）俯冲带研究为国家发展战略提供重要科学支撑

国务院2016年印发的《“十三五”国家科技创新规划》面向2030年“深度”布局，要构筑国家先发优势，围绕“深空、深海、深地、深蓝”，发展保障国家安全和战略利益的技术体系。其中，前“三深”与地球科学密切相关，就是要探明宇宙天体、地球海洋、地球深部可利用资源，为国家的资源安全提供重要保障。人类社会的生存和发展离不开矿产资源、宜居环境。俯冲带研究与国家的深空、深海、深地战略密切相关。这是因为俯冲带是重要的洋陆转换带，不仅是大洋、大陆物质进入地球深部地幔甚至核幔边界的重要位置，而且也是矿产与油气资源形成、火山喷发和地震活动的主要场所，与人类的生存和发展密切相关。因此，对俯冲带的深入研究，可以揭示地球深部物质组成、性质以及深部动力学与物质循环过程，从而为揭示矿产与油气资源形成背景、火山喷发、地震活动机理提供重要证据，为资源勘探、火山与地震灾害预防提供理论指导，为国家发展战略需求和社会经济发展提供重要科学支撑。

第四节 板块俯冲带研究关系国民经济与国家安全的战略布局

俯冲带作为稀有、稀土、稀散等金属矿产的主要聚集地，在国民经济发展与国防安全方面都具有重要的战略意义，在解决“卡脖子”矿产资源问题、保障国家战略安全等方面具有重要作用。另外，俯冲带广泛发育地震活动和火山活动，这也深刻影响了人们生活和生命财产的安全。

一、俯冲带相关矿床保障了国民经济的发展

俯冲带从构造上讲属于造山带的一部分，中国是全球唯一三大造山系均有分布的国家，包括中亚古生代－中生代造山带、特提斯古生代－新生代造山带和环太平洋中生代－新生代造山带。三大造山系产出了大量的多种类型矿床，与中国较具特色的华南陆内矿产和华北克拉通前寒武纪矿产一起，成为保障我国国民经济发展的资源支撑。从全球角度来看，与俯冲带直接相关的矿床类型和矿产资源更是对人类近现代经济和工业的发展起到了举足轻重的作用。

中华人民共和国成立以来，随着中国整体工业体系的建立和一系列五年计划的连续实施，国民经济对矿产资源的需求不断攀升，矿产资源受到的重视程度不断提高。各种大宗矿产（如铜、铁）、贵金属矿产（如金、银）和稀有金属矿产（如锂、铍）等，在不同时期都成为地质研究和勘查工作的重点，在国家科技项目资助体系中占据了重要位置。特别是近年来，科学技术部实施的各类 973 计划和“深地资源勘查开采”国家重点研发项目，国家自然科学基金委员会实施的大量重点研究项目和“战略性关键金属超常富集成矿动力学”重大研究计划，以及自然资源部实施的公益科研项目等，都对中国重要矿产资源的科学研究和勘查应用进行了持续的强有力支持。其中，对中亚（新疆－兴蒙）、特提斯、环太平洋（东南沿海、东北东部）三大造山系中与俯冲带相关矿床的支持占有非常重要的地位。随着中国“一带一路”倡

议的推进，与境外俯冲带和相关成矿带的综合对比工作也得到了国家不同层面的大力支持，并在近年来取得了显著成效。

俯冲带相关矿产资源的勘探和开发，促进并保障了中国现代化工业生产与经济的飞速发展，其中最为显著的实例是中国铜矿资源在近年取得的突破性进展。数据显示，中国铜金属查明资源量自 2006 年的约 7000 万 t 迅速提升到 2018 年的 11 400 万 t，增长了 62.86%。这些铜矿储量的增多主要来自青藏 - 三江和新疆 - 兴蒙等重要造山带，并且大多与俯冲带成矿作用密切相关。需要指出，中国正处于高质量发展时期，目前已有的铜矿资源量不能满足工业化需求。但这些俯冲带铜矿资源的重大发现，无疑极大地缓解了中国在重要金属资源上的紧缺局面，一定程度上保障了国民经济的高速发展。

二、俯冲带地质灾害是国家安全和国家战略布局的重要考量

中国大陆被多个俯冲带或俯冲碰撞带所包围，东边为西太平洋板块俯冲带，东南边为菲律宾海板块俯冲带，西南边为印度板块与欧亚板块的俯冲碰撞带。板块的俯冲及碰撞在中国及其周边地区产生了一系列地震和火山活动，如何防范和应对这些地震和火山活动对于中国的国家安全和国家战略布局具有重要的影响。

板块的俯冲运动导致地下介质动力应变累积，容易形成大地震，造成严重的地震灾害及其次生灾害。西太平洋板块俯冲所产生的强震主要分布于日本岛弧地区。西南边的印度板块与欧亚板块的俯冲碰撞，以及东南边菲律宾海板块与欧亚板块的俯冲碰撞，都有可能对中国安全造成严重影响。

在印度板块与欧亚板块的俯冲碰撞前缘喜马拉雅碰撞带，经常会发生超过 7 级的强震，如 2015 年的 7.8 级尼泊尔大地震和 1950 年的西藏察隅 8.6 级大地震。虽然西藏地区人口相对稀少，西藏察隅 8.6 级大地震仍然造成地表建筑物的强烈破坏和 3000 多人的死亡，并造成一些重大的次生地质灾害，如山体滑坡，甚至山体崩塌堵塞雅鲁藏布江。由于西藏地区有中国重要的军事设施和基地、水电站、铁路和公路运输要道，所以要尤其关注印度板块俯冲碰撞所造成的强震对这些基础设施的破坏。这些基础设施建设的选址要特别注意避开强震风险区和可能的次生灾害发生区域。

菲律宾海板块与欧亚板块的俯冲碰撞每年都在台湾地区造成了一系列的中强地震。例如，1999 年的台湾集集 7.6 级地震，造成 2000 多人死亡，经济损失达 92 亿美元，同时强震也波及福建、浙江、广东等沿海地区。此外，由于南海块体向菲律宾海之下俯冲，在中国南海东缘的菲律宾海域也时常发生

中强地震；虽然离中国大陆距离较远，但要十分警惕俯冲带地震引起海啸而造成次生灾害。由于东南沿海是中国经济最发达的地区，有大量的城市群和密集的人口，同时还有中国重要的基础设施，如大亚湾、漳州、福清等地核电站，以及密集的城际高铁网和沿海海底电缆设施，所以要密切关注由菲律宾海板块俯冲所导致的中强地震灾害及其可能引起的海啸等次生灾害。在东南沿海地区构建高效、稳定、可靠的地震预警及海啸预警网尤其重要。

除地震外，俯冲带也是火山喷发密集发生的场所。由于俯冲带的岩浆通常比较富含挥发分（一般以水和 CO_2 为主），岩浆在上升过程中会发生挥发分饱和出溶，气泡生长甚至会导致岩浆房破裂，形成对人类和环境危害巨大的气驱爆发式火山喷发。1991 年 6 月，位于欧亚板块和菲律宾海板块边界的皮纳图博（Pinatubo）火山喷发是 20 世纪最大的两次火山喷发之一，导致地球进入两年的火山冬天。中国东北地区近 1000 年来多次发生较大规模的火山喷发。这些火山虽位于大陆内部，但一般认为与西太平洋板块俯冲之间存在密切成因联系。长白山天池和黑龙江五大连池等火山重新活化对中国东北地区安全造成的风险需要予以重视。中国西南地区的腾冲火山也与印度板块东向俯冲到缅甸下方密切相关，需要密切关注其活动性。因此，构建长白山天池、五大连池、腾冲等活火山的有效的监测预警网对预防火山灾害具有十分重要的意义。

第二章 发展规律与研究特点

第一节 板块俯冲带的学科定义与内涵

俯冲带通常指一个板块俯冲于另一个板块之下的构造带，亦称消减带或贝尼奥夫带。根据俯冲板块的地质学属性，一般将俯冲带划分成两大类：①大洋俯冲带，又称 B 型俯冲带，是大洋岩石圈板块俯冲的产物（图 2-1），在太平洋周边造山带之下最为典型；②大陆俯冲带，又称 A 型俯冲带，是大陆岩石圈板块俯冲的产物（图 2-2），在西阿尔卑斯造山带和大别－苏鲁造山带之下最为典型。

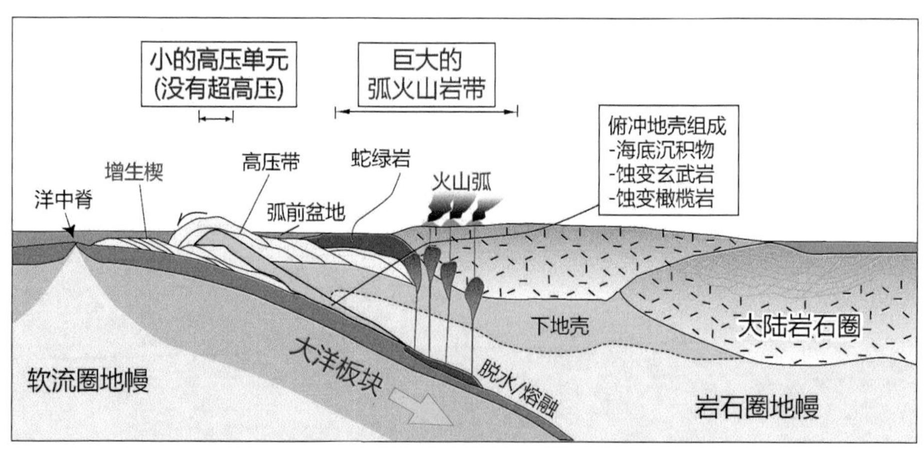

图 2-1 大洋俯冲带结构及其常见产物示意图（修改自 Liou et al.，2009）
岩石圈地幔之上的都是大陆地壳

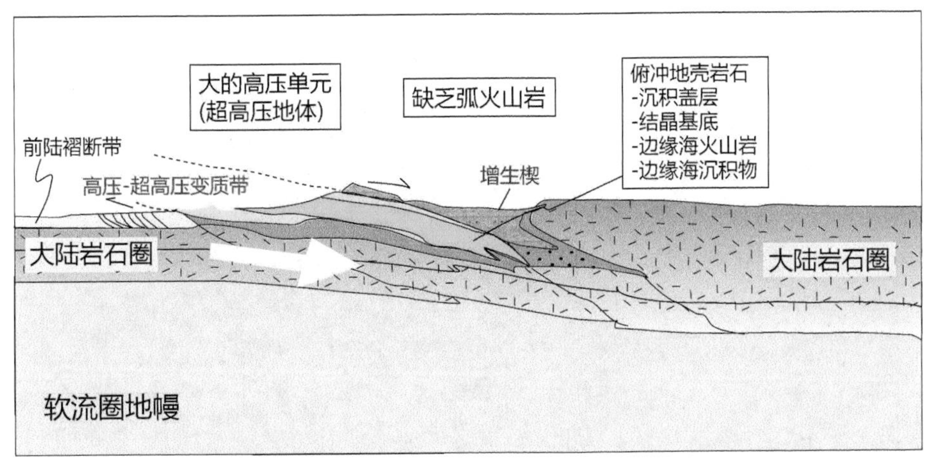

图 2-2　大陆俯冲带结构及其常见产物示意图（修改自 Liou et al.，2009）

大洋俯冲带还可进一步划分为两个亚类（Stern，2002）：一是马里亚纳俯冲带（图 2-3），属于弧后扩张的高角度洋－洋俯冲带，在西太平洋较为常见；二是安第斯俯冲带（图 2-4），属于弧后挤压的低角度洋－陆俯冲带，在东太平洋较为常见。在这两者之间的有日本型俯冲带，为出现弧后扩张的高角度俯冲带。洋－洋俯冲带以发育海沟、火山弧和弧后盆地为特征，但是增生楔很小或者缺乏。在洋－陆俯冲带之上，虽然发育有火山弧和弧后收缩，但是缺乏弧后盆地，不过增生楔较大且普遍发育（Zheng，2021b）。

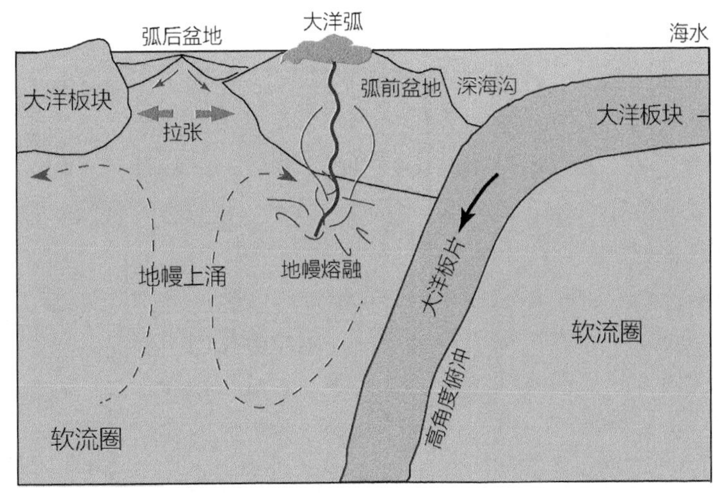

图 2-3　洋－洋俯冲带结构示意图（以西太平洋马里亚纳俯冲带为例）

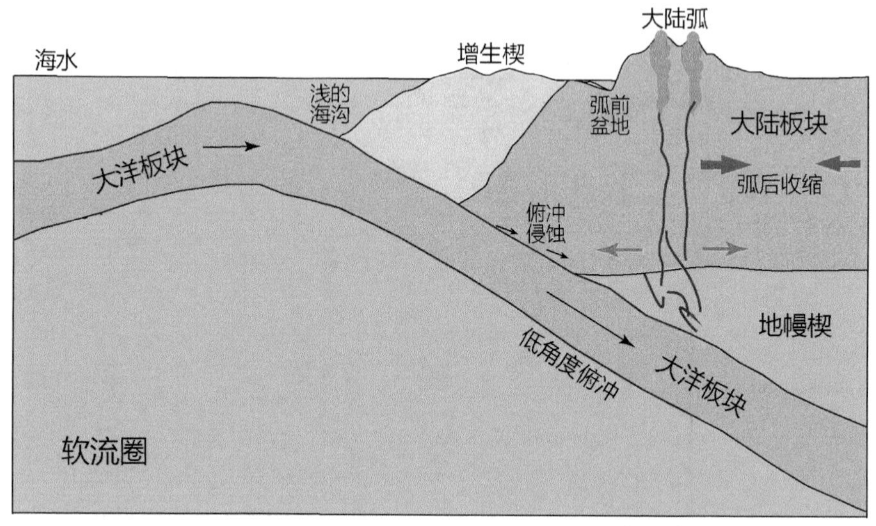

图 2-4　洋－陆俯冲带结构示意图（以东太平洋安第斯俯冲带为例）

大陆俯冲带又称为大陆碰撞带，也可进一步划分为两个亚类（郑永飞等，2009）：一是喜马拉雅型碰撞带，是一个古老大陆边缘俯冲到一个年轻大陆边缘弧之下所形成的陆－弧碰撞带；二是大别－苏鲁型碰撞带，是一个古老大陆边缘俯冲到一个更为古老大陆边缘之下所形成的陆－陆碰撞带。虽然在文献中常见将大陆俯冲与大陆碰撞互用的情况，但是它们在深度定义上是有所不同的。大陆碰撞是相对于地壳深度而言的，而大陆俯冲则是相对于岩石圈地幔深度而言的（Zheng，2021b）。

近半个世纪以来，世界各国科学家对俯冲带开展了广泛深入研究，俯冲带研究已进入相对独立的学科发展阶段。根据其研究内涵，俯冲带科学的基本定义可表述为：研究俯冲带结构、组成、过程、产物及其资源、环境、灾害、生命效应等相关内容的学科。俯冲带科学与其他学科领域的紧密联系体现在以下几个方面。

1）俯冲带科学涉及高压物质科学和材料科学。俯冲带是发生俯冲作用的板块边缘部位，主要由以下部分组成：俯冲板片向下弯曲形成的海沟或前陆盆地；因板片俯冲而刮削下来的弧前增生楔或沉积楔形体；俯冲板片之上的地幔楔；板片俯冲到一定深度因部分熔融而形成的火山岩浆弧，以及与火山岩浆弧伴生的双变质带或高压－超高压变质带等。大洋或大陆岩石圈在俯冲带进入地幔，到一定深度会被深部地幔熔融同化甚至消亡。俯冲隧道中物质组成在高压环境下发生相变，不同阶段形成不同产物，如地质流体、熔体、岩浆、矿床、各种高压变质矿物和变形构造等。研究这些物质组成，不但涉及传统的岩石学和地球化学，也涉及纳米尺度的材料科学和高压物质科学。

2）俯冲带科学涉及广泛的地球物理科学。利用地震层析成像、大地电磁、大地热流、重力、地磁学等地球物理学科的方法体系，并结合高温高压实验和计算地球动力学等，不仅可以观测到俯冲带现今的地表地质过程，还可以探知俯冲带的深部结构特征、物质组成，为了解俯冲带在地质历史时期的演化过程提供关键依据。例如，地震层析成像方法能够为识别俯冲板片的深部形态、探讨地幔楔的物理化学性质、理解岩浆弧火山的起源以及相关的地球动力学过程提供约束。俯冲板片形态不仅与板块间的相对汇聚速率、仰冲板块的运动方向和绝对速度、俯冲板片的年龄以及是否存在外来地体有关，还受到地幔楔构造特性的影响（Zheng and Chen，2016）。正是这些差异，造成了全球不同俯冲带的不同俯冲形态。此外，俯冲带地球物理探测与构造地质学、地球动力学模拟的结合，还可以揭示岩石圈地幔的年代结构，为未来地幔动力学、地幔结构演变研究提供可检验的窗口。

3）俯冲带科学是生命科学与地球科学的结合点。由于俯冲带是地球内部最冷的部分，而生物虽然在温度大于150℃时不能存活，俯冲带几乎肯定与最深（最高压力）的生物圈相关。海沟系统是地球上大气圈、水圈、岩石圈、生物圈乃至下地幔之间相互作用最为活跃的场所，是研究这些圈层之间物质交换和能量交换过程，揭示地球系统运行和演变、极端环境下暗生命生存机理的关键地区（李三忠等，2020）。

4）俯冲带科学是联系海洋科学与固体地球科学的纽带。俯冲带输入物质的组成与水圈的泥沙输运密切相关，而物质组成又与水圈所处纬度相关。在深时时期的海陆格局重大变化过程中，要恢复进入俯冲带的物质组成，既需要海洋科学与固体地球科学的紧密结合，也需要海底科学调查与数值模拟的结合。

5）俯冲带科学是探索地球动力学的基石。在现代板块构造体制下，俯冲带早期演化阶段具有低的地热梯度，因此冷俯冲挤压是产生阿尔卑斯型变质作用的必要条件，而在晚期演化阶段由于俯冲板片回卷产生了高的地热梯度（Zheng，2019）。过去将俯冲板片脱水交代地幔楔作为其部分熔融形成弧岩浆的物理化学机制，后来将地幔楔降压加热作为其部分熔融形成弧岩浆的构造机制。现今俯冲带的地热梯度还受地幔楔温度的影响，而地幔楔温度变化又受控于俯冲板片的年龄。在太古宙和古元古代时期，地幔温度普遍比显生宙地幔高 100～250℃，板片边缘变形并大多处于韧性状态，板片以暖俯冲为主（Zheng and Zhao，2020）。这将导致俯冲板片易于脱水熔融产生含水的长英质熔体，由此交代上覆地幔楔使其相对富集不相容元素（包括水和碳）。地幔楔熔融可形成大洋弧拉斑玄武岩和大陆弧钙碱性安山岩，在弧后地区常出现巴肯型高温－超高温变质岩。因此，以这些变质岩和岩浆岩为对象，结合俯冲带温压结构和动力体制随时间演化的性质，采用计算地球动力学模拟，能够为解决板块构造起始－夭折－再起始以及体制转变等重大问题提供关键约束。

6）俯冲带科学是建立地球系统科学的关键。俯冲带是地球各圈层物质的混合器，涉及俯冲板片的上覆沉积物、洋壳和大洋岩石圈地幔，以及俯冲板片来源的流体与上覆地幔楔物质的相互作用。这些过程促进了俯冲带上盘钙碱性火山岩或侵入体、矿床和新生陆壳的形成。因此，从地球系统科学理念出发，俯冲带是地球最为显著的多圈层相互作用的场所，是探索影响地球宜居环境演变和地质灾害发生机制的理想研究对象，这使得俯冲带科学成为多学科交叉建立地球系统科学的关键。例如，20 世纪 90 年代提出的俯冲工厂概念，强调了俯冲带物质（如沉积物、流体）的再循环过程。深入揭示俯冲工厂再循环的过程、行为、归宿和效应，以及各种再循环物质的通量和平衡问题，就需要根据地球系统科学概念，综合多学科成果进行集成创新。为此，特别需要关注俯冲再循环的两个基本研究领域：一是再循环过程，即岩石、沉积物和流体通过俯冲带所涉及的物理、热力、矿物学和地球化学过程；二是再循环通量，即固体、流体、元素和溶解物质，通过俯冲带的通量和途径。

第二节　板块俯冲带的研究历史

一、国际板块俯冲带研究历史

俯冲带是汇聚板块边缘发生相互作用的构造带，是岩石圈再循环进入地幔的地方，是约 55 000 km 汇聚边缘板块从地表向地幔深部的延伸。板块在俯冲带处的重力下沉为板块运动提供了动力，是产生地壳变形与岛弧岩浆作用的直接原因。俯冲带地质过程为人类社会发展提供了丰富的矿产资源和能源，同时也产生了严重的自然灾害，直接或间接地影响了气候变化，进而对地球的宜居性产生重要影响。自 20 世纪 50 年代以来，俯冲带一直是很多国家地质学、地球物理学和地球化学领域科学家研究的热点，经历了 70 多年的发展历程，可归纳为如下三个发展阶段。

（1）20 世纪 50～60 年代，为俯冲带研究的早期探索阶段

Amstutz（1951）最早引入俯冲（subduction）一词，意指一段岩石圈下沉的过程。随后，很多研究发现环太平洋地区岩浆作用与深部地震密切相关，并有明显规律性变化。由 Wadati（1935）发现、经 Benioff（1954）研究证实，在日本岛弧及其他地区地震带的震源深度从洋向陆出现由浅到深的变化规律，由此勾画出俯冲带的几何轮廓。Kuno（1959）对玄武岩岩石化学成分的研究发现，从大洋向大陆方向玄武岩的碱度升高，进而提出幔源岩浆成分与深度有关，由此激发了一系列地幔橄榄岩体系的实验研究（如 Turner and Verhoogen，1960；Kushiro，1968）。Miyashiro（1961）发现日本列岛和环太平洋地区变质带成对分布，低温 / 高压型变质带分布在大洋一侧，而高温 / 低压型变质带分布在大陆一侧，据此提出了变质相系和双变质带的概念，为板块俯冲提供了变质作用的证据。此外，Coats（1962）通过对阿留申群岛岩浆类型和地壳结构研究，提出了很有预见性的俯冲带轮廓。但是，直到 20 世纪 60 年代末期板块构造理论提出之前，科学家并没有充分认识这些现象之间的内在联系及其整体科学意义。

（2）20 世纪 70 年代至 90 年代中期，为俯冲带研究的重要发展阶段

伴随板块构造理论的出现，俯冲带研究成为地球科学中的热点，在各相关领域都取得了明显进展。例如，Isacks 和 Molnar（1969）首先发现在全球范围内中、深源地震都分布在俯冲板块中，并且震源机制与俯冲深度之间有明显变化规律；White 等（1970）正式将“subduction”一词引入板块构造研究；Veith（1977）发现千岛群岛下部存在双层地震带，其上、下层可能分别与俯冲板片中的榴辉岩和蛇纹石脱水有关（Hasegawa et al.，1978）。Molnar 等（1979）首先建立了地震带分布、板块汇聚速率和海沟处板片年龄之间的定量关系。

通过对环太平洋俯冲带的综合研究，Uyeda 和 Kanamori（1979）把大洋俯冲带分为两个端元类型：智利型和马里亚纳型。Jarrard（1986）依据应变强度，把俯冲带分成 7 个等级，第一级表现为强烈伸展，如马里亚纳型，第七级为强烈挤压，如智利型。Shreve 和 Cloos（1986）提出了俯冲隧道的概念，认为在俯冲板块与上覆板块之间一般存在自由空间，其中来自上、下板块的物质混杂在一起，并可以发生向下或向上流动（Cloos and Shreve，

1988a，1988b），从而很好地解释了在板块汇聚过程中超高压岩石发生折返与增生楔的形成机理。

大洋俯冲带之上弧岩浆作用研究一直备受关注。Ringwood（1975）基于地质和地球物理观测与实验岩石学研究，提出了经典的岛弧玄武岩浆形成模式（图 2-5），强调俯冲板片脱水是驱动地幔楔部分熔融的主要因素。后来的研究将该模式推广到所有大洋俯冲带，包括板片熔体与地幔楔橄榄岩反应、受改造的地幔物质上升降压熔融产生玄武质和安山质熔体，并在镁铁质岩浆喷出之前经历复杂的分异、混合和地壳混染等过程。

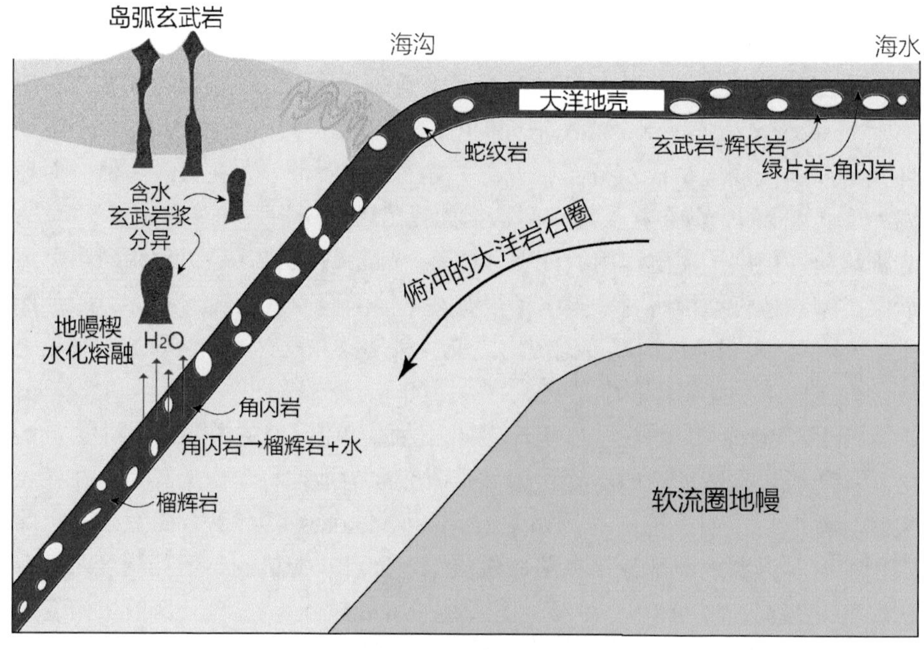

图 2-5　俯冲大洋板片脱水引起地幔楔熔融产生岛弧玄武岩浆的经典模式（修改自 Ringwood，1975）

该模式假设，板片流体通过交代地幔楔橄榄岩使其固相线降低，从而引起水化熔融

Kay（1978）首先在阿留申群岛发现俯冲地壳熔融形成的埃达克岩，这类岩石强烈亏损重稀土元素，并具有很高的 Sr/Y 值，表明它们是与榴辉岩平衡的熔体（Defant and Drummond，1990），支持俯冲板片熔融的认识。但是，系统的实验岩石学研究表明，含水地幔橄榄岩部分熔融可以形成各种玄武岩和安山岩（Tatsumi，1981；Umino and Kushiro，1989），不需要加入板片熔体。Gill（1981）总结发现，汇聚边界的岩浆作用主要发生在俯冲板片距地表 80～160 km 深度，因此现在一般将弧下深度定义为 80～160 km（Zheng，

2019）。

与洋壳俯冲有关的变质作用研究主要集中在环太平洋地区和阿尔卑斯－喜马拉雅造山带，如 Ernst（1988）指出美国西部弗朗西斯科蓝片岩与西阿尔卑斯蓝片岩有着不同的 *P-T* 轨迹，反映折返机制或俯冲带其他动力学条件不同。20 世纪 80 年代中期以来，随着陆壳变质岩中柯石英和微粒金刚石的发现（Chopin，1984；Smith，1984；Sobolev and Shatsky，1990；Xu et al.，1992），人们开始认识到大陆地壳可以俯冲到大于 100 km 的地幔深度，并又折返回来，大陆地壳俯冲带的研究成为地球科学中的又一热点（Chopin，2003；Zheng，2012）。

随着地质和地球物理观测资料的积累，从 20 世纪 90 年代开始学界对俯冲带综合数值模拟研究取得了实质性进展，从而可以定量评价影响俯冲带动力学过程的各种物理参数（如板块年龄、汇聚速率、俯冲角度等）之间的关系，确定俯冲带的温压结构（如 Kirby et al.，1991；Peacock，1991）。

（3）20 世纪 90 年代中期至今，为俯冲带研究的全面发展阶段

对俯冲带研究的各个领域包括地球物理学、地质学、地球化学和地球动力学等都取得了很大进展。以 1994 年在加利福尼亚州圣卡塔利娜岛召开的俯冲带专题研讨会为标志（Bebout et al.，1996），俯冲带研究开始走向多学科联合。根据 Bebout 等（2018）的总结，概述过去 25 年俯冲带科学的研究进展如下。

1）地震学的发展很好地解决了俯冲板块中地震构造与力学机制（Rondenay et al.，2008；Shillington et al.，2015），特别是流体对俯冲边缘力学机制和地震成因的影响（Saffer and Tobin，2011）；尤其是利用地震层析成像技术，分析俯冲板片的深部状态（Zhao et al.，1994，1997；van der Hilst et al.，1997），证明有些俯冲板片停留在 660 km 的不连续面，有的达到了幔－核边界。

2）通过连续 GPS 监测，发现了俯冲带慢滑动事件，以及与之有关的一系列地震学和大地测量学现象（Gomberg and the Cascadia and Reyond Working Group，2010）。

3）认识了俯冲剥蚀的程度与意义。发现在大多数汇聚板块边缘的上部板块中发生物质丢失，即剥蚀，只有少数俯冲带上部板块出现物质增加，形成增生楔（Scholl and von Huene，2007）。

4）基于实验岩石学和相平衡模拟研究，定量阐述了俯冲沉积物、基性岩和超基性岩在俯冲不同阶段发生的变质反应和流体行为（Schmidt and Poli，

1998，2003，2014；Hacker，2008）。

5）由于微量分析技术和模拟计算能力的发展，能够较精确地测定矿物及熔体包裹体的挥发分和微量元素成分，进一步限定和模拟俯冲带化学循环过程（Plank，2005；Frezzotti et al.，2011；Bebout，2014；Zheng，2019），解释弧型火山喷发的原因等（Wallace，2005；Zellmer et al.，2015）；尤其是加深了对俯冲带中超临界流体性质和物质运输能力的理解（Manning，2004；Hermann et al.，2006；郑永飞等，2016），发现名义上无水矿物可以储存大量的水。

6）随着技术设备的进步，开展了一系列有关俯冲带的实际探测与观测，如探测深海沟（Cui et al.，2013；Okumura et al.，2016）、水下火山弧和弧后盆地火山以及有关的地热系统和火山碎屑沉积等（Baker et al.，2008；Dziak et al.，2015）。

7）通过对古俯冲带高压－超高压变质岩的研究，揭示出大部分洋壳俯冲折返的榴辉岩和部分陆壳型榴辉岩都经历了以形成硬柱石为特征的冷俯冲过程（Wei and Clarke，2011；Faryad and Cuthbert，2020）；明确了大陆深俯冲的极限深度可以达到斯石英稳定域（Liu L et al.，2007，2018）；揭示出在地球演化过程中俯冲样式的规律性变化（Stern，2002；Holder et al.，2019）。此外，古俯冲带变质岩温度压力条件的研究对理解现代俯冲带的温压结构提供了限定（Penniston-Dorland et al.，2015；Zheng，2021c）。

8）随着各种观测资料的积累和计算能力的提升，对俯冲带温压结构模拟研究得到了很大发展。Peacock 和 Wang（1999）首先建立了日本岛东北部和西南部发生冷俯冲和热俯冲的温压结构模型，很好地解释了两个地区岩浆和地震作用的差异。近年来，又提出了一系列适合现代全球俯冲系统的温压模型（Syracuse et al.，2010；van Keken et al.，2011）。

9）俯冲工厂研究，汇聚板块边缘的俯冲系统可比拟为一个工厂（图 2-6），俯冲的大洋板块是输入工厂的原料；从弧前区逸出的流体、泥火山物质以及蛇纹岩底辟体，在弧与弧后区形成的岩浆，以及生成的矿床等是工厂的产品；俯冲板片所发生的变质反应、脱水和熔融等过程则是这个工厂内部的工艺流程（Eiler，2003）。俯冲工厂研究的核心科学问题是俯冲再循环，主要研究再循环过程和再循环通量两个基本领域。俯冲作用改造了俯冲板块和上覆板块，紧密联系着地表和地球内部的各个圈层，俯冲工厂在地球系统的运行和演化中占据着中心位置。

由于地球上的重大地质灾害以及大量矿产资源和能源都与俯冲带有关，很多西方国家都非常关注俯冲带科学综合研究。例如，美国国家科

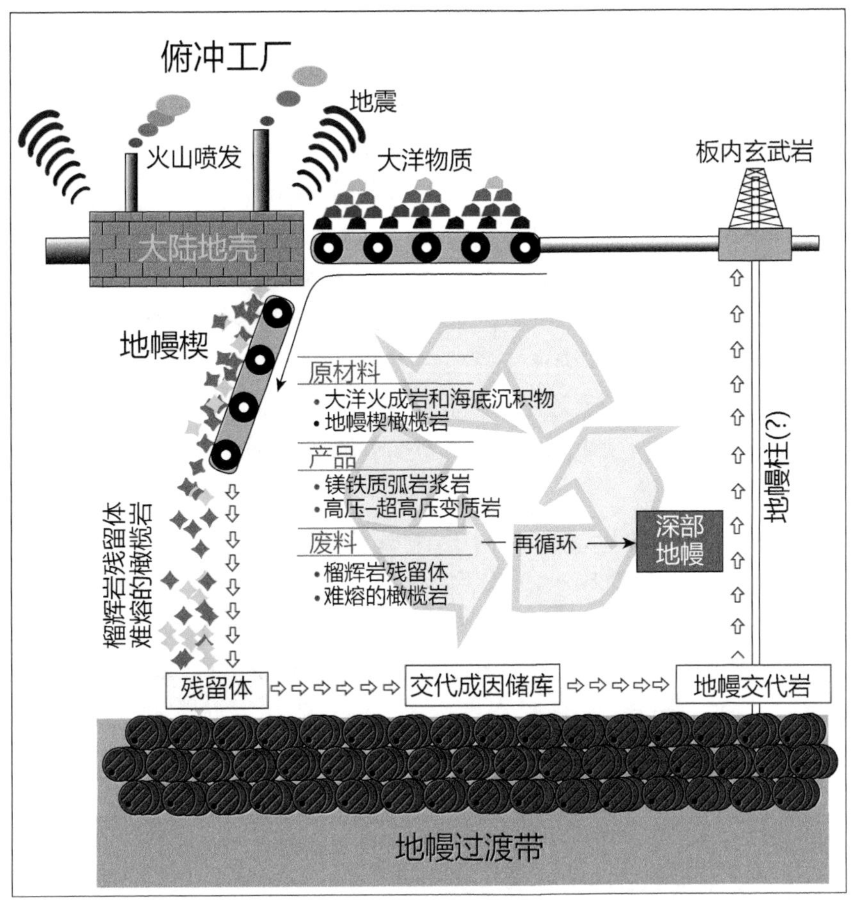

图 2-6 俯冲工厂材料加工流程示意图（修改自 Zheng and Zhao，2017）

学基金会（NSF）会启动了大陆边缘计划（MARGINS）和地学棱镜计划（GeoPRISMS），前者研究焦点是俯冲工厂和孕震带，后者主要瞄准俯冲物质循环和变形。美国从 2018 年起实施了一个更大的“俯冲带观测计划”（Subduction Zone Observatory），旨在揭示俯冲带边缘短期和长期演化，建立更广泛的科学网络，培养地球物理学、地质学、地球化学和地球动力学领域的综合型人才。日本、新西兰、中－南美洲、欧盟和东南亚等国家、地区或组织也先后启动了类似的研究计划，如欧盟启动了“聚焦板块之间”（Zooming in between plates）计划，旨在研究俯冲过程与地质灾害，建立综合研究和人才培养平台。显然，俯冲带研究仍然是 21 世纪国际地球科学的前沿与热点。

二、国内板块俯冲带研究历史

对板块俯冲带的研究建立在板块构造理论框架之下。20 世纪 60 年代末提出的板块构造理论用球面上的刚性块体旋转来解释板块间的相对运动，通过大洋的打开和闭合建立了洋盆演化与大陆构造演化的联系，掀起了对板块俯冲带研究的第一波浪潮（McKenzie and Parker，1967；Morgan，1968；Le Pichon，1968；Wilson，1968）。中国老一辈地质学先驱如尹赞勋、李春昱、傅承义、常承法、郭令智等在 70 年代初就向国内学者系统介绍了板块构造理论，指出验证这一理论首先需要在大陆内部寻找古缝合带，确定人类赖以生存的大陆是由大洋板块消亡、陆块逐步拼贴形成的。因此，中国板块俯冲带的研究历史也是中国大地构造格架逐渐清晰、构造演化历史逐步细化的过程。

传统观念认为，中国大陆主要由华北、华南、塔里木三大古老稳定陆块和环绕它们的活动带组成。1960 年中国科学院组织了首次青藏高原综合科学考察，以中国科学院地质研究所常承法为代表的一批中国地质学家在研究基础极为薄弱的青藏高原识别出若干消亡的大洋俯冲带和大陆碰撞带，提出青藏高原是由多个陆块和岛弧在古生代、中生代和古近纪相继向亚洲拼合增生形成的，块体之间存在消亡大洋的证据——蛇绿岩（常承法和郑锡澜，1973）。1980 年常承法在青藏高原国际科学研讨会上做了“青藏高原的大地构造演化”报告，受到国际同行的高度评价，立即吸引了法国、英国、美国的顶尖地球科学家来到中国开展青藏高原国际合作研究。

国际合作项目的开展推动了中国科学家快速融入国际地学舞台。1980～1983 年开展了中 - 法合作项目“喜马拉雅地质构造与地壳上地幔的形成和演化”研究（肖序常等，1988），1985～1986 年开展了中 - 英合作项目“拉萨 - 格尔木地质综合考察”。1986～1995 年中国地质科学院、武汉地质学院、长春地质学院、中国科学院地质研究所、中国科学院地球物理研究所等 8 家单位，联合完成了亚东 - 格尔木 - 额济纳旗的地学大断面（吴功建等，1991；王泽九等，1995）。1991～1992 年中国地震局与美国合作开展了 PASSCAL 实验，在拉萨 - 格尔木青藏公路沿线布设了 11 台宽频带天然地震仪，进行天然地震观测（丁志峰和曾融生，1996）。“国际合作青藏高原及喜马拉雅深部探测”（International Deep Profiling of Tibet and the Himalayas，INDEPTH）计划在 1992 ～2000 年进行了三期，通过地震反射、天然地震观测、大地电磁测深等手段，揭示了喜马拉雅造山带、雅鲁藏布江缝合带和班公－怒江缝合带的地壳波速结构、电性特征和深部过程（Zhao et al.，1993；Yuan et al.，1997；Wei et al.，2001；赵文津等，2008）。这些重要成果验证了常承法等一批中国板块构造先驱的奠基性工作，已成为关于喜马拉雅和青藏高原形成演化的所有构造模型的基础。此外，南京大学郭令智团队于20世纪70年代发现了扬子地块与华夏地块俯冲碰撞所形成的元古宙沟弧盆体系，突破了板块构造限于显生宙的观念（郭令智等，1980），开创了我国研究地球早期板块构造机制的先河。

在20世纪末，地质矿产部组织开展了对中国超基性－基性岩的普查，寻找与蛇绿岩相关的豆荚状铬铁矿。中国地质科学院地质研究所、成都地质矿产研究所和中国地质科学院地质力学研究所与西藏自治区地质矿产局合作，于1979年完成了雅鲁藏布江缝合带罗布莎地区铬铁矿综合研究，罗布莎矿区成为中国最大的铬铁矿床。中国地质科学院地质研究所方青松、白文吉带领的金刚石组通过人工重砂分选，在罗布莎岩体和位于班公－怒江缝合带的东巧岩体中发现了金刚石（Bai et al.，1993）。之后杨经绥团队在全球蛇绿岩地幔橄榄岩和铬铁矿中陆续发现了金刚石、柯石英等超高压矿物以及一些新矿物，使蛇绿岩型金刚石与深部地幔动力学的联系成为国际研究热点（杨经绥等，2013）。

传统的板块构造理论认为，低密度的大陆岩石圈在浮力作用下会长期“漂浮”在软流圈之上，保留多期大陆聚合与裂解的地质记录。但是，在意大利西阿尔卑斯山（Chopin，1984）和挪威西片麻岩省（Smith，1984）发现含柯石英的超高压变质岩表明，陆壳物质可以俯冲到至少80 km的地幔深度（压力＞2.8 GPa）并折返到地壳层位。这一发现打开了认识造山带演化、壳幔物质循环与地幔动力学的新窗口，迅速掀起了板块俯冲带研究的第二波浪潮（Chopin，2003；Liou et al.，2004；Zheng，2012）。

以中国地质科学院地质研究所许志琴和安徽省地质矿产局徐树桐为代表的一批地质学家很快在大别山发现了柯石英（Xu，1987；Okay et al.，1989）和金刚石（Xu et al.，1992）。中国学者通过积极的国际合作，迅速成长为国际上超高压变质带研究的主力军和领跑者。通过30多年的研究发现：华南陆块的巨量陆壳物质曾在早三叠世俯冲至华北板块之下并快速折返，使大别－苏鲁造山带成为全球规模最大、保存最好的超高压变质带（Cong et al.，1995；Zheng，2012）。大别－苏鲁超高压变质岩的俯冲深度可能超过200 km（Ye et al.，2000），并具有过剩氩（Li et al.，1994）、氧同位素负异常（Yui et al.，1995；Zheng et al.，1996）、钕同位素正异常（Jahn et al.，1996）等特点。1995 年在大别山成功召开了第三届国际榴辉岩野外会议，扩大了中国学者的国际影响力。由许志琴任首席科学家的中国大陆科学钻探工程（Chinese Continental Scientific Drilling Project，CCSD），2000 年立项，2005 年钻探完工，2007 年验收，是国际大陆科学钻探计划中首个以研究“大陆板块汇聚边界的地幔动力学”为目标的科学钻，位于江苏东海县 5158 m 深的 CCSD 主孔岩心广泛存在柯石英，连续的岩心记录和地球物理剖面为揭示大陆物质深俯冲和折返的过程、机制和壳幔相互作用提供了重要信息（Yang J S et al.，2003；Liu F L et al.，2007；Xu et al.，2009；Zheng et al.，2009）。

全球已发现了 30 余条含柯石英、微粒金刚石或超硅石榴子石的超高压变质带（Liou et al., 2009, 2014；Zheng, 2012）。中国地质学家通过开展地质学、地球化学、地球物理学和地球动力学的系统研究，尤其是锆石 U-Pb 定年技术的应用，在中国大陆厘定了 9 条高压－超高压变质带：①大别－苏鲁超高压变质带（240～225 Ma）（Xu，1987；Okay et al.，1989；Xu et al.，1992）；②西南天山超高压变质带（320～315 Ma）（Zhang et al.，2002；Gao et al.，2009；Wei et al.，2009）；③柴北缘超高压变质带（450～420 Ma）（Yang et al., 2001；Song et al., 2003）；④阿尔金超高压变质带（509～475 Ma）（Zhang et al.，2001；Liu L et al.，2002，2007）；⑤北秦岭超高压变质带（514～485 Ma）（Yang J S et al.，2003；Liu et al.，2016）；⑥祁连高压变质带（510～460 Ma）（Song et al.，2006）；⑦西藏羌塘高压变质带（243～217 Ma）（Li et al.，2006）；⑧西藏松多高压变质带（257～267 Ma）（杨经绥等，2006）；⑨喜马拉雅高压－超高压变质带（55～44 Ma）（Kaneko et al.，2003；Wang Y H et al.，2017）。这些国际瞩目的研究成果揭示了显生宙以来洋－陆俯冲、陆－弧俯冲、陆－陆俯冲的差异性，为进行全球板块重建提供了重要基础。

大陆深俯冲是板块汇聚过程中的常见现象，因此汇聚板块边缘是岩石圈发生破坏的场所。对大陆地壳增长曲线的构建显示，太古宙以来的大陆地壳体积趋于稳定，表明俯冲消亡的陆壳和新生陆壳基本达到平衡。新生大陆地壳增加最显著的两种环境是大火成岩省和大陆边缘弧，板块俯冲过程与大陆弧地壳生长之间的关系是 20 世纪末国际地学研究的前沿和热点。大陆边缘弧常含有附近大陆地壳再循环的地球化学信息，而大量新生地壳的形成需要地幔物质的参与，最明显的特征就是花岗岩的初始 Nd 同位素组成接近亏损地幔值，这在中亚造山带非常突出。对天山造山带的详细研究（Zhang et al.，2002）业已揭示出古生代大洋俯冲证据；根据中国北方造山带花岗岩 Nd 同位素特征，发现这里是显生宙地壳增生最为显著的地区。

综上所述，中国大陆漫长复杂的地质演化历史使其成为研究古板块俯冲带的理想实验室，中国板块俯冲带研究紧跟国际潮流，借助岩石学、地球化学和地球物理探测的新技术、新手段，取得了系列成果，国际影响力显著提升，为发展板块构造理论提供了重要支撑。

第三节　本学科的发展规律和特点

一、学科发展动力

中国在板块俯冲带研究领域具有独特的地域优势。中国俯冲带的类型多样，形成时间跨度大，出露岩石规模大、类型丰富。这些俯冲带既有世界最典型的大洋俯冲带（如西南天山）、大陆俯冲带（如大别－苏鲁造山带），也有经历大洋俯冲、弧陆碰撞或陆陆碰撞形成的复合造山带（如喜马拉雅－青藏高原造山带）。西南天山是世界上最大的大洋俯冲超高压俯冲带，而大别－苏鲁造山带是世界最大的大陆深俯冲超高压变质带，中亚造山带是世界上最典型的显生宙陆壳生长的构造带，在中央造山带中发现多种矿物出溶结构以及斯石英副象，指示这些陆壳岩石经历了大于 300 km 的超深俯冲而后返回地表。固体地球科学的发展历史表明，对世界上典型地质区域、典型样品的全面剖析是认识地球运行规律的重要步骤。对俯冲带的研究更是如此。中国在俯冲带领域的独特地域优势为俯冲带学科的发展奠定了坚实基础。

板块俯冲带的发展是由自身发展规律决定的。板块俯冲带作为地球科学研究的重要内容，在 20 世纪 80 年代之后得到快速发展。由于俯冲带深刻影响着地球各个圈层系统的演化，因此固体地球科学的发展要求板块俯冲带这个学科在各个方向飞速发展。1984 年国外学者在大陆表壳岩石中发现柯石英这个超高压变质矿物后，大陆深俯冲和超高压变质成为研究热点。这也同时使整个俯冲带的研究焕然一新。中国科学家迅速抓住这个时机，首先在大别－苏鲁造山带发现了柯石英和金刚石，而后在整个中央造山带的多个造山带中陆续发现了柯石英、金刚石、矿物出溶结构、白云石分解为菱镁矿结构、斯石英假象和副象等，并通过大规模取样、在锆石中寻找柯石英包裹体等方法，确定了超高压变质岩的出露规模。特别是，2001 年在江苏连云港东海毛北村启动了中国大陆科学钻探工程，这是国际上在大陆造山带打下的最深科学钻，极大地推动了大陆动力学的发展。通过地质年代学、矿物温压计和视剖面相图模拟等手段创新，进一步确定了俯冲带变质 *P-T-t* 轨迹，勾勒出造山带的构造演化历史；岩石学、地球化学和锆石学等方面的进步和交叉，进一步揭示了原岩属性、流体活动与元素迁移、壳幔相互作用、岩浆岩形成、矿床成因等俯冲带关键问题。同时，地质学、地球物理学和地球化学等学科的交叉融合，对俯冲带结构、过程和产物的认识起到了极大的推动作用。这一点充分体现在对喜马拉雅 - 青藏高原造山带的研究中。

板块俯冲带的发展极大地得益于技术进步的推动作用。近年来，随着热力学数值模拟、微区原位微量元素和同位素分析、高精度同位素分析的技术进步，对俯冲带岩石经历的变质时限、变质温度压力条件、元素迁移和同位素分馏行为有了更精确的认识。特别是原位离子探针 U-Pb 定年和稳定同位素（H-O-S-Si 等）分析，结合岩石变质温度压力限定，可以把变质时代和变质条件精确对应，为揭示变质岩记录的造山带构造演化提供关键信息。高温高压实验技术的进步，推动了对俯冲带流体性质、地震成因、元素迁移和同位素分馏行为的认识。特别是超临界地质流体的识别和认识，对理解俯冲带元素迁移和岩浆形成演化提供了新视角。地球动力学数值模拟的发展，为揭示更接近地质实际的地质过程提供了可能，在认识俯冲起始、俯冲和折返过程等俯冲带动力学过程起到了重要作用。地球物理学手段的进步在确定俯冲带地震成因、内部结构、板片脱水行为和板片动力学过程方面具有极大的推动作用。

板块俯冲带的发展是中国经济社会发展的现实需要。经济社会的飞速发展对资源能源的需求极大增加，特别是矿产资源和化石能源。稀有金属、稀散金属、稀土金属等关键金属对国家经济发展、国防安全起着关键作用，需求量也逐年增加。俯冲带是这些关键金属产出的重要构造环境，因此对俯冲带的研究，特别是俯冲带如何形成矿产资源、如何勘查找矿势在必行。俯冲带还是重要的地震活动带，对人民生命健康、国家军事、核电 / 水电站等布局具有重大影响。因此，这也推动着俯冲带研究向前发展。

板块俯冲带的发展与国际科学发展密切相关。国际上对俯冲带的研究一直非常重视，并且开展了诸多重大研究计划，如美国国家科学基金会启动的大陆边缘计划（MARGINS）、地学棱镜计划（GeoPRISMS）和 2018 年启动的“俯冲带观测计划”（Subduction Zone Observatory），旨在揭示俯冲带边缘短期和长期演化，建立更广泛的科学网络，培养地球物理学、地质学、地球化学和地球动力学领域的综合型人才。日本和欧洲也先后启动了类似的研究计划。这在一定程度上刺激了国内俯冲带的研究。近年来国内俯冲带研究迎头赶上，推动了“特提斯地球动力系统”重大研究计划和“西太平洋地球系统多圈层相互作用”重大研究计划，以及诸多与俯冲带有关的重大项目。从20世纪80年代开始，中国在大别-苏鲁造山带及其西延的秦岭-柴北缘造山带、喜马拉雅造山带及其北延的特提斯俯冲带等领域的国际合作逐渐加强，通过人员、技术和思想的广泛交流，极大地推动了国内板块俯冲带研究走向深入，迈进国际舞台。

板块俯冲带的发展离不开稳定的人才队伍、充足的资金支持和高水平的平台建设。幸运的是，中国自20世纪80年代开始就非常重视板块俯冲带的发展。例如，科学技术部从20世纪末开始，先后设立了4个973项目支持大陆俯冲带的研究：1999～2003年“大陆深俯冲作用”；2003～2007年“大陆板块会聚边界的地幔动力学与现代地壳作用”；2009～2014年“深俯冲地壳的化学变化与差异折返”；2015～2019年“大陆俯冲带壳幔相互作用”。2017年起，国家自然科学基金委员会又先后实施了“特提斯地球动力系统”重大研究计划和“西太平洋地球系统多圈层相互作用”重大研究计划。这些重大项目的实施，培养和稳定了一支由中国科学院院士领衔、以优秀中青年科学家为主体的俯冲带研究的重要人才队伍，极大提升了国际影响力，在大陆深俯冲、壳幔相互作用、印度-亚洲大陆碰撞过程和效应等一些研究领域已经走入世界前列。同时，以国家重点实验室和部属重点实验室为主体的重大科研平台，部署了俯冲带研究所需的尖端仪器设备，建立了前沿岩石地球化学分析技术、数值模拟和解译技术等，极大地推动了俯冲带研究的发展。

二、人才培养和人才队伍特点

在国家的大力支持下，目前俯冲带研究领域的人才呈蓬勃发展态势，人才队伍呈现稳定化、年轻化、现代化和国际化的特征。

一是队伍稳定化。20世纪80年代以来，随着国家经济实力逐步增大，对地球科学研究的投入也不断增加。在此形势下，一批板块俯冲带相关的重大项目逐步实施，培养和稳定了一支高水平的研究队伍。这支队伍以中国科学院院士和富有活力的中青年教授领头，以青年人才为骨干，结构合理、年富力强、交叉融合性强，在俯冲带研究中形成强有力的人才基础和梯队。

二是队伍年轻化。年轻人才是科学研究的生力军。随着中国研究基础的加强，仪器设备和分析技术的提升，学术思想的活跃，近年来在俯冲带研究领域不断有年轻人才踊跃出现，逐渐承担起重任，为俯冲带研究保持了新鲜的活力。通过自主培养、和国外顶尖科研机构联合培养等方式，培养了大批优秀青年人才。由于国内学术条件和学术环境的改善，通过国家自然科学基金委员会“海外优秀青年基金”等项目，引进归国的青年人才也越来越多。

三是队伍现代化。这主要体现在研究平台通过购置和研发重大仪器设备，建立和发展最新的分析技术和研究方法，使得俯冲带领域的研究人员在研究手段和工作方法上跻身世界先进水平，在部分领域达到世界领先。例如，结合薄片原位尺度副矿物 U-Pb 定年、微量元素分析和视剖面相平衡温度压力计算模拟，研究人员可以对造山带变质岩经历的 *P-T-t* 轨迹给出精确限定，恢复了造山带的构造演化历史。同时，通过对锆石 U-Pb 定年、矿物包裹体确定及矿物包裹体的温度压力限定，研究人员可以识别出多个新的超高压变质带，并对超高压变质岩出露的规模给出了很好的限定。通过自我创新、国际交流合作，研究人员研究思路也更加现代化，在俯冲带岩浆作用、大陆俯冲隧道等方面取得了原创性理论创新成果，在大陆超深俯冲、大陆碰撞成矿、俯冲带化学地球动力学、喜马拉雅 - 青藏高原造山带结构和碰撞过程等方面取得了诸多原创性发现。同时，在国家重大项目引导下，研究人员在地质学、地球物理学、地球化学等方向学科交叉越来越多、越来越深入，通过协同攻关、融合创新，极大地推动了对俯冲带结构、过程和产物的认识。

四是队伍国际化。近年来，俯冲带领域国际化合作进一步加强，研究视野进一步国际化，不仅研究中国的俯冲带，还走向世界，对国际上著名的俯冲带（如西阿尔卑斯造山带、波希米亚造山带、挪威西片麻岩省等）进行调查和研究，开展多方位多层面的国际合作，让世界了解中国学者在俯冲带领域的付出、力量和创新。特别是 2009 年中国学者在中国西宁成功组织召开第八届国际榴辉岩会议，并组织对北祁连、柴北缘造山带开展了野外考察。这极大提升了中国在俯冲带研究的国际影响力。同时，通过联合培养博士研究生、设立国家自然科学基金委员会双边合作研究项目等方式，提升了青年人才的国际化视野和学术水平，促进了俯冲带研究进一步走向深入。

三、学科交叉状况

俯冲带是一个非常复杂的体系。俯冲带研究呈现高度学科交叉的特征。

一是与地球物理学的交叉。在喜马拉雅 - 青藏高原造山带、秦岭 - 桐柏 - 大别 - 苏鲁造山带等的研究中，地球物理学发挥了重要作用。它揭示了俯冲带的几何结构、地质结构和温压结构，限定了俯冲带的地震特征、板片形态和俯冲过程。例如，根据地球物理综合剖面对北秦岭 - 桐柏造山带的岩石圈结构进行分析发现存在两期造山作用的结构特点，并将其分别归属于加里东期和印支期造山作用的结构残余（袁学诚等，2008）；部分学者通过地震层析成像和航磁特点，揭示了该造山带的几何结构特征，认为在北秦岭（Sun et al.，2015）或华北南缘（He and Zheng，2018）存在高速异常体，代表了消减的原特提斯洋残片。然而该造山带的相对平整连续的莫霍面表明造山带山根已经垮塌，结合区域岩浆岩和混合岩化特征，推测该垮塌过程应与早白垩世伸展作用相关（He and Zheng，2018）。

二是与地质学（构造地质学、矿物岩石矿床学）的交叉。用俯冲带的思路来研究大地构造及成岩成矿问题，不仅拓宽了俯冲带研究的视野，还丰富了俯冲带研究的活力。例如，阿尔金造山带研究通过矿物岩石学与构造地质学的交叉，率先确定阿尔金南、北两条高压带，论证提出阿尔金早古生代是一条经历板块俯冲 - 碰撞地质演化的造山带，确定阿尔金造山带北部地块、北阿尔金蛇绿混杂岩带、中阿尔金地块以及南阿尔金俯冲碰撞杂岩带四个构造单元的划分（许志琴等，1999），为认识青藏高原北部早期构造格局及演化提供了坚实的理论依据。与地质科学的学科交叉，推动了秦岭 - 桐柏造山带造山单元的详细划分和构造演化特点。通过研究秦岭造山带的岩石学特征和构造变形特点，将秦岭 - 桐柏造山带划分出宽坪群、二郎坪群、北秦岭群、南秦岭群等构造单元，在北秦岭单元发现含柯石英和金刚石等超高压矿物的超高压榴辉岩相岩石（Yang J S et al.，2003），并利用俯冲带的研究思路建立了古生代增生造山带和中生代碰撞造山带两大造山作用的动力学过程。在红安 - 大别 - 苏鲁造山带，与地质学科的交叉确定了造山单元的详细划分和构造演化特点。采用激光拉曼和电子探针的综合分析方法，在大别 - 苏鲁超高压变质带各类强退变质岩石的锆石中普遍发现以柯石英为代表的超高压矿物包裹体及其矿物组合，借以证明大别 - 苏鲁地体曾发生巨量陆壳物质深俯冲 - 超高压变质，进一步确认大别 - 苏鲁地体是一条大于 2000 km 的世界上规模最大的超高压变质带。采用电子探针分析、配合传统温压计和相平衡模拟的综合研究手段，准确建立超高压变质带中各种岩石特别是强退变质岩石不同阶段的 *P-T* 条件，发现不同变质带存在差异性变质演化。利用俯冲带的研究思路建立了三叠纪碰撞造山带和侏罗 - 白垩纪碰撞后再造的动力学过程。与矿床学交叉，将大别造山带沙坪沟钼矿成矿作用与俯冲带过程联系起来。

三是与地球化学的交叉。俯冲带的流体过程对壳幔物质循环具有重要意义，特别是对俯冲带流体中元素活动和同位素组成的研究，对理解地球壳幔演化具有重要意义。地球化学分析手段的进步，特别是微区原位地球化学分析和高精度同位素分析技术的发展，在俯冲带同位素年代学、化学地球动力学等方面取得显著成果。例如，通过与地球化学的交叉研究，确定了祁连－柴北缘造山带蛇绿岩的类型和形成环境，厘定出大洋中脊蛇绿岩、弧后盆地蛇绿岩和洋底高原蛇绿岩等，通过放射性同位素 Sr-Nd-Pb 研究，与全球蛇绿岩对比，初步确定了祁连山蛇绿岩的分区。确定了祁连山两条不同类型的弧火山岩带，北侧为大陆边缘弧，南侧为洋内弧，可以与现代西太平洋岛弧对比。通过同位素年代学的研究，确定了俯冲带形成和发展过程，演绎了从大洋俯冲到大陆碰撞的造山带旋回。通过俯冲带与矿床资源研究的交叉，确定了祁连－柴北缘俯冲碰撞过程的矿床成因类型、成矿规律和找矿方向。同样地，在其他造山带，与地球化学的交叉，准确限定了变质岩的原岩属性、不同变质阶段的时限、流体活动时限、地球化学分异行为、造山带岩浆岩成因、岩浆运移及演化过程、成矿过程等，极大地推动了俯冲带研究的精确性和定量化。

四是与实验岩石学和数值模拟的交叉。俯冲带变质作用、流体活动、部分熔融、元素迁移、成矿作用、地震成因等，都需要确定不同温度、压力、氧逸度和水活度条件下物相的性质与组成。实验岩石学作为一种模拟手段，在这些方面发挥着独特作用。以西天山超高压变质带为例，开展了榴辉岩化碳酸岩的实验岩石学研究工作：在岩相学观察基础上，利用高温高压实验成功地实现了生成碳氢化合物的变质反应，从而提出了在冷俯冲带中大量水流体存在的条件下，碳酸岩矿物可以通过歧化反应形成非生物成因的甲烷气，为板块俯冲带作为“工厂”生产出非生物成因的碳氢化合物提供了理论和实验证明，对进一步开展深部非生物成因的油气勘查具有重要指导意义。该成果曾被选入国际“深部碳观测计划”（Deep Carbon Observatory，DCO）的新闻通讯报道中，并以“在合适的条件下，俯冲带工厂可以生产出碳氢化合物”为题进行了深入报道；同时，该项成果还被美国科学促进会（AAAS）选为能“重新改写教科书中有关化石能源新技术”的 15 篇代表性成果之一。另外，在天山俯冲带的几何结构和温压结构岩石学研究的基础上，利用数值模拟的方法，提出了洋壳深俯冲超高压变质榴辉岩两阶段抬升折返机制。在南阿尔金造山带，通过矿物学、岩石学与高温高压实验研究交叉，相互验证，研究确定江尕勒萨依泥质片麻岩石英颗粒中大量蓝晶石＋尖晶石＋金红石针柱状体是先存富铝铁斯石英降压出溶的产物，以及榴辉岩绿辉石和石榴子石中长棒状多晶石英集合体是斯石英副象的证据，确定其中部分岩石的俯冲深度大于 300 km，为识别和研究大陆超深俯冲做出了突出贡献。同时，与实验岩石学交叉，对来自大别造山带含柯石英的片麻岩开展了压力达 24 GPa、温度达 1800 ℃的高温高压实验，探讨了深俯冲陆壳的命运；对天然榴辉岩的部分熔融高温高压实验，为俯冲陆壳发生部分熔融的条件和机制提供了限定；通过榴辉岩高温高压流变学研究，为探讨俯冲带地球动力学过程和机制提供了可靠支撑；通过“湿”榴辉岩的高温高压流变学实验，提出名义上无水矿物脱水致裂是俯冲带深部地震发生的可能机制。

四、成果转化态势

板块俯冲带研究成果的应用及其转化主要体现在三个方面。

1）俯冲带研究需要结合详细的区域调查和地质工作，这些基础数据能极大推动相关学科的发展。例如，西北大学研究团队近 30 年来对阿尔金造山带构造演化以及高压－超高压变质作用的研究成果被 2004 年出版的青藏高原北部 1:4 000 000 及 1:250 000 苏吾什杰幅、石棉矿幅、瓦石峡幅、阿尔金山幅、且末一级电站幅等地质图间接引用，同时被《造山的高原——青藏高原的地体拼合、碰撞造山及隆升机制》（许志琴等）、《塔里木盆地板块构造与大陆动力学》（贾承造等）、《中国蛇绿岩》（张旗、周国庆）、《青藏高原北部前寒武纪地质初探》（陆松年）、《昆仑山及邻区地质》（李荣社等）等专著收录，为中国西部塔里木盆地及青藏高原北缘大地构造等研究提供了重要支撑。

2）板块俯冲带的研究推动了矿产资源的勘查和发现，直接服务于人类社会的发展。作为全球成矿模式推动找矿勘查的经典实例——斑岩型铜矿，即为俯冲带成矿类型中最为突出的代表。斑岩型铜矿成矿模式自 20 世纪 70 年代建立以来，经过了近半个世纪的不断发展和完善，已经成为全球最为重要和成熟的成矿模式。斑岩型铜矿系列产出了全球 70% 的铜以及大量的金和钼等金属资源，在全球众多成矿类型中具有举足轻重的地位。进入 21 世纪以来，斑岩型铜矿成矿机制研究工作仍然是全球矿床学研究的热点，并且不断有重大研究成果的产出，如新建立的“碱性斑岩体铜金成矿模型”，由中国学者在青藏高原冈底斯成矿带斑岩型矿床研究基础上提出的“大陆碰撞造山带斑岩成矿模型”，以及利用大数据揭示全球斑岩型矿床大规模成矿规律和控制因素等。这些最新研究成果直接推动了相关找矿勘查的进展，如中国冈底斯斑岩型成矿带的重大勘查发现，加拿大太平洋东岸碱性斑岩型铜矿的持续发现，等等。除了斑岩型铜矿系列外，与俯冲带直接相关的其他成矿类型，如 VMS 型、夕卡岩型、浅成低温型等，其在矿床地质和成矿机制方面的研究成果都已经对相关找矿勘查起到关键的推动作用，成为地质科研成果进行勘查应用转化的范例。此外，在承接中国地质调查局西安地质调查中心“西昆仑—阿尔金成矿带主要含矿地层层序、时代及构造环境研究”项目过程中，对南阿尔金俯冲碰撞杂岩带中相关富矿地层开展了一系列研究工作，确定了南阿尔金迪木那里克铁矿为622～510 Ma形成的火山沉积型铁矿，与南阿尔金洋盆发育的早期裂谷盆地火山沉积作用相关，而长沙沟铁矿则是南阿尔金深俯冲板片断离后引发的同折返幔源岩浆作用的产物；确定北阿尔金喀腊大湾铁矿为521～506 Ma形成的夕卡岩型铁矿，其成矿与俯冲带火山活动引起的接触变质相关。这些矿床成因的初步研究对后续阿尔金造山带找矿具有指导意义。

3）板块俯冲带的地震、火山等自然灾害频发，因此板块俯冲带的研究对于防灾减灾具有重要意义。对俯冲带结构、过程和效应的深入认识将十分有助于人们掌握地震活动和火山喷发的规律，做好防灾减灾工作，保障国家安全。利用地震学、大地测量学、地质学和地球动力学紧密结合的多学科交叉研究方法，对俯冲带地震灾害及其次生地质灾害的研究，有利于深入理解俯冲带地震的孕育与发震机制、破裂过程、灾害特征等工作，促进地震预警系统以及海啸预警系统的建设。同时，俯冲带地震动力学的研究工作，有利于对灾害相关的风险进行有效评估，尤其是针对国防安全设施和国家大型基础设施等进行强震动计算和灾害评估。此外，俯冲带的火山活动和岩浆喷发的研究，有利于对活动岩浆房参数和状态进行监测和分析，进而为评估火山喷发前景提供关键依据，并更好地对火山岩浆活动及关联地震活动进行预测和预警。因此，板块俯冲带作为地震和火山灾害的集中爆发地，其研究水平决定了相应的防灾减灾能力。

五、研究组织形式

板块俯冲带的研究具有立足基础、面向需求、多学科交叉等特征，因此研究组织形式应该具有多样性，从而最大限度地发挥特长，共同促进该学科领域的发展。

1）自由探索。板块俯冲带研究主要属于基础研究领域，应该鼓励自由探索。因此，这方面的研究以个人申请项目为主。这种自由探索有助于充分发挥个体科研人员的主观能动性，发现学科新前沿，揭示新现象和新规律，推动学科发展。为进一步引导原始创新和重大发现，建议在国家自然科学基金委员会、中国科学院和科学技术部等竞争性项目中，加强对俯冲带原创项目的引导设置，增加支持额度。

2）联合攻关。板块俯冲带学科是高度交叉的研究，涉及地球物理学、地质学、地球化学、高温高压实验、地球动力学模拟等，需要各学科协同攻关。并且，板块俯冲带研究也与国家战略需求密切相关，因此需要进行一定的集成研究。因此，需要设置一些大型项目，围绕特定关键科学或技术问题进行联合攻关。近年来在俯冲带研究中，也体现了重大项目由院士牵头，多家单位合作，多学科联合攻关的态势。例如，在中国中央造山带的研究方面，中国地质科学院、中国科学技术大学、北京大学、吉林大学、西北大学、中国地质大学（武汉）、中国地质大学（北京）、南京大学等国内优势科研院所自 20 世纪 80 年代以来不同程度参与，近年来更是在重大项目上密切合作，为全面理解大洋俯冲带和大陆俯冲带变质历史、流体活动、壳幔相互作用、俯冲带岩浆作用等提供了坚强支撑。

3）平台建设。为了推动俯冲带研究的稳步发展，以及在前沿科学及应用领域等特定方向进行强化，建议围绕俯冲带研究，设立类似基金委员会的科学中心，进一步建设高水平研究队伍，稳定产出高水平成果，并通过各种方式加强国际合作交流，持续提升国际影响力。

六、资助管理模式

作为基础研究领域的重要部分，俯冲带研究在国家自然科学基金委员会青年项目、面上项目和重大项目等，以及科学技术部 973 计划，教育部、中国科学院的很多资助上得到了充分体现。因此，通过布置竞争性研究课题，增加竞争性课题的支持额度，积极发挥个体优势，对板块俯冲带学科发展极为有利。

作为高度交叉的学科，板块俯冲带研究涉及地球物理学、地质学、地球化学、高温高压实验和地球动力学模拟等，需要各学科协同攻关。因此，在资助体系上也适时布置了一些大型项目，围绕某一关键科学问题进行共同攻关研究，如科学技术部国家科技重大专项、国家自然科学基金委员会重大研究计划和重大项目等。板块俯冲带的研究也与国家战略需求密切相关，因此需要进行集成攻关。设置持续高强度资助的体系，如国家自然科学基金委员会基础科学中心，也是很好的举措。根据俯冲带研究特点，应加大支持重大基础设施建设，包括重大仪器设备、前沿分析方法和技术、数据处理软件等。

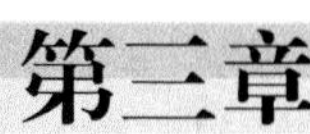

第三章 发展现状与发展态势

第一节　本学科主要研究领域的发展状况与趋势

一、板块俯冲带类型

（一）大洋俯冲带

一般意义上的俯冲带即指大洋俯冲带，是地表物质向地球内部传输的纽带，也是板块构造理论的核心环节之一。蚀变的大洋地壳，连同上覆沉积物及下伏的岩石圈地幔构成俯冲板片的主体，由海沟处进入俯冲带，板片在俯冲带深部经历高压－超高压变质作用发生脱水形成弧岩浆作用，脱水后的板片进一步俯冲到深部地幔完成地球表层－内部的物质循环。

1. 研究现状和趋势

（1）大洋俯冲带的起始及驱动力

洋壳俯冲如何开始及其驱动力问题是俯冲带研究的热点和基础。一般来说，新生的大洋岩石圈随着时间的推移逐渐变冷变重，这为俯冲持续提供了动力来源。俯冲起始的机制包含自发俯冲和诱发俯冲两种模式（Stern，2004）：自发俯冲起始的动力来源于板块自身的负浮力，如大洋转换断层或被动大陆边缘处由于重力不稳定引发的岩石圈坍塌；诱发俯冲起始的动力则来源于远场对先存薄弱带的作用。区分两种方式的根本标志是从俯冲板片发生埋藏到上覆板片开始拉张的时间间隔：自发俯冲起始时，俯冲板片向下俯冲，上覆板片同时拉张以平衡区域动力；而诱发俯冲起始时，俯冲板片被迫下插累积到一定时间后，上覆板片才发生拉张。大洋俯冲作用一旦起始，驱动力就变得非常重要（陈凌等，2020）。一种观点认为，大洋俯冲受控于岩石圈之下的地幔对流系统，特别是起源于核幔边界的地幔柱作用于板块底部，促使大陆裂解，并驱动板块运动（Wilson，1963，1973；Morgan，1971）。另一种观点认为，大洋板块在自身的重力作用下发生俯冲，俯冲板片经历高压变质之后密度大于周围地幔，自身的负浮力对与之相连的板块产生拖曳作用促使持续俯冲（Anderson，2001；Conrad and Lithgow-Bertelloni，2004）。后者为目前地球科学界较为普遍接受的观点。当大洋岩石圈俯冲殆尽时，后面的陆块与上盘板块发生碰撞，将会导致两种结果：一是比较轻的大陆岩石圈因浮力太大难以俯冲，俯冲的大洋板片发生断离造成俯冲终止（Davies and von Blanckenburg，1995）；二是大陆岩石圈在深部大洋岩石圈的拖曳下持续俯冲，大洋俯冲带逐渐转换成大陆俯冲带（Zheng，2021b）。

（2）大洋俯冲带的结构组成、分类及演化

大洋俯冲带体系的结构组成包括俯冲板片、海沟-增生楔体系、地幔楔-岛弧体系及相关的弧前盆地、弧后盆地等（Stern，2002）。俯冲板片由上至下包括沉积物、洋壳、岩石圈地幔。沉积物包含前陆及远洋的陆源、硅质和碳酸盐质沉积物，全球范围俯冲沉积物的平均成分与大陆上地壳一致。洋壳是俯冲板片中最重要的部分，包含上部的洋中脊玄武岩以及下部的席状岩墙和辉长岩。洋壳之下是蛇纹石化的岩石圈地幔。海沟是俯冲板片消减进入地球内部的入口，是位于汇聚板块之间的狭长条带，记录了全球俯冲带的长度，也是地球表面最深的地方，如马里亚纳海沟。从俯冲板片表层刮削下来的深海沉积物及洋壳碎片在海沟处堆积，形成了混杂堆积的增生楔。一旦板块俯冲到地幔深度，在俯冲板片和上盘板块之间就形成了由岩石圈地幔和软流圈地幔组成的三角形地幔楔。地幔楔受板片流体交代发生部分熔融，最终导致岩浆作用形成岛弧。由于汇聚板块之间的挤压作用，在岛弧与海沟之间形成弧前盆地，在岛弧远离俯冲带的方向形成弧后盆地。因此，平行展布的沟-弧-盆体系是鉴别古大洋俯冲带的有效标志之一（Stern，2002）。

根据汇聚板块的性质不同，可以把大洋俯冲带分成洋-洋俯冲带和洋-陆俯冲带两种。前者为一个大洋板块俯冲到另一个大洋板块之下，在西太平洋边缘形成典型的马里亚纳型俯冲带（图 2-3）；后者为大洋板块俯冲到大陆岩石圈之下，在东太平洋边缘形成典型的安第斯型俯冲带（图 2-4）。二者的差异不仅表现在上覆地幔楔的不同，而且存在弧后动力体制和增生楔大小等方面的差异（郑永飞等，2016）。

根据俯冲角度大小，一般将大洋俯冲带分成低角度（缓）俯冲（<30°）、中角度（正常）俯冲（40°～50°）和高角度（陡）俯冲（>70°）三种类型。在现代汇聚板块边缘，大多数大洋俯冲带都是正常俯冲和陡俯冲，只有大约10% 属于缓俯冲，基本上位于太平洋周边。一般来说，低角度俯冲导致弧后收缩（图 2-4），高角度俯冲引起弧后拉张（图 2-3）。

在板块俯冲的不同阶段，大洋俯冲带的形态可以发生变化。快的汇聚速率导致低角度俯冲和海沟正向前进，而慢的汇聚速率导致高角度俯冲和海沟反向后退。低角度俯冲为俯冲板片变质脱水和部分熔融提供了适当的温度压力条件，产生了同俯冲弧火山岩的水化和交代地幔源区（郑永飞等，2016）。高角度俯冲为后撤板片部分熔融提供了合适的温度压力条件，后撤板片产生的熔体交代地幔楔，形成了俯冲后板内火山岩的交代地幔源区。当板片进入软流圈地幔顶部，还可能发生平板俯冲，即大洋板片在大陆岩石圈之下的水平运动现象，多出现在太平洋周边地区。全球地震层析成像显示大洋板片的平板俯冲也可能发生在地幔过渡带中，板片在地幔过渡带平躺、滞留从而在其上部形成“大地幔楔”（Maruyama et al.，2009）。此外，当洋壳俯冲接近尾声发生陆－陆碰撞，或者板片俯冲被其他过程严重阻碍时，深部高密度俯冲板片会在重力的作用下对浅部板片产生一个巨大的拖曳力，从而引发大洋岩石圈在薄弱处断离。

根据汇聚板块边缘地热梯度的不同，可以将俯冲带分为三种类型（Zheng,2019）：①冷俯冲带（<11℃ /km）；②暖俯冲带（11～30℃ /km）；③热俯冲带（>30℃ /km）。根据地热梯度划分的三种类型俯冲带产生了三种类型的变质相系（图 3-1），分别为阿尔卑斯型、巴罗型和巴肯型。沿着小于11℃ /km 的低地热梯度俯冲会形成冷的或者超冷的俯冲带，对应的进变质相系在 *P-T* 轨迹上从阿尔卑斯型高压蓝片岩相到高压－超高压榴辉岩相，常见的含水矿物是硬柱石、多硅白云母和黝帘石。沿地热梯度 11～30℃ /km 俯冲的暖俯冲带，对应的进变质相系在 *P-T* 轨迹上从巴罗型低压绿片岩相经过中压角闪岩相到高压麻粒岩相，铝硅酸盐矿物为蓝晶石，常见的含水矿物为绿泥石、绿帘石和角闪石。沿地热梯度大于 30℃ /km 发生的低压变质作用，对应的进变质相系在 *P-T* 轨迹上为巴肯型角闪岩相到麻粒岩相，铝硅酸盐矿物为红柱石或夕线石，在麻粒岩中缺少含水矿物。

（3）大洋俯冲带的内部过程及产物

俯冲板片在进入俯冲带之前一般遭受了广泛的洋底热液蚀变。蚀变的洋壳形成大量的含水矿物和碳酸盐矿物，是俯冲板片中水和 CO_2 的重要来源。

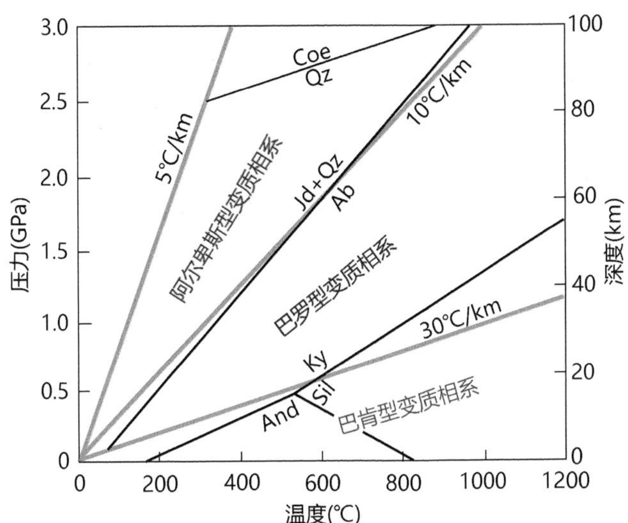

图 3-1　汇聚板块边缘区域变质相系与变质热梯度关系相图（修改自 Zheng and Chen，2017）
重要的相转换边界包括 SiO_2 和 Al_2O_3 的同质多相转变、钠长石分解成硬玉 + 石英的反应。矿物缩写：Ab- 钠长石；And- 红柱石；Coe- 柯石英；Jd- 硬玉；Ky- 蓝晶石；Sil- 夕线石；Qz- 石英

大洋岩石圈的冷却下沉以及在海沟入口处的弯曲都会造成大量的深大断裂，海水沿着断裂进入岩石圈地幔发生蛇纹石化（Ranero et al.，2003），蛇纹石化岩石圈地幔的厚度目前尚无统一的认识，但是被认为是俯冲板片中水的重要储库之一。

板片进入俯冲带之后，随着温度压力的增加，进入俯冲带的板片会在低地热梯度下经历阿尔卑斯型变质作用、在中等地热梯度下经历巴罗型变质作用（Zheng，2021c），发生一系列与之相关的岩石相变、矿物相转变、变质反应和流体作用。例如，洋壳物质在俯冲带的弧前深度（40～80 km）经历蓝片岩相到榴辉岩相高压变质作用，在 80～160 km 的弧下深度经历榴辉岩相超高压变质作用。基性变质岩中常见的含水矿物为角闪石、硬柱石、多硅白云母和黝帘石。俯冲带温压结构控制了俯冲岩石中含水矿物的稳定性，决定了板片脱水的位置，进而影响了弧岩浆作用的有无及其发生的位置（郑永飞等，2016）。板片脱水是俯冲带深部最引人注目的过程（Schmidt and Poli，1998）。一方面，脱水作用使俯冲板片密度变大，自身产生的负浮力拖曳着板片不断俯冲，扮演着俯冲驱动力的角色；另一方面，俯冲板片脱水释放的流体交代上覆地幔楔，诱发地幔楔发生部分熔融，最终导致岩浆作用形成岛弧（Tatsumi，1986；McCulloch and Gamble，1991）。板片脱水诱导的俯冲带 - 岛弧体系是俯冲带实现元素迁移、物质循环的主要途径（图 3-2）。例如，沉积物为俯冲板片贡献了重要的微量元素，如K、Sr、Ba、Th等，在岛弧玄武岩中发现这些不相容元素的富集证明了俯冲沉积物的贡献（Plank and Langmuir，1993）。俯冲带另一个产物就是诱发中深源地震，环太平洋俯冲带与地震带的耦合分布显示出二者的因果关系。俯冲板片变质脱水引起的流体超压促使岩石发生脆性破裂，可能是诱发俯冲带地震的重要方式（Hacker et al.，2003）。俯冲板片深俯冲携带壳源物质进入更深部地幔，一方面可以导致地幔的不均一性，另一方面实现了地球地表－内部的物质循环。

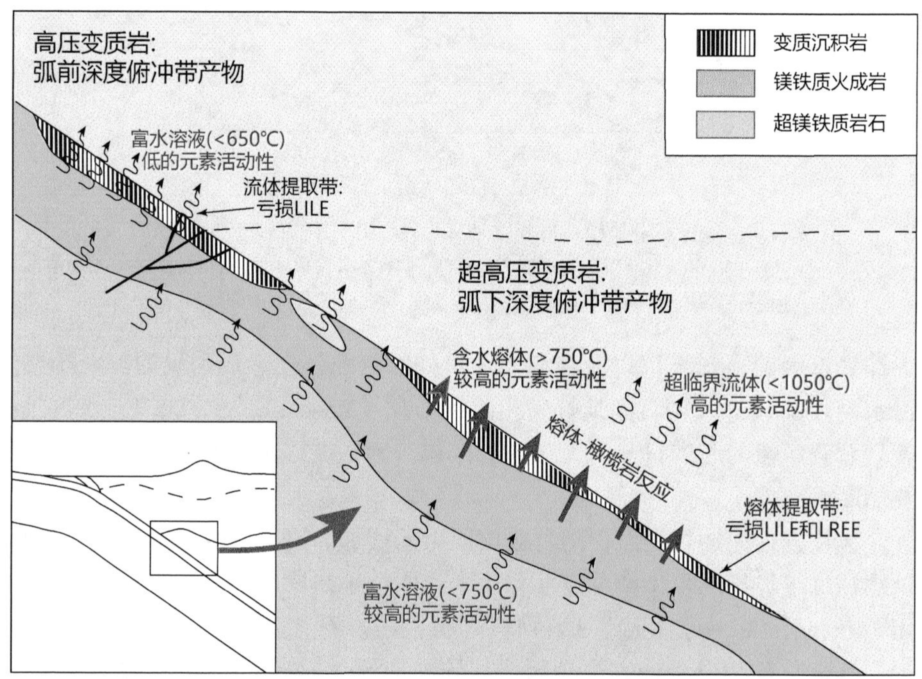

图3-2　弧前和弧下深度俯冲带变质作用析出不同类型的流体活动性元素交代地幔楔（引自Zheng，2019）

左下角图指示的是俯冲带的结构，框中是板片－地幔楔界面

2. 关键科学问题

（1）大洋俯冲的起始机制

地球可能是太阳系中唯一存在板块构造活动的星球，板块的运动和演化几乎合理解释了地壳中记录的和地球深部图像所展示的所有地质现象，并从根本上控制了大气圈和海洋的演化。尽管板块构造行为和几何学的许多方面都已清楚，但人们仍对一些重要问题缺乏共识。例如，地球上的板块构造是

何时启动的？为什么板块构造仅出现在地球而非其他星球上？另外，对具体某个俯冲带的起始机制也不清楚，现代俯冲带为什么广泛存在于环太平洋周围而非其他位置？俯冲带起始的关键控制因素是什么？俯冲带的启动涉及哪些具体地质过程？这些都是未来俯冲带研究的关键科学问题。

（2）俯冲带控制的地震和火山

全球地震带与俯冲带分布的高度重叠显示二者具有密切的成因联系。中深源地震（震源 70～700 km 深度）多发于俯冲带，但目前对中深源地震的诱发机理依然不清楚。主流观点认为俯冲板片变质脱水释放流体，流体超压促使岩石发生脆性破裂是诱发俯冲带地震的主要方式（Hacker et al.，2003）。但也有研究强调在俯冲带温压条件下，岩石应该发生韧性而非脆性变形，认为俯冲带局部薄弱面发生剪切热不稳定性以及亚稳态橄榄石相变，是诱发俯冲带地震的主要原因（John et al.，2009）。除了含水矿物脱水，在温度升高和相变条件下，名义上无水矿物中的大量结构水析出，可以形成类似含水矿物的脱水反应。因此，名义上无水矿物中的结构水脱水致裂是一种可能的深源地震机制（Zhang et al.，2004）。对俯冲带深源地震发震机理的认识则更为模糊。俯冲带控制着岛弧岩浆作用，但弧下地幔楔发生的是直接加水熔融（单阶段过程），还是需要额外的热源驱动地幔楔部分熔融（两阶段过程）？大洋俯冲带内有没有板片熔融发生？板片熔体对弧岩浆成分的贡献比例是多少？俯冲带过程是否直接控制火山喷发的启动、持续时间、规模和强度？这些都有待深入探究。

（3）俯冲带控制的关键元素循环

俯冲带控制着全球的物质、挥发分等关键元素的循环。地质历史上关键元素强烈的再分配与矿物和生物多样性的爆发、大氧化事件、海洋缺氧事件，以及相关的大气成分的变化密切相关，并影响着气候和生命的演化历史。例如，碳、氢、氮、硫等元素是创造宜居环境的生物关键元素，深入理解俯冲带内部这些关键元素循环是如何进行的，俯冲变质 - 岩浆系统中氟、氯等卤素是如何分配到熔体和流体中的，以及金属成矿元素是如何在俯冲带内循环、如何在岛弧系统中富集的，是今后俯冲带研究的关键科学问题。

（二）大陆俯冲带

1. 研究现状

按照传统的板块构造学说，大陆地壳由于密度较低，不可能俯冲到高密度的地幔中。然而，20 世纪 80 年代地球科学家分别在西阿尔卑斯和挪威西部的变质表壳岩石中，发现了超高压变质矿物柯石英（Chopin，1984；Smith，1984），之后又分别在哈萨克斯坦科克切塔夫（Kokchetav）地块和中国大别山的变质表壳岩石中（Sobolev and Shatsky，1990；Xu et al.，1992）发现了金刚石，证明大陆地壳曾俯冲到大于120 km的地幔深度发生超高压变质（图3-3），然后折返至地表。这些发现改变了传统的地球动力学观念，在国际上引发了大陆深俯冲和超高压变质作用的研究热潮，开启了大陆深俯冲作用研究的序幕。

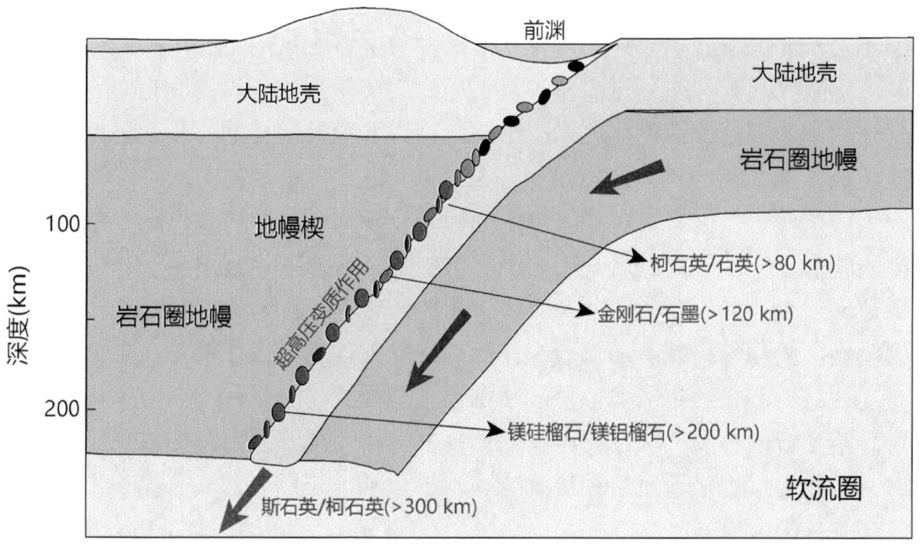

图 3-3　大陆地壳俯冲深度的矿物学标志（引自郑永飞等，2015）

30多年来，大陆深俯冲作用的研究已成为国际固体地球科学界最活跃的前沿领域之一。地球科学家通过深入的研究，相继在全球至少30个碰撞造山带中发现了超高压变质岩及大陆深俯冲作用存在的证据，其中包括中国的大别-苏鲁、北秦岭、柴北缘、阿尔金、西南天山和东昆仑6个地区，完善和丰富了板块构造理论，极大地推动了人们对于大陆俯冲带相关的汇聚板块边缘造山过程的认知（Zheng et al.，2011；Zheng，2012；郑永飞和陈伊翔，2019）。迄今，国际上关于大陆深俯冲作用研究突出的进展有以下几个方面。

（1）陆壳岩石可深俯冲到斯石英稳定域的地幔深度

对大陆俯冲深度的研究，继柯石英和金刚石之后，近年来取得了两次重要突破。其一，超硅石榴子石等特征矿物的出溶结构指示俯冲深度超过200 km，如挪威西部、希腊以及中国的苏鲁、北秦岭、柴北缘和南阿尔金等地的榴辉岩、石榴橄榄岩和片麻岩等（van Roermund and Drury，1998；Ye et al.，2000；Mposkos and Kostopoulos，2001；Liu et al.，2005；Song et al.，2005）。其二，在南阿尔金泥质片麻岩和榴辉岩中发现先存斯石英的出溶结构和斯石英副象（Liu L et al.，2007，2018），以及挪威西部片麻岩区的石榴橄榄岩中石榴子石出溶辉石（Spengler et al.，2006）和大别超高压石榴橄榄岩的普通辉石出溶斜顽辉石（Liu X W et al.，2007）等，这些研究把陆壳俯冲 / 折返深度由柯石英稳定域推进到斯石英稳定域的地幔深度（＞300 km）。因此，有必要将其定义为超深俯冲，特指俯冲到斯石英稳定域温压条件对应的地幔深度（＞300 km）的超高压变质岩石。

（2）精确确定地壳深俯冲和折返不同变质阶段的变质年龄

得益于测试技术平台（包括 SHRIMP、SIMS 和 MC-ICPMS）的不断进步与完善，尤其是微区原位（包括薄片尺度）定年技术的日渐成熟，目前能够实现对高压 / 超高压岩石峰期变质、退变质时代和原岩形成时代的精确测定。锆石是超高压变质岩中的常见副矿物，通过对其阴极发光图像、不同微区矿物包裹体及其组合、Ti 含量温度计和微量元素配分的分析，已实现对不同变质阶段事件的精确定年。结合对榍石、独居石和金红石等其他副矿物 U-Th-Pb 以及含钾矿物的 Ar/Ar 定年，可以清晰查明造山带各个阶段的构造热演化历史。此外，石榴子石在变质过程中通常保存复杂的生长和扩散环带，通过对每个石榴子石环带进行精细 Sm-Nd 和 Lu-Hf 同位素定年，可揭示更精细的造山过程。高精度定年结果表明，世界范围内陆壳深俯冲和折返主要发生于显生宙，时间跨度上从新元古代晚期到新生代。与大洋地壳俯冲 / 折返的持续时间相比，大陆地壳俯冲 / 折返的持续时间总体上相对较短。此外，大的超高压变质地体在地幔深度的居留时间相对较长（一般在 15 ± 2 Ma），而小的超高压变质地体在地幔深度的居留时间相对较短（Kylander-Clark et al.，2012；郑永飞等，2013）。

（3）超高压变质岩 *P-T-t* 演化轨迹重建

大陆深俯冲超高压变质岩 *P-T-t* 演化轨迹的重建对认识造山带的演化过程具有重要意义。目前，常用的方法是基于岩石学观察研究的反演以及相平衡模拟计算获得的视剖面图的正演。通过这些方法的综合应用，并结合对不同变质阶段的定年结果，已成功地对很多造山带中超高压变质岩进行了精确的 *P-T-t* 演化轨迹重建。这些研究结果表明，俯冲带中超高压岩的 *P-T* 轨迹均为顺时针型，进一步还可区分为“近发卡状”降温降压型、近等温降压型和先增温后降压型三种类型，反映了不同造山带或同一造山带中不同类型超

高压变质岩折返 *P-T* 路径上的差异。与洋壳俯冲带超高压变质岩相比，大陆俯冲带超高压变质岩一般先后经历了近等温降压和先增温后降压的折返过程（Zheng，2021d），结果受到不同程度的叠加改造，看上去好像属于中温型甚至高温型，峰期变质温度达到 600～900℃甚至 900～1000 ℃（张立飞，2007），但是该温度含有叠加变质的组分。另外，近年来数值模拟正演也再现了大陆俯冲带中超高压岩石顺时针型的 *P-T* 演化轨迹（Sizova et al.，2012；李忠海，2014），但是在高温 / 低压下受到叠加变质的超高压岩石可能会出现逆时针型 *P-T* 演化轨迹（Zheng and Chen，2017）。

（4）大陆深俯冲发生的必要条件

前人研究表明，低密度陆壳物质深俯冲的必要条件是前期俯冲的高密度大洋板块的重力拖曳或牵引。虽然这一认识尚未被地球物理观察直接证实，但在一些造山带如阿尔卑斯、波希米亚、红安和柴北缘等高压 - 超高压地体中发现了俯冲变质洋壳和陆壳岩石的共存（Zhou et al.，2015；Zhang et al.，2017），而且越来越多的动力学数值模拟研究也再现了这一过程（李忠海，2014）。数值模拟研究还进一步表明，单向高角度陡俯冲是引发陆壳深俯冲和超高压变质作用最主要的途径，而陆壳低角度的缓俯冲一般只形成高压岩石，缺乏超高压变质岩石（李忠海，2014）。这一结果与前人关于喜马拉雅西构造结和中喜马拉雅地区的地质观察研究非常吻合。同理，可以推断陆壳超深俯冲以及牵引陆壳发生超深俯冲的先期洋壳也可能是高角度的陡俯冲。

（5）大陆俯冲带中超高压变质岩的折返机制与路径

高压 - 超高压岩石的折返机制一直是大陆深俯冲研究的核心科学问题之一，是理解俯冲碰撞造山带动力学演化的关键。过去 30 多年来，前人基于地质观察、实验和数值模拟结果，提出了十余种超高压变质岩的折返模型（李忠海，2014；刘贻灿和张成伟，2020；张立飞和王杨，2020）。根据综合折返的路径不同，可将超高压变质岩的折返区分为岩片沿俯冲隧道逆冲折返和片麻岩以穹隆状侵出折返这两种基本方式（郑永飞等，2009；Zheng，2021d）。基于数值模拟研究提出的底辟折返模型，是指深俯冲的地壳物质在弧下深度以底辟柱体的形式穿过上覆地幔楔而折返到地表（Hall and Kincaid，2001；Gerya and Stöckhert，2006；Sizova et al.，2012）。但是，该模型要求地幔楔发生张性裂隙以便超高压变质岩片垂直上升通过，这点对于冷的刚性上覆岩石圈板块是难以实现的（Liu M Q et al.，2017）。

俯冲隧道模型最早是针对大洋俯冲隧道提出的（Shreve and Cloos，1986；Cloos and Shreve，1988a，1988b），近年来被扩展到大陆俯冲带研究领域（Gerya et al.，2002；郑永飞等，2013；张建新，2020）。该模型解释了不同变质级别的变质岩构造混杂在一起的地质现象，并得到数值模型的验证（Beaumont et al.，2009；李忠海，2014；Liu M Q et al.，2017）。在俯冲隧道模型中，深俯冲大陆板片的行为受控于两种驱动力的竞争结果：一种是俯冲的低密度地壳物质所导致的向上的浮力；另一种是俯冲大洋板片向下的牵引力。当俯冲隧道中地壳物质的强度不足以承受其浮力所导致的差异动力时，就会在俯冲板片中发生应变集中并形成逆冲断层，进而导致物质的拆离和折返。因此，深俯冲地壳的逆冲折返发生在大陆碰撞的晚期阶段（郑永飞等，2013）。

穹隆式侵出模型指地壳根部岩石在经受部分熔融后，长英质熔体裹带熔融残留体沿减薄的岩石圈上升，在地表形成麻粒岩－混合岩－花岗岩穹隆。该模型的重要特征是超高压变质岩经受高温－超高温的变质叠加，一般发生在大陆碰撞之后（Zheng，2021d）。

（6）大陆俯冲带中存在显著流体活动

大陆俯冲带中流体的存在不仅会促进变质反应进程、诱导岩石体系发生部分熔融，还会使俯冲板块内部发生显著的流体迁移，引起元素活动和同位素变化（Bebout，2007；Zheng，2019）。近年来研究（Zheng，2009）发现，一些名义上无水的超高压矿物（如石榴子石、绿辉石和金红石等）含有微量结构水，证明深俯冲陆壳可以携带浅部流体进入深部地幔。同时，在超高压变质岩及其包裹的脉体中发现了超临界流体出现的岩石学证据（Hermann et al.，2006；Zheng et al.，2011），这种流体对通常不活动元素（如重稀土元素和高场强元素）具有很强的溶解和迁移能力。造山带中高压－超高压岩石还可以发生显著部分熔融作用。人们在超高压变质的长英质岩石中发现了混合岩化现象和薄片尺度的部分熔融结构、在榴辉岩中发现了以石英和长石为主的晶体包裹体，这些都指示深俯冲陆壳岩石在折返过程中发生了不同程度的部分熔融（Chen Y X et al.，2017）。超高压岩石在地幔深度的部分熔融会显著降低岩石黏滞度，促进甚至诱导深俯冲陆壳岩石从地幔深度的快速折返（Wallis et al.，2005；Labrousse et al.，2011）。

2. 发展趋势和关键科学问题

1）地壳深俯冲特征指示矿物的保存机制及其识别，实验模拟超高压特征指示矿物成分变化与 P-T 条件的对应关系及其对矿物出溶的控制作用。

2）大陆地壳在什么情况下才能发生深 / 超深俯冲，其组成、性质、规模如何；精细的地质观察研究与数值模拟密切结合、相互印证，定量再现和评估大陆俯冲带中超高压变质岩石形成与折返的地球动力学过程；深俯冲陆壳板片发生弱化拆离的控制因素及其对折返过程和路径的影响；不同同位素体系定年所得年龄与超高压岩石不同变质阶段 *P-T* 条件关系的准确制约。

3）深俯冲陆壳岩石发生部分熔融的时限、熔融机制、空间范围及其对折返过程的关系；大陆俯冲板片不同深度来源流体的物理和化学性质，俯冲带不同深度壳幔相互作用的实验研究和数值模拟；俯冲带不同深度交代作用的橄榄岩和壳幔相互作用的岩浆岩记录。

4）大陆深/超深俯冲及其折返作用对碰撞造山带的形成与演化过程究竟有哪些限定；对碰撞造山带的组成、结构、构造形态又有哪些影响。

（三）俯冲工厂与俯冲隧道过程

1. 研究现状和趋势

俯冲带是地球上许多重要地质过程发生的场所，其中最重要的过程就是促使地壳物质循环进入地幔、新的地壳物质的产生及大陆地壳的形成，这一过程被形象地称为“俯冲工厂”（Eiler，2003）。这一概念最初在 20 世纪末由美国国家科学基金会的重大研究计划大陆边缘计划（MARGINS）所提出，而后很多学者开始从事相关研究（Hacker et al.，2003；Tatsumi，2005；van Keken et al.，2011）。岛弧火山作用是俯冲工厂的主要产物，岛弧玄武岩是研究俯冲带温压结构和地球化学分异的重要对象。

俯冲工厂原材料有俯冲的大洋沉积物、大洋岩石圈及上覆的地幔楔等；产品包括弧岩浆岩、新生大陆地壳、高压-超高压变质岩、矿床及地震活动等；脱水、交代、变质、变形及部分熔融作用是工厂内部的加工工艺流程，而加工后的残留物是指被化学改造（脱水、变质、交代等）的俯冲洋壳和沉积物，以及最终拆沉到地幔的残留镁铁质弧下地壳等，这些残留物被运移和储存到深部地幔，并可作为地幔柱的原材料参与地幔循环（图 2-6）。

具体来说，洋壳沉积物、蚀变洋壳和大洋岩石圈地幔等原材料随着俯冲作用进入工厂，然后会发生一系列变质反应，形成高压-超高压变质岩，并出现脱挥发分乃至部分熔融，从而导致俯冲板块发生明显的地球化学分异（图 3-4）。这一地球化学分异作用通过流体交代活动影响上覆地幔楔，受到流体交代的地幔楔发生部分熔融形成初始弧岩浆，这些岩浆发生分异进一步形成具有大陆地壳成分特征的安山质岩浆岩，并伴随大规模物质运移、元素分异作用及大规模成矿作用；同时也诱发了地震和火山活动，释放了碳、氮、硫和水等挥发分，影响地球大气环境。改造后（变质、脱水及部分熔融）的残留俯冲物质通过进一步俯冲循环到地幔深处，弧岩浆分异后的残留镁铁质弧下地壳也可由于重力作用拆沉到深部地幔，这些进入地幔的物质参与深部地幔循环，影响地幔的组成和流变性质，导致地幔的化学和流变学不均一性（Zheng，2019）。因此，俯冲工厂从原料到产品的“生产过程”形成了复杂的地球动力学循环体系，是地球上大气圈、水圈、岩石圈和深部地幔相互作用及物质和能量交换的场所。

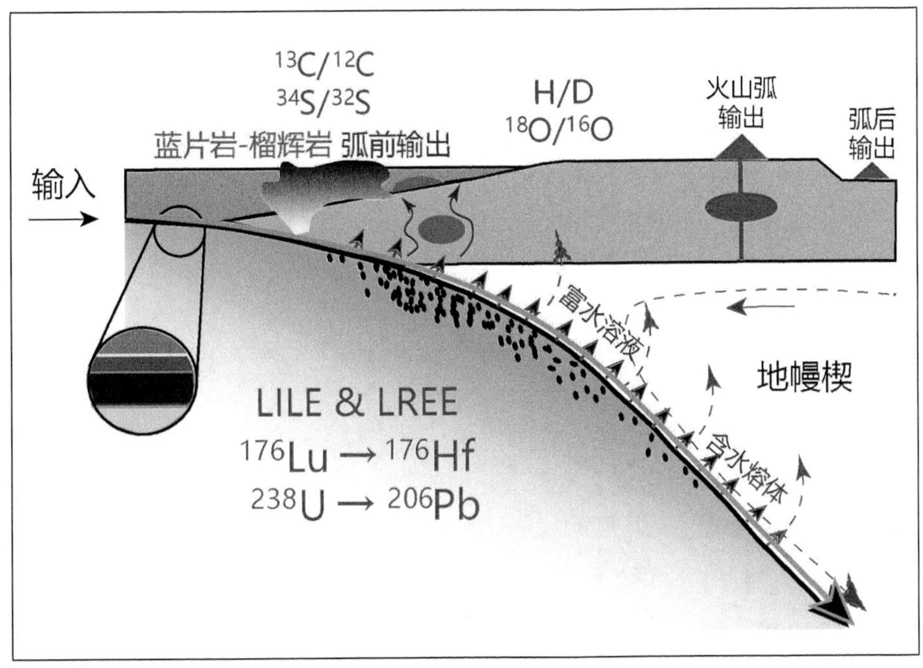

图 3-4　大洋俯冲带地球化学传输俯冲工厂输入输出物质示意图

俯冲工厂可分为活动大陆边缘俯冲体系和洋内俯冲体系。相对于前者，洋内俯冲体系曾被认为是研究俯冲工厂更为理想的场所（Eiler，2003）。它所形成的大洋弧直接建立在薄的基性洋壳基础上，不受大陆物质混染的影响，更能直接研究新生大陆地壳的生长机制。典型的现代俯冲工厂实例是西太平洋的伊豆 - 小笠原 - 马里亚纳（IBM）洋内俯冲体系，近 20 年来已成为全球地球科学家关注的焦点。近年来，中国科学家也通过国际合作和组织科考团对这一全球著名的俯冲工厂——IBM 洋内俯冲体系进行多学科的联合考察和研究。

对现代汇聚板块边界的直接观察和研究可以获得正在进行的俯冲工厂过程的重要信息。然而，由于俯冲工厂过程主要发生在俯冲带深部，深海钻探难以触及，受分辨率所限的地球物理探测也难以精确获得俯冲工厂过程的细节。在造山带中出露有古俯冲工厂过程不同阶段的物质。虽然受后期改造和破坏，但这些物质记录了俯冲工厂生产和加工全过程的信息，对古俯冲带物质和结构的精细解剖是重建古俯冲工厂过程的重要手段，加深了对俯冲带深部过程的认识。

俯冲工厂过程是板块构造体系的重要环节，在地球板块构造启动时就开始运转。因此，古俯冲工厂过程的研究也涉及地球板块构造启动和初始大陆地壳形成机制等重大科学问题。此外，在俯冲工厂过程中，在原材料上以往研究更强调了大洋物质，但越来越多的工作显示大陆物质作为原材料在俯冲工厂过程中也起重要作用。这些大陆物质既包括陆源碎屑沉积岩，也包括来自俯冲带上盘构造剥蚀的大陆基底，还包括深俯冲的大陆岩石圈。近 30 年的研究显示，大陆岩石圈在大洋岩石圈的拖曳之下，可以俯冲到地幔深部甚至过渡带，并参与地幔循环，引起地幔的不均一性。俯冲工厂过程中大陆地壳的生长和返回至地幔两种相反的过程维持了地球动力学系统中洋 - 陆物质通量的平衡。

俯冲隧道是指汇聚板块边界上下板片之间具有独立运动学特征的相对窄的薄弱带，主要由低密度、低黏度、高度剪切变形的基质（沉积物或蛇纹岩）和少量高密度、高黏度的块体所组成，具有典型混杂岩特征，并可向下和向上流动（回流），形成隧道环流（图 3-5）。俯冲隧道的概念模型最初是基于对现代大洋俯冲带的地质和地球物理观察及数值模拟而建立的（Shreve and Cloos，1986; Cloos and Shreve，1988a，1988b），模型设计的最大深度为 30 km。随着数值模拟技术的发展和对古俯冲带研究的深入，一些学者把俯冲隧道模型延伸到俯冲带 80～100 km 及更大深度，并由此来探讨俯冲带深部动力学环境及古俯冲带高压 - 超高压岩石的折返机制。

近年来，俯冲隧道模型也运用到了大陆俯冲带中，用于解释大陆俯冲过程中的流体作用、壳幔相互作用及超高压陆壳岩石的折返机制，透视大陆俯冲构造过程及其产物，大陆俯冲隧道是对俯冲隧道模型的拓展（Gerya et al.，2002; Guillot et al.，2009; Butler et al.，2013；郑永飞等，2013；Ring et al.，2020；张建新，2020）。然而，由于大陆俯冲带的俯冲物质组成、上下板块性质及板块 - 地幔楔之间俯冲界面性质等方面与大洋俯冲带不同，大陆俯冲隧

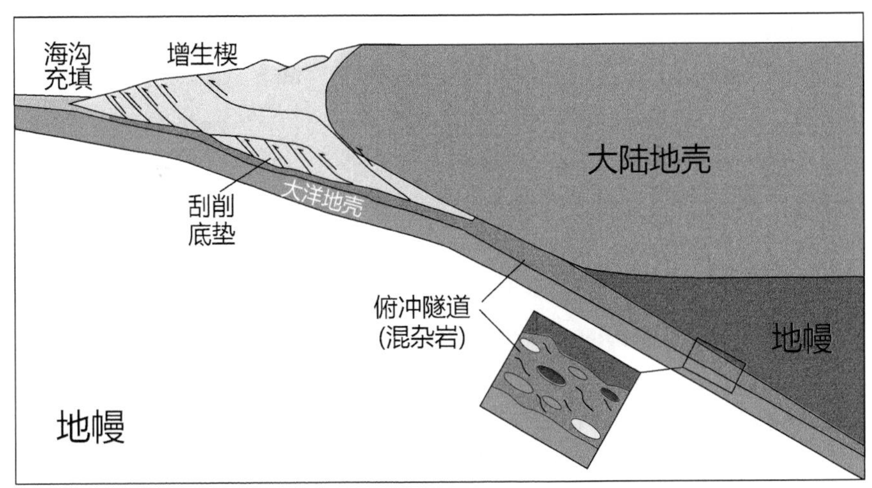

图 3-5　俯冲隧道内部物质流动示意图（修改自郑永飞等，2013）

道在组成、结构及动力学特征等方面也与大洋俯冲隧道存在一定差异。总体上，无论是大洋俯冲带还是大陆俯冲隧道，均是板块界面相互作用以及实现地球表层与内部之间物质和能量交换的重要场所，俯冲隧道过程从某种程度上就是上下板片相互作用的过程，也是地球表层与内部之间相互作用的纽带（郑永飞等，2013）。相互作用过程包括变质脱水、部分熔融、流体活动、元素迁移、韧性变形及地震活动等。因此，俯冲隧道可看作整个俯冲工厂体系的一个“加工车间”，俯冲隧道过程既是俯冲工厂过程中的重要环节（如俯冲隧道内脱水及部分熔融作用形成的流体对上覆地幔楔的改造），同时还具有相对独立的物质运动和循环体系。

2. 关键科学问题

相对于对整个俯冲工厂过程在宏观上已有较清晰的认识而言，人们对俯冲隧道过程的了解还很有限，并存在很多争议。其存在的科学问题和挑战主要表现在以下几方面。

（1）俯冲隧道的几何结构及随深度的变化

在几何学上，俯冲隧道等同于上下板片之间具有一定宽度的界面，是上下板片之间复杂的剪切带。然而，受分辨率所限，地球物理方法对现代俯冲带数千米以下的俯冲界面的结构还不能准确识别（Sage et al.，2006）。同样，由于目前的方法所限，数值模拟也不能精确地模拟俯冲隧道更细微的结构特征。由于后期的叠加改造和俯冲隧道随时间的演化，古俯冲隧道的结构也难以完整保存。因此，现有的研究手段还不能详细解剖俯冲隧道的结构。

一些学者认为俯冲板片和上覆地幔楔之间的最大解耦深度是俯冲隧道的最大深度。当大于这个深度时，俯冲板片和上覆地幔楔之间黏滞耦合在一起，代表了俯冲隧道消失，并驱使弧下热的软流圈流动（Wada and Wang，2009；Abers et al.，2017）。对于大洋俯冲带，这个深度在 80 km 左右（Syracuse et al.，2010），大致与通过折返的洋壳榴辉岩的岩石学研究所提出的榴辉岩的“不归点”一致（Agard et al.，2018）。然而，少量古俯冲带中折返的洋壳性质的榴辉岩具有超高压性质，其所代表的古俯冲隧道的深度可能大于 100 km。此外，这些认识主要针对大洋俯冲隧道，对大陆俯冲隧道的最大深度还没有一个明确的认识，如 Gerya 等（2008）的数字模拟显示大陆俯冲隧道可延伸到 150 km，郑永飞等（2013）认为大陆俯冲隧道至少可延伸到 200 km。

（2）俯冲隧道的变形机制及流变学性质

位于板块界面的俯冲隧道是经历强烈变形作用的高应变带，俯冲隧道中物质的弱化是俯冲地壳拆离和折返的先决条件。然而，起主导作用的弱化机制仍然存在争议。俯冲隧道物质的流变学是以位错蠕变为主还是以溶解－沉淀蠕变或扩散蠕变为主？其变形的动力条件和应变速率是否存在差异？俯冲隧道的变形是非常复杂和不均匀的，受俯冲隧道的部位、地热梯度、隧道内混杂岩中块体－基质组成和比例、矿物相变以及变形分解作用等多种因素所制约，并随俯冲隧道的时空演变而发生变化。例如，能干性相对强的块体和相对弱的基质变形机制存在明显差异；随俯冲深度变化，一些层状硅酸盐矿物（绿泥石、滑石、蛇纹石等）向链状硅酸盐矿物（角闪石、辉石等）转变，物质流变学强度增加，从而影响了整个俯冲隧道混杂岩的流变学性质和流动规模（Penniston-Dorland et al.，2018）。

（3）俯冲隧道与地震

汇聚板块边界和俯冲带是地球上地震活动的主要地带。从理论上讲，地震活动主要不应发生在俯冲隧道内，以低黏度变沉积岩和蛇纹岩为特征的俯冲隧道的性质决定了其内部以无震蠕动为主。然而，一些折返的古俯冲带岩石保留有古地震的重要标志，如高压角砾岩和榴辉岩相假玄武玻璃（Angiboust et al.，2012；Menant et al.，2018），这些古地震是发生在古俯冲隧道内还是在俯冲的大洋（或大陆）板块内部？另外，俯冲隧道内地震样式主要取决于隧道内混杂岩块体的大小及所含比例，隧道内地震与相对能干的块体的脆性变形有关，而基质则表现出以无震的流动为特征，俯冲隧道中混杂岩的变形分解为无震和地震滑动（Yin et al.，2018）。正是由于俯冲隧道流变学的不均一性，隧道内物质的变形可同时表现为脆性破裂和黏滞剪切，显示出重复性的摩擦和黏滞应变循环，这与地球物理观察和模拟计算所获得的俯冲带中慢地震特征一致。

（4）俯冲隧道与高压－超高压岩石折返

俯冲隧道模型很好地解释了一些古俯冲带特别是大洋和大陆俯冲带中高压－超高压变质岩石的折返（郑永飞等，2013；张建新，2020）。然而，是否俯冲隧道模型适合所有的高压－超高压变质岩石的折返？或者是否所有的高压－超高压岩石的折返都与俯冲隧道有关？俯冲隧道岩石具有典型混杂岩特征，隧道内块体和基质以及块体之间的变质峰期条件及 *P-T* 轨迹存在明显差异，并表现出不连续性特征。但是在一些古俯冲带中，高压－超高压变质岩石显示出几千米至数十千米均匀或连续的变质作用特征。总体上，俯冲隧道模型能合理解释具有混杂岩特征的高压－超高压地体的折返，但难以解释大而连续的且规模巨大的含有片麻岩穹隆的高压－超高压地体的折返。

（5）俯冲隧道的物质运移过程（运动学）

数值模拟显示，俯冲隧道的物质运移以平行隧道的运动为主，但也存在垂直向上的运动（Gerya et al.，2008；李忠海等，2015）。对于大洋俯冲隧道，其物质的折返路径以平行于隧道的回流为特征，而在大陆俯冲隧道中，由于放射性生热和黏滞加热等因素，可发生部分熔融，形成热的俯冲隧道。这些高温部分熔融且具有正浮力的高温超高压岩石打破了俯冲隧道的平衡，可垂直向上穿切上覆地幔楔（Gerya et al.，2008；Hacker et al.，2011），底辟或上侵到上覆岩石圈的中下地壳。然而，近年来的研究显示，洋壳俯冲隧道的混杂岩也可能垂直底辟抬升进入地幔楔，并运移到弧岩浆的源区，并发生部分熔融形成弧岩浆岩（Codillo et al.，2018）。数值模拟显示隧道中低黏度物质可发生环流（Gerya et al.，2002）。古俯冲隧道中折返的高压变质岩也记录了非常复杂的 *P-T* 轨迹及压力循环，反映了俯冲隧道内多期俯冲－折返循环（Blanco-Quintero et al.，2011；Li J L et al.，2016）。

总之，对于俯冲工厂和俯冲隧道，人们能见到的只是部分原材料和加工后的产品，对于复杂的加工工艺过程不能直接观察到，需要地质、地球物理、地球化学、物理和数值模拟及对古俯冲工厂和俯冲隧道进行综合研究，来揭示俯冲工厂及俯冲隧道的内部运作过程，最终重建俯冲系统的循环和地球动力学过程。

二、俯冲带壳幔物质循环

（一）俯冲带变质作用与流体活动

随着温度压力的增加，进入俯冲带的板片会经历高压－超高压变质作用，发生一系列的岩石相变、变质反应和流体作用。这些变质反应控制了俯冲板片的浮力和向下拖曳机制，并可以解释流体的形成、释放和运移过程，来自俯冲板片的流体交代上覆地幔楔橄榄岩，最终导致复杂的岛弧岩浆作用（Tatsumi，1986；McCulloch and Gamble，1991）。

1. 研究现状和趋势

（1）俯冲带变质作用

基性岩构成俯冲洋壳主体，其变质作用及脱水反应是理解俯冲带各种地质过程的最关键因素。基性岩在俯冲带低地热梯度条件下一般经历沸石相、葡萄石－绿纤石相、绿片岩相、蓝片岩相到榴辉岩相，岩石相变由一系列的脱水反应所控制（Schmidt and Poli，1998）。在前两个阶段，沸石、绿纤石和葡萄石是主要含水矿物，全岩含水量为8%～9%。高温高压实验研究显示，从很低级变质开始，变质基性岩中主要含水矿物包括绿泥石、绿帘石、硬柱石、角闪石、硬绿泥石和白云母等。绿泥石含水量很高（达12%），在温度超过600 ℃时，绿泥石完全分解转变成石榴子石。硬柱石含水量也很高（达11%），稳定于高压低温区域。变质基性岩中角闪石的温度上限受全岩成分影响很大，压力上限为～2.4 GPa。绿帘石稳定于3 GPa以下的400～750 ℃的中温区域。白云母的稳定域很大，其压力上限可达到9 GPa以上的斯石英稳定域（魏春景和郑永飞，2020）。

俯冲板片沉积岩包括硅质岩、泥质岩和杂砂岩等，在泥质岩和杂砂岩中出现的主要含水矿物为叶蜡石、绿泥石、硬绿泥石、纤柱石、十字石、白云母和黑云母等（魏春景和郑永飞，2020）。有关变质沉积岩亚固相线条件下的实验资料相对较少。对于泥质岩成分体系，叶蜡石仅出现于非常富铝的泥质岩中，温度上限低于450 ℃。硬绿泥石的稳定上限也主要受温度控制，最高不超过600 ℃。绿泥石在1.5 GPa以下，主要受温度控制，温度上限约为650 ℃；而在1.5 GPa以上，绿泥石的稳定上限明显受压力控制，可达到3GPa。黑云母为低压矿物，随着压力升高会转化为白云母（Vielzeuf and Schmidt，2001）。白云母的稳定域非常大，压力上限可到9～10 GPa。

俯冲板片橄榄岩中主要含水矿物为叶蛇纹石、绿泥石、滑石，高压时出现A相、10Å相、E相、斜硅镁石和含水瓦兹利石（魏春景和郑永飞，2020）。叶蛇纹石出现于小于700 ℃、小于7 GPa的*P-T*范围，叶蛇纹石升温可转变为绿泥石和10Å相，升压可转变为A相。绿泥石的压力上限约为5 GPa，温度上限约为780℃。滑石局限于低压中温条件，介于叶蛇纹石与角闪石稳定域之间，压力小于1.5 GPa，但在单矿物滑石脉中，滑石可稳定至4.5～5.0 GPa（Pawley and Wood,1995）。在富Ca和Na的二辉橄榄岩或地幔岩中，会出现钙质角闪石，其稳定*P-T*范围为500～1100 ℃，小于3 GPa（Fumagalli and Poli，2005）。如果岩石中有K_2O，会形成金云母和钾－碱镁闪石，这是地幔楔橄榄岩受到流体交代作用的主要特征。金云母的温度上限不超过1150 ℃，压力不超过9 GPa，随着压力增加，金云母转变为钾－碱镁闪石（Konzett and Ulmer，1999）。

（2）俯冲带流体活动

在不同地热梯度下，俯冲洋壳脱水量在不同深度有显著差别。在暖－热俯冲带中，完全水化的洋壳岩石在小于80 km的弧前深度释放了大约2/3的水；而在冷和非常冷的俯冲带中，在弧前深度脱水量大概只有1/3。在80～160 km的弧下深度，在暖到热俯冲带中洋壳大概还有1%的水；而在冷到超冷的俯冲带中还有大约4%的水储存在含水矿物硬柱石、黝帘石、硬绿泥石、滑石和多硅白云母中（郑永飞等，2016）。主要含水矿物硬柱石在弧下深度的分解造成冷洋壳在这个范围内的脱水量非常可观，强烈的脱水造成上覆地幔楔的大规模水化，最终导致弧岩浆作用（Tatsumi，1986；McCulloch and Gamble，1991）。在硬柱石消失之后，变质基性岩中的含水矿物只有多硅白云母，由于大多数基性岩贫钾，其含水量很少（＜0.1%）（魏春景和郑永飞，2020）。但在超高压条件下，名义上的无水矿物中还可以含有一定量的水（Zheng，2009）。只有在极少数的热俯冲带，基性岩才能穿过湿固相线，发生部分熔融形成钾质花岗岩和奥长花岗质熔体（Hernádez-Uribe et al.，2019）。

泥质沉积物在俯冲到弧下深度持续升温过程中先后发生绿泥石和硬绿泥石脱水，在温度升至550～600 ℃铁镁质含水矿物消失后，含水矿物为多硅白云母，岩石含水量取决于岩石K_2O含量，对平均成分泥质岩其含水量可达到1.85 wt%①（魏春景和郑永飞,2020）。大多数冷俯冲带的地热梯度不会穿过泥质岩的固相线，但在较热的俯冲带中，俯冲沉积物会在80～160 km深度发生部分熔融，形成富钾花岗质熔体（魏春景和郑永飞，2020）。

超基性岩在俯冲过程中的流体行为取决于其初始水化程度。大多数俯冲

① wt%表示质量百分数，下同。

超基性岩的水化程度都很低（<25%），远不能达到饱和状态，因此不会在弧前俯冲阶段发生明显脱水。其实，即使超基性岩完全水化，变成蛇纹石和绿泥石，它们也不会在弧前俯冲阶段分解脱水，超基性岩明显脱水发生在弧下俯冲阶段（Ulmer and Trommsdorff，1995）。在平均和较热俯冲带中水化橄榄岩会在 120～180 km 的弧下深度先后发生叶蛇纹石、绿泥石和 10Å 相脱水；在冷俯冲带中，部分叶蛇纹石在约 160 km 深处转变为绿泥石发生部分脱水；在非常冷的俯冲带中，在约 220 km 深度叶蛇纹石转变为 A 相发生部分脱水。因此，在俯冲过程中超基性岩脱水主要发生在 120～220 km 的地幔深度，在持续俯冲至地幔过渡带前不会发生明显脱水（魏春景和郑永飞，2020）。由于俯冲板片内部温度总是低于板片表面，洋壳俯冲过程中不会引起俯冲板片橄榄岩层发生部分熔融。

（3）俯冲带流体类型

俯冲带的流体根据其所含水和溶质比例的不同，大致可以分为富水溶液、含水熔体和超临界流体三种类型（图 3-6）。富水溶液一般密度低，表现为以水及挥发分为主，只溶解有限的各种元素的离子（Manning，2004）。这种富水溶液可以在很大的压力范围内存在（1～6 GPa），但是其产生的温度一般低于 650 ℃。自然界岩石中富水溶液的直接记录为流体包裹体（Xiao et al.，2000）。高压 - 超高压洋壳榴辉岩中流体包裹体的研究显示，俯冲带流体中的溶质以 NaCl、$MgCl_2$、$CaCl_2$ 为主，含少量的 KCl，盐度在 3 wt%～50 wt%。除了卤化物，富水溶液中的其他溶质包含 Si、Al、Ti、碳酸盐、硫化物 / 硫酸盐等。此外，流体包裹体中也含一定量的 N_2、CO_2 或 CH_4（Bebout and Penniston-Dorland，2016）。

在热俯冲带，当板片温度穿过洋壳岩石的湿固相线时，洋壳岩石在水饱和条件下发生熔融，形成含水熔体，最低熔融温度约为 650 ℃。含水熔体比富水溶液密度高，一般溶解有 1%～15% 的水。然而在水不饱和条件下，角闪石和黑云母最高可稳定至约 900 ℃和 2.5 GPa，其上发生脱水熔融；多硅白云母在更高的温压条件下发生脱水熔融。在低于 1 GPa 压力下，洋壳熔融产生富钠的长英质熔体，而在高于 3 GPa 压力下则产生富钾的熔体（Schmidt and Poli，2014）。一些天然超高压岩石样品中的晶体和玻璃包裹体，被认为是含水熔体的岩石学记录（Hermann et al.，2013）。这些固体包裹体显示与岛弧岩浆类似的微量元素特征，即富集大离子亲石元素和轻稀土元素，亏损高场强元素。

在弧下深度，当温压条件超过俯冲岩石体系的第二临界端点时，流体中

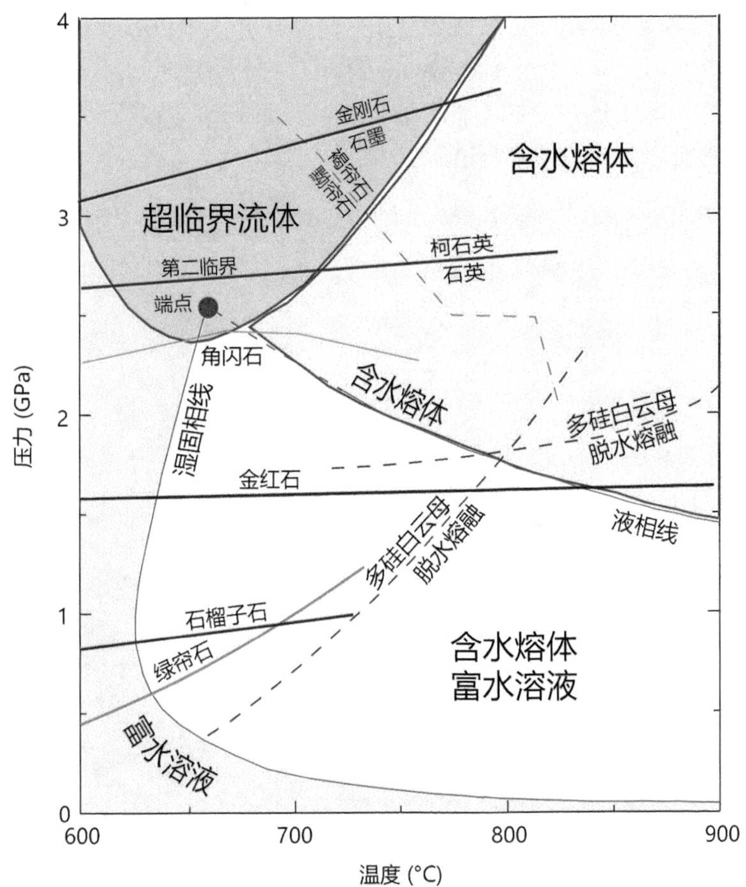

图 3-6　俯冲带三种类型流体形成域与花岗质地壳岩石脱水熔融温度压力关系示意图（修改自 Xia et al.，2010）

硅酸盐或硅酸盐组分和硅酸盐熔体中水的溶解度逐渐增加，形成溶质含量在 30 wt%～70 wt%、介于富水溶液和含水熔体之间的中间成分流体（Hermann et al.，2006；Ni et al.，2017）。对于俯冲带流体来说，超临界流体被用于描述富水溶液与含水熔体的完全混溶相，其形成会导致在先前相中不溶的许多组分具有更大的溶解度，大大提高了流体运移溶质的能力。

（4）俯冲带流体－岩石相互作用与元素迁移

俯冲带流体不仅可以迁移易溶于水的大离子亲石元素，也可以运移重稀土元素、高场强元素和过渡族金属元素等（图 3-7）。实验岩石学表明，一些挥发性组分（如卤素 F 和 Cl、轻元素 B 以及 CO_2 等）的加入可能会明显提高重稀土元素、高场强元素和过渡族金属元素在流体中的溶解度和活动性

（Guo et al.，2019）。此外，自然界高压－超高压脉体中可以含有碳酸盐矿物、硫化物、磷灰石、云母等矿物，说明流体中含有大量的 C、N、S 和卤族元素等挥发分，俯冲带流体对挥发性元素的运移在解析俯冲板片－岛弧物质循环方面扮演着重要的角色（Li et al.，2020）。

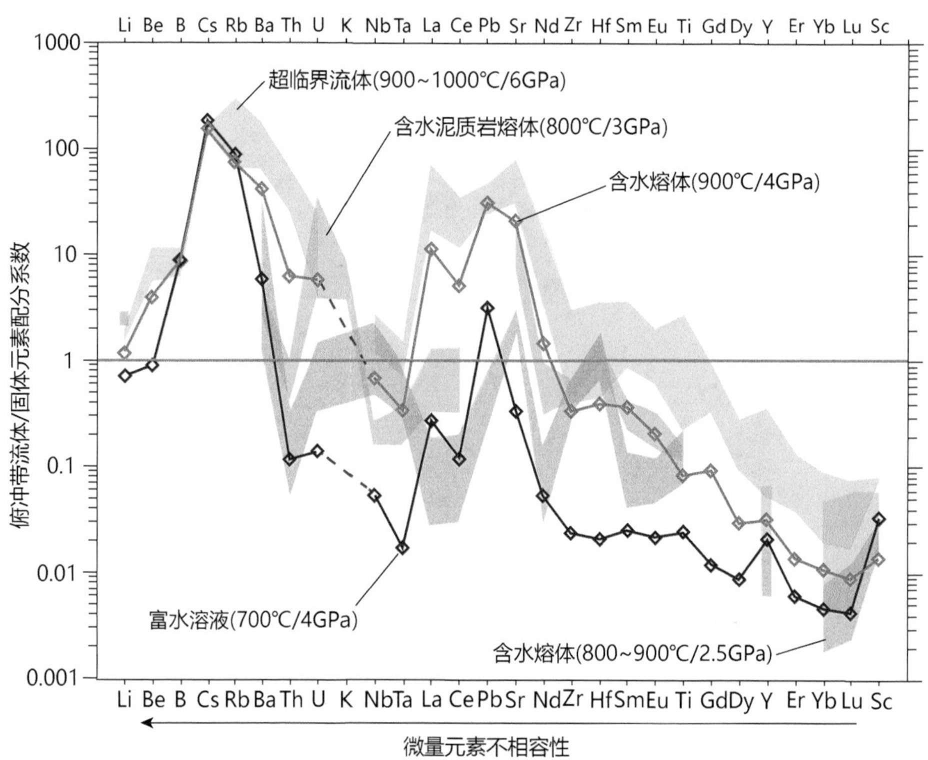

图 3-7　俯冲带三种类型流体与固体之间的微量元素配分系数大小随其不相容性变化规律图解（修改自 Zheng，2019）

弧岩浆岩和交代地幔岩的地球化学特征指示，板片来源的俯冲带流体携带大量溶质组分（如大离子亲石元素和轻稀土元素）进入了地幔楔（Plank and Langmuir，1993；Ryan et al.，1995）。然而相当多的研究发现，俯冲板片中含水矿物（如硬柱石和蓝闪石）脱水分解释放的流体具有非常低的溶质组分含量。流体在运移过程中会与俯冲隧道内各类岩石通过溶解－沉淀机制进行物质交换，从而不断改变岩石和流体自身的化学组成。这表明，俯冲板片释放的低溶质组分富水溶液在迁移上升过程中与周围岩石发生了水 / 岩相互作用，导致大量组分从围岩进入流体，并伴随这些流体最终迁移进入岛弧地幔源区（Bebout and Penniston-Dorland，2016）。

高压－超高压变质岩石中发育的高压变质脉体代表了俯冲带流体曾经的运移通道和最终结晶产物，而脉体与其寄主岩石之间的反应带则记录了成脉流体与岩石相互作用的直接信息（John et al.，2012; Guo et al.，2019; Li et al.，2020）。不同类型脉体及相关主岩的详细地球化学研究证实，洋壳俯冲释放的流体主量元素成分富 Si、Na 和 Ca；而微量元素成分复杂多样，以富集大离子亲石元素（如 Cs、Rb、K、Ba、Sr 和 Pb 等）、亏损高场强元素（如 Nb、Ta、Zr、Hf 和 Ti）为特征。

2. 关键科学问题

（1）俯冲带变质作用精细过程解析

俯冲带变质过程中的相变对板片脱水、俯冲带地震和弧岩浆作用具有重要意义。矿物相转变涉及一系列的变质反应，但是目前的研究多来自岩相学观察等定性研究，尚缺乏热力学模拟和高温高压实验对变质反应精细过程的定量厘定（李继磊，2020）。俯冲带变质相转变涉及哪些具体的变质反应？这些变质反应的温压范围和全岩控制因素是什么？相变速率和时间尺度如何？相变导致的板片岩石密度增加、体积减小、孔隙度和黏滞性的改变如何？对这些物理化学参数的精确限定是理解俯冲带结构演化、俯冲动力学、俯冲带地震诱因的关键因素。

（2）俯冲带流体活动的时间尺度

俯冲带是地球内部流体活动最为活跃的场所之一。流体－岩石相互作用是地球动力学过程的基本组成部分，地质流体通过化学反应、物质转移、能量转移与构造变形联系起来，控制了岩石圈的物理化学性质。尽管流体－岩石相互作用对地壳的动态演化至关重要，但是由于缺乏合适的计时器，且传统的放射成因测年技术提供了绝对年龄而不是持续时间，它们的不确定性太大，无法揭示快速和短暂的地质过程，所以流体－岩石相互作用的时间尺度基本上没有受到精确的约束。近年来通过地球化学模拟获得的流体－岩石反应时间非常短暂，只有短短的几十年到几百年的尺度（John et al.，2012; Beinlich et al.，2020），这对于传统的以百万年计的地质事件来说异常短暂。此外，实验室模拟流体－岩石相互作用过程仍然是一个难题。

（3）超临界流体的特征及自然界存在形式

高温高压实验和理论模拟研究都显示，俯冲带变质岩 *P-T* 路径可以通过体系的湿固相线，在第二临界端点之上的压力下形成超临界流体（Kessel et al.，2005；Hermann et al.，2006；Zheng et al.，2011）。虽然在天然样品中观察到一些可能暗示超临界流体作用的现象（肖益林等，2020），但是目前确切的超临界流体的岩石学记录依然比较少见。大洋俯冲带岩石中尚未发现能够指示超临界流体的证据，能够确切指示超临界地质流体存在的地质地球化学证据仍很缺乏（Zheng，2019），还没有建立超临界地质流体存在的标志性识别指标（Ni et al.，2017）。

（4）俯冲带 C-H-N-S 等挥发分的循环与演化

地球上的挥发分对地球形成与圈层分异、岩浆演化与氧化还原状态、金属成矿元素富集以及全球气候变化等诸多地质过程至关重要。大洋俯冲板片可以携带大量挥发分进入地幔深度，而岛弧岩浆喷发过程中释放了大量的 CO_2、N_2、SO_2、H_2S 等气体，暗示俯冲板片释放了大量挥发分，并通过弧岩浆作用的方式又循环回地表。厘定这些气体的释放通量对理解大气圈演化、提高地球对人类的宜居性具有重要作用。目前，对这些挥发性元素在俯冲板片中的存在形式、变质演化以及脱挥发分机制（变质作用、流体交代或熔融作用）的认识还非常有限。要厘清俯冲带对碳、氢、氮、硫等挥发分的循环机制，首先要对洋壳俯冲带变质岩中这些元素的俯冲形式和演化关系开展综合研究。

（二）俯冲带壳幔相互作用

自地球早期发生壳幔分异以来，壳幔之间的相互作用就开始主导着地球系统的物质循环和能量交换过程。特别是在地球内部动力体制转变为“板块构造”主控之后，这些循环和交换过程进一步导致不同圈层的分异和成熟，并使得地球逐渐成为太阳系中“宜居”的类地行星。如果说俯冲带过程是板块构造的核心，那么俯冲带壳幔相互作用则是俯冲带过程的关键。

1. 研究现状

俯冲带的基本结构由俯冲板片、上盘岩石圈和软流圈地幔楔构成（Stern，2002；Zheng and Chen，2016）。俯冲隧道是俯冲带板片界面之间发生壳幔相互作用最为强烈的位置（图 3-7），由来自俯冲板片和地幔楔的物质经构造混杂而成（郑永飞等，2013）。俯冲板片和地幔楔之间在俯冲隧道中发生的相互作用，成为俯冲带系统内物质循环的主要方式。这些相互作用具体表现为俯冲板片与地幔楔之间强烈且复杂的物理混杂作用、强烈的壳幔物质交换和流体活动，诱发汇聚板块边缘大规模的构造、岩浆、变质和成矿作用，并导致对流地幔组成和状态的进一步演变（Zheng，2019；Schmidt and Poli，2014）。

俯冲带在几何结构、温度场和流动场上的差异，会造成地幔楔物理结构和化学属性的不同，进而产生俯冲板片和上覆地幔楔之间的不同的相互作用（Stern，2002；Zheng and Chen，2016）。俯冲角度的平与陡、上覆板片的厚与薄、俯冲板片的冷与热、俯冲速率的快与慢、上覆板片流变学强度的大与小等，都会导致地幔楔几何结构的不同，从而产生不同类型与程度的板片脱水和熔融过程、壳幔相互作用、弧岩浆作用以及地幔楔运动过程等（Schmidt and Poli，2014；Zheng，2019）。同时，根据俯冲板片属性的不同，俯冲带可划分为洋－洋、洋－陆和陆－陆俯冲带，其中壳幔相互作用分为两种类型：大洋板片与地幔楔相互作用，以及大陆板片与地幔楔相互作用。

大洋岩石圈通常包括沉积物、蚀变洋壳和地幔橄榄岩。大洋岩石圈板片随着俯冲深度增加和温度升高发生一系列脱挥发分过程，具体包括从脱孔隙水到岩石进变质脱挥发分，再到高压/超高压条件下释放流体或高温下发生熔融。随后，这些流体在俯冲隧道内以弥散流和通道流的方式运移，并与俯冲隧道内地壳或地幔的组分发生复杂相互作用（Manning，2004；Schmidt and Poli，2014；郑永飞等，2016）。最后，形成的流体和低密度混杂体向上迁移，引起地幔楔的富化，甚至诱发部分熔融，形成弧岩浆，并最终演化为新生地壳（Spandler and Pirard，2013；Kelemen et al.，2014）。经历富化作用后的地幔和俯冲残余的板片，被卷入对流地幔后，成为板内岩浆的重要源区（Hofmann，1997；郑永飞等，2018）。

与大洋俯冲带相比，大陆俯冲带具有相对较低的温度，并且俯冲陆壳具有相对较少的易挥发组分，使得在大陆俯冲过程中相对缺乏流体和岩浆活动（Zheng，2012；Hermann and Rubatto，2014）。然而，大量的岩石学、地球化学研究和实验模拟表明，在深俯冲的大陆地壳内发育规模较小但类型丰富的流体活动并伴随流体－岩石相互作用（Zheng and Hermann，2014；Frezzotti and Ferrando，2015）。此外，大陆俯冲带在弧前和弧后深度可能具有和大洋俯冲带类似规模的流体活动，深俯冲大陆板块折返过程中也存在显著的熔体活动（Zheng and Chen，2016）。同时，大陆俯冲隧道内不同深度来源的地幔楔岩石也记录着复杂的熔体活动和壳幔相互作用，如对中国大别－苏鲁、柴北缘以及意大利 Ulten、挪威西片麻岩省等造山带中地幔橄榄岩的研究，都取得了丰富的成果（郑建平等，2019）。如何利用造山带内地幔橄榄岩深入揭示俯冲带壳幔相互作用，一直是研究俯冲带物质循环过程的重点和难点。

2. 发展趋势

俯冲带壳幔相互作用可总结为物理和化学两种形式。俯冲带壳幔物理相互作用主要表现为：①不同岩石通过机械混合形成混杂岩带，表现为几何结构上呈现为塑性变形的构造混杂岩；②充填于地幔楔橄榄岩内因剪切所发育裂隙的各种脉岩，它们是不同岩性构造混杂过程中所形成的流体活动的结晶产物，如各种辉石－角闪岩脉体、金云母脉、蛇纹石－滑石脉等；③岩石和矿物的塑性为主的变形，如高度塑性揉皱、拉伸甚至局部互层化、片理化和矿物的形变、位错和滑移特征等，反映了地幔楔的物理化学条件、含水性和构造环境；④地幔楔橄榄岩的水化交代过程影响了岩石的密度、磁性、波速等物理性质，并对应造山带区域上的地球物理航磁、重力和波速异常现象（郑建平等，2019）。

俯冲带壳幔化学相互作用主要体现为壳幔元素交换和物质循环，通过流体与地幔楔岩石发生化学反应来实现（Zheng and Hermann，2014）。在不同温压条件下的硅酸盐岩石－水体系中，俯冲带内的流体可划分为富水溶液、含水熔体和超临界流体（Hermann et al.，2006；Zheng et al.，2011；Ni et al.，2017）。此外，低密度 C-H-O 流体、碳酸盐流体和硫化物熔体等其他成分的流体或熔体，也在俯冲带壳幔物质交换循环中扮演重要角色（Schmidt and Poli，2014）。当俯冲隧道的温压条件达到俯冲板片内含水矿物的脱挥发分条件后，能够产生不同性质的流体或熔体。这些流体相对富硅，在化学上与地幔岩石不平衡，从而会与俯冲隧道内裹挟的地幔楔碎片，主要是橄榄岩块，发生不同方式和程度的相互作用。研究显示，在俯冲带环境下流体－橄榄岩相互作用的方式有两类（Zheng，2012；O'Reilly and Griffin，2013）：一类是隐性交代作用，是流体与橄榄岩发生化学反应但还没有生成新生矿物；另一类是显性交代作用，流体与橄榄岩反应形成新生矿物，其中均伴有流体的产生及其对地幔楔橄榄岩的溶解－运移－沉淀作用。这些作用会产生如下效应：①造成地幔楔橄榄岩内不相容元素和挥发分的富集；②形成新生的无水/含水硅酸盐矿物、碳酸盐矿物以及伴生硫酸盐矿物/硫化物、单质和氧化物等；③富硅熔体与橄榄岩反应，诱发贫硅矿物（橄榄石）的溶解和富硅矿物（辉石等）的结晶，形成橄榄岩内的辉石岩脉体；④导致在橄榄岩矿物内保留流体包裹体及其与矿物反应后的产物。

在大洋型俯冲带橄榄岩中，来自软流圈对流地幔楔的硅酸盐熔体占据主导，与岩石圈地幔楔反应形成纯橄榄岩、铬铁矿岩、辉石岩、辉长－辉绿岩脉等；而来自俯冲大洋板片的流体以富水溶液为主，主要沿着俯冲隧道内裂隙或剪切带等薄弱带交代上覆岩石圈地幔楔，在俯冲型蛇绿岩底部形成含水矿物聚集脉体。在大陆型俯冲带橄榄岩中，来自俯冲陆壳的硅酸盐熔体、碳酸盐熔体和富水溶液占据主导，与大陆岩石圈地幔楔反应形成多种交代岩石或交代矿物；而来自对流地幔楔的贫硅熔体，却很罕见。产生这些差异的根本原因在于俯冲带的物理几何结构、温度场结构和动力学过程的不同（郑永飞等，2016）。

大洋型和大陆型俯冲带壳幔相互作用的差异，可以部分解释豆荚状铬铁矿体仅在蛇绿岩内发育，而极少存在于大陆岩石圈地幔楔的现象（郑建平等，2019），即前者主控因素是软流圈来源的熔体（富 Cr），而后者是俯冲陆壳来源的小比例流体（贫 Cr）。同时，已有证据表明大陆俯冲带可能存在先期俯冲大洋板片来源流体对大陆岩石圈地幔楔的交代作用，这些先前交代形成的蛇纹石化橄榄岩可能释放流体反向交代俯冲大陆板片（如阿尔卑斯的白片岩）（Chen et al.，2016；Xiong et al.，2022）。这些俯冲带壳幔物理作用过程限制着造山带和地幔楔的空间几何结构以及构造组合，同时也约束了壳幔化学相互作用所发生的物理空间。宏观和微观的变形作用对俯冲带壳幔界面的变质过程、流体活动以及交代作用，均起着关键的“催化”作用。更重要的是，岩石和矿物的物理性质是进行地质学、地球化学与地球物理学交叉的重要纽带，能够为岩石圈演化、俯冲带地幔动力学和地震观测数据的解读提供理论支撑。因此，需要大力开展对俯冲带地质体橄榄岩，特别是对不同蚀变程度的橄榄岩进行矿物岩石特征的对比研究，无疑是俯冲带壳幔相互作用的研究新趋势。

3. 关键研究对象

造山带中的地质体橄榄岩是记录俯冲带壳幔相互作用最直接的对象。这些橄榄岩多以汇聚板块边缘地质体橄榄岩的形式产出，主要分为两类极端情况：一类是反映大洋俯冲过程的俯冲带型蛇绿岩；另一类是记录大陆俯冲过程的地幔楔型造山带橄榄岩。俯冲带型蛇绿岩的主要特点是在岩石学上显示玄武岩 - 辉长岩 - 方辉橄榄岩 - 二辉橄榄岩这样一个典型的地幔熔融和分离结晶层序，其中地幔橄榄岩显示难熔属性和明显的流体交代特征，但是该层序的上层岩石或洋壳（如玄武岩和玻安岩等），在微量元素地球化学特征上显示岛弧属性（Bodinier and Godard，2014；Zheng and Chen，2016）。正是这些地球化学特征变化记录着大洋俯冲带壳幔相互作用。全球典型的俯冲带型蛇绿岩出现在塞浦路斯的特罗多斯（Troodos）、法国的新喀里多尼亚、中国雅鲁藏布江带的罗布莎和泽当等（郑建平等，2019）。

与蛇绿岩不同，造山带橄榄岩分为地幔楔来源的 M 型和俯冲大陆岩石圈来源的 C 型（Bodinier and Godard，2014），均记录了地幔楔与大陆板片的相互作用过程。M 型造山带橄榄岩主要经历与大陆边缘有关的汇聚或裂解等动力学过程及其相关的岩石物理和化学性质演变，但少量蛇绿岩经历俯冲带过程叠加也可形成造山带橄榄岩（Bodinier and Godard，2014）。按照变质相和压力条件的不同，造山带橄榄岩可分为三类：①高压 / 超高压型，如挪威西片麻岩省和大别－苏鲁带中的石榴子石相橄榄岩地质体；②中压型，主体尖晶石相，如摩洛哥的 Beni Bousera 和西班牙的 Ronda；③低压型，浅部伸展体制下的尖晶石相和斜长石相橄榄岩地质体，如意大利阿尔卑斯的 Lanzo 和 External Ligurides。同时，这三类造山带橄榄岩可形成于板块汇聚造山过程（如高压 / 超高压型），或产生于大陆边缘裂解抬升过程（如中－低压型）。壳源 C 型造山带橄榄岩包括挪威西片麻岩省的 Svartberget 橄榄岩、中国阿尔金造山带南部的橄榄岩以及大别造山带的碧溪岭橄榄岩等，它们均是早先侵入到大陆深部地壳的超基性－基性堆晶岩体，后伴随大陆俯冲－折返而形成的石榴子石相造山带橄榄岩（郑建平等，2019）。通过系统研究不同深度和构造背景下的地幔楔样品，能够进一步揭示俯冲带壳幔界面的物质循环、能量交换等一系列细节过程。

4. 存在问题与展望

俯冲带壳幔相互作用的本质是两种构造单元（俯冲地壳和地幔楔）在机械作用（物理过程）主导下发生复杂的化学相互作用，核心过程是俯冲带流体和地幔楔岩石在多种物理化学条件下发生相互作用。俯冲带大洋型橄榄岩记录了洋－洋俯冲启动、发展、成熟甚至消亡阶段复杂的流体交代、熔－岩和水－岩相互作用、变形变质过程、金属成矿作用（如铬铁矿成因）以及壳幔物质交换动力学过程；俯冲带大陆型橄榄岩则记录了从洋－陆俯冲到陆－陆俯冲、碰撞、折返阶段的强烈物理变形变质过程、多种性质的流体交代作用以及高度复杂的壳幔物质循环过程。

如何揭示这些过程，需要未来对俯冲带地质体橄榄岩开展进一步的交叉学科研究，具体包括：①利用微区原位地球化学和显微构造学等综合分析手段约束交代过程与变质、变形的时空关系；②精细识别在俯冲带内流体的起源与性质，甚至在微米至纳米尺度对流体活动的产物进行系统研究；③结合热力学模拟计算恢复俯冲带壳幔相互作用的温度、压力、氧逸度、流体组分等物理化学条件；④利用地幔楔岩石属性及演化约束汇聚板块边缘的金属成矿作用；⑤结合地质学、地球化学、数值模拟、高温高压实验和岩石物性资料，对特定构造区的岩石圈演化开展跨学科研究。

（三）俯冲带镁铁质岩浆作用

1. 研究现状和发展趋势

板块俯冲带是连接地壳和地幔的纽带，是联系地表与地球深部的通道，岩石圈的俯冲是地壳与地幔之间物质和能量交换的重要过程，又被称为俯冲工厂（图 3-4），也是深入理解化学地球动力学的关键（Stern，2002；Zheng，2012，2019；Zheng and Chen，2016；Zheng et al.，2020）。俯冲带镁铁质岩浆作用研究对于了解大洋俯冲带的发生和发展、壳幔相互作用、大陆地壳增生等方面意义重大，一直是地球科学研究的热点。国际上俯冲带镁铁质岩浆作用研究突出的进展有以下几个方面。

（1）俯冲带结构和弧岩浆类型

大洋俯冲带之上发育不同类型的镁铁质弧岩浆作用。根据上覆板块性质的不同，可以将大洋俯冲带分为两大类：洋－洋俯冲带和洋－陆俯冲带（Stern，2002；Gill，2010）。洋－洋俯冲带以发育大洋弧（洋内岛弧）玄武岩和弧后盆地玄武岩为特征（图 3-8），在西太平洋最为典型，其中的镁铁质岩石包括高镁玄武岩、玄武安山岩、高铝安山岩、玻安岩和赞岐岩。洋－陆俯冲带以发育大陆边缘弧（大陆弧）安山岩为特征（也有少量玄武岩和流纹岩），基本上不发育弧后盆地玄武岩，在东太平洋边缘最为典型。根据上覆板块现在的构造应力场，可以进一步划分为挤压型安第斯山陆缘弧和伸展型科迪勒拉陆缘弧，其中安第斯山陆缘弧岩浆作用以大型中酸性岩基和火山岩为主，科迪勒拉陆缘弧的岩浆作用以双峰式火山岩为特征。

（2）俯冲起始及洋内弧镁铁质岩浆作用

俯冲起始的研究是板块构造理论研究的前沿课题（吴福元等，2019）。现代 IBM 岛弧岩浆作用研究基本或者初步确定了洋－洋俯冲带形成过程，并提出了不同的初始俯冲诱发机制（Stern，2004；Stern and Gerya，2018；Niu et al.，2001；Arculus et al.，2019），俯冲相关的（SSZ 型）蛇绿岩被认为是代表特提斯构造域沟－弧－盆体系初始俯冲阶段的产物（Ishizuka et al.，2014）。

俯冲带流体对弧岩浆成分和微量元素的控制：作为重要的元素搬运介质

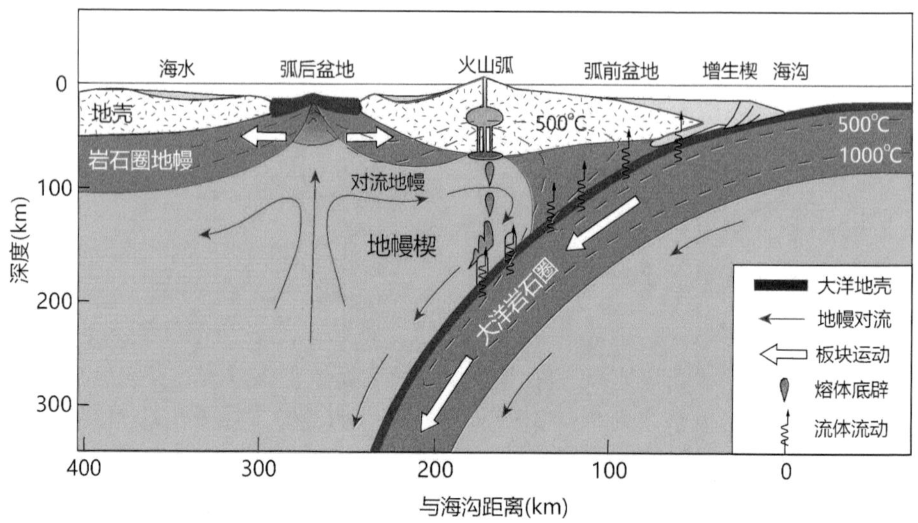

图 3-8　西太平洋俯冲带之上大洋弧和弧后盆地岩浆作用发育示意图
（修改自 Stern，2002）

的流体可分为富水溶液、含水熔体和超临界流体。岩浆中的流体含量对大离子亲石元素和轻稀土元素的富集有重要影响。实验岩石学证明，流体搬运微量元素的能力随温度升高而增加，随压力增加而增加，所产生的超临界流体会搬运除重稀土元素、Y 和 Sc 以外大部分的微量元素（Kessel et al.，2005）。

（3）俯冲板片熔融机制和壳幔相互作用

通过实验和热力学模拟，学者分析了俯冲板片熔融的动力学机制，确定埃达克岩和赞岐岩的形成与板片熔融及其与地幔楔相互作用的关系（Defant and Drummond，1990；Tatsumi，2006；Wei and Clarke，2011），建立了埃达克岩和赞岐岩与太古宙 TTG（奥长花岗岩－英云闪长岩－花岗闪长岩）的成因联系，以及太古宙初始大陆地壳的形成机制（Rapp et al.，2003；Martin et al.，2005）。

（4）俯冲带镁铁质岩浆的氧逸度

氧逸度控制挥发性元素（C-H-N-S）的存在形式、稳定性与迁移和循环行为，并且与地球内部的物理化学性质密切相关（Arculus，1985；Ballhaus，1993；Frost and McCammon，2008）。俯冲带氧逸度研究进展包括：①岛弧岩浆岩及地幔楔具有比洋中脊岩浆更高的氧逸度，与俯冲带有密切关系（Arculus，1985；Kelley and Cottrell，2009）；②建立了一系列氧逸度与温度压力的计算体系及不同俯冲带环境下的氧逸度特征；③深化了对氧化还原熔融机制（redox melting mechanism）的认识，厘清弧岩浆的成因（Parkinson and Arculus，1999）；④初步确定了氧逸度与微量元素分配及其行为、Fe^{3+} 含量、S 通量以及金属成矿作用之间的关系（Evans and Tomkins，2011）。

（5）弧镁铁质岩浆作用与大陆地壳增生

平均大陆地壳成分是安山质到英安质的（Rudnick and Gao，2003；Hacker et al.，2011），俯冲带岩浆作用是理解大陆地壳生长机制和过程的关键（Gazel et al.，2015）。随着俯冲的持续进行，岛弧地壳的性质开始发生由中基性向安山质新生大陆地壳的转化，这种转化对于地质历史上的大陆地壳生长具有重要的意义（Kelemen et al.，2003）。

（6）洋脊俯冲的镁铁质岩浆作用

洋脊俯冲广泛存在于地质历史时期的增生造山带和现代的俯冲带（如太平洋东岸），是洋中脊与俯冲带相互作用的场所。在俯冲过程中洋中脊两侧的岩石圈发生撕裂形成板片窗，软流圈地幔上涌可以造成很强的弧前热异常，并形成类似洋中脊玄武岩的富铌玄武岩，同时俯冲的洋壳板片发生熔融形成埃达克岩，二者构成双峰式火山岩组合，典型例子为墨西哥的维斯凯诺（Vizcaino）半岛（Aguillon-Robles et al.，2001）。在加勒比地区，俯冲的洋中脊板片熔融形成埃达克岩，而上涌的太平洋地幔穿过地幔楔发生熔融形成弧后偏碱性镁铁质火山岩（Abratis and Wörner，2001）。

（7）陆缘弧后伸展及拆沉的镁铁质岩浆作用

安第斯山挤压型陆缘弧由于有厚的大陆岩石圈，地幔楔熔融的镁铁质岩浆很难直接到达地表，而表现为以中酸性为主的岩浆活动。但俯冲带弧后伸展和大陆克拉通岩石圈拆沉也是陆缘弧常见的现象（如北美的科迪勒拉陆缘弧），也是壳幔物质交换和陆缘弧镁铁质岩浆作用产生的重要机制。关于弧后伸展的构造机制，目前最为流行的是俯冲板片回卷模型（Dewey，1980）。弧后伸展可以引起弧后位置岩石圈的减薄，在洋－洋俯冲带之上形成裂谷火山盆地，其中可包含与洋中脊玄武岩成分类似的火山岩。根据科迪勒拉盆岭省的岩石圈结构特征，Bird（1979）提出了岩石圈拆沉模型，并被广泛应用于解释喜马拉雅造山带以及华北岩石圈减薄和破坏的机制（Gao et al.，2004；Yang et al.，2008）。岩石圈拆沉造成软流圈地幔上升到浅部并发生降压熔融，形成以碱性玄武岩为主的镁铁质岩浆，并沿张性正断层上升到地表。同时，上涌的软流圈地幔使陆壳发生熔融产生酸性火山岩，与镁铁质岩浆构成双峰式火山岩组合。

（8）洋底高原与俯冲带碰撞的岩浆作用

无震海岭或大洋高原广泛存在于现代海洋的大洋盆地之上，一般认为是地幔柱活动形成的大规模岩浆活动产物。由于其以海山链或洋底高原形式存在，地形远高于大洋盆地，因此常常会与俯冲带或岛弧发生碰撞，这是现代大洋板块动力学系统中十分普遍且非常重要的一种地质过程，与俯冲带结构变化、岛弧岩浆作用及物质循环、大陆地壳生长与演化、上覆板块变形或地形沉降/隆升以及金属成矿关系非常密切，已成为当前国际地质研究的热点之一。一般认为，洋底高原与俯冲带碰撞可以造成俯冲带阻塞，是初始俯冲或新的俯冲带形成的重要机制之一（Stern，2004）。祁连造山带拉脊山地区洋内弧的形成与洋底高原碰撞有密切关系（Song et al.，2017；Yang et al.，2019），同时造成大量洋底高原火山岩在大陆边缘堆积，因此是大陆地壳增生的重要方式之一。

2. 关键科学问题

1）地球早期俯冲带镁铁质岩浆作用的确定及其特征。太古宙古俯冲带的识别是确定板块构造起源的关键。根据太古宙地球热流值的特征，分析俯冲带结构和物理化学条件、俯冲板片与地幔楔相互作用、俯冲带镁铁质岩浆岩岩石学和地球化学特征。

2）镁铁质岩浆成因机制。俯冲板片流体与地幔楔交代作用及其俯冲带镁铁质岩浆成因，俯冲板片如何发生脱流体和熔融，这些流体和熔体的迁移机制及其与地幔楔的交代过程需要精细的岩石学和地球化学研究。

3）构造背景和俯冲带结构。不同构造环境下，造成俯冲带结构变化的主要原因是什么；俯冲带起始和后撤过程的动力学过程及岩浆作用有哪些关键标志，如何识别；大陆弧和洋内弧岩浆作用的异同性以及洋内弧形成过程和机制；沟-弧-盆体系形成机制和动力学过程。

4）俯冲带镁铁质岩浆与壳幔相互作用。俯冲带与地幔楔之间的物质交换和相互作用如何造成地幔楔不均一性；洋内弧镁铁质岩浆作用的岩石组合与大陆地壳生长的关系；俯冲带岩浆作用氧逸度变化及其与俯冲带的关系；洋底高原的俯冲/碰撞效应及其与弧岩浆作用关系；俯冲带岩浆作用与大型-超大型矿床的成因联系。

三、俯冲带岩浆作用与大陆地壳增生

（一）弧后盆地岩浆岩

1. 弧后盆地及其岩浆作用

弧后盆地是指岛弧靠近大陆一侧的深海盆地，是在陡倾俯冲下海沟向大洋方向迁移、上覆岩石圈扩张而成，与海沟、岛弧一起组成了沟 - 弧 - 盆体系（Taylor，1995；Taylor and Martinez，2003；Christie et al.，2006）。弧后盆地具有清晰的洋中脊构造、部分有海山分布，其岩浆作用的主要产物是玄武岩，被称为弧后盆地玄武岩（back arc basin basalts，BABB）。弧后盆地玄武岩在成分上具有两种类型：早期的具有岛弧玄武岩特点，源区为地幔楔；晚期的与洋中脊玄武岩相似，源区为软流圈地幔，在岩性上以拉斑玄武岩为主（Taylor，1995）。

全球弧后盆地主要集中分布于西太平洋和地中海地区，约占全球弧后盆地面积的 75%（Taylor，1995；Christie et al.，2006）。西太平洋地区共分布 20 余个大小不等的弧后盆地。一系列国际大洋钻探计划的开展，使得西太平洋地区的弧后盆地研究程度高。该地区的弧后盆地数量多，在时空分布上具有代表性，在基底属性上具有多样性，是研究弧后盆地成因和弧后盆地岩浆作用的理想对象。

根据弧后盆地扩张中心相对于岛弧的位置，可以将弧后盆地按扩张类型分为三类（Martinez and Taylor，2006）。第一类盆地是由弧后大洋岩石圈发生裂解形成的弧后盆地。该类弧后盆地形成的玄武岩在靠近岛弧这一侧具有明显的岛弧岩浆的信号，然而在相反的方向上岛弧岩浆的信号是微弱的，而洋中脊岩浆的信号显著增强。日本海海盆和冲绳海槽即为这种类型的弧后盆地。第二类盆地是由岛弧岩石圈中央发生裂解形成的弧后盆地，新生盆地落在了活动弧与残留弧之间，此类弧后盆地形成的早期玄武岩具有岛弧岩浆的信号，而晚期玄武岩具有洋中脊岩浆的信号。劳海盆和菲律宾海盆即为该类型的弧后盆地。第三类盆地是在岛弧前部拉张形成的弧后盆地，是由俯冲板片回卷与仰冲盘岩石圈前进相结合所导致的弧前边缘的伸展减薄裂解形成的。该类弧后盆地玄武岩普遍存在俯冲地壳物质的影响，这种类型的弧后盆地可见于伊豆 - 小笠原裂谷。

弧后盆地在时间上也存在分阶段演化的特征，可以概括为四个阶段（Taylor，1995）。①初始张裂（initial rifting）阶段，岛弧、弧后或者弧前开始裂开形成裂谷，裂谷内岩浆作用密集，这些岩浆具有弧岩浆的特点。典型例子如新西兰的陶波盆地和冲绳海槽的北端。②持续伸展（continued stretching）阶段，盆地被持续加宽，形成更大的沉降，大量岩浆喷出于盆地，形成不对称的分布。③初始扩张（initial spreading）阶段，岩浆扩张中心形成，扩张过程与全球开放大洋接近。典型例子如冲绳海槽的南部和哈维盆地等。④成熟扩张（mature spreading）阶段，形成成熟的海底扩张中心，产出与全球洋中脊玄武岩相似的洋壳。典型案例如马努斯海盆、马里亚纳海槽中部和北斐济海盆等。

2. 弧后盆地玄武岩研究现状和趋势

（1）地幔源区组成的多样性

Pearce 和 Stern（2006）通过微量元素比值发现弧后盆地玄武岩的源区存在三种不同来源的物质，分别为地幔楔中的初始物质、俯冲板片释放的流体、外来的地幔物质。地幔楔是俯冲板片上盘呈现楔状区域。地幔楔自身在地球化学上存在高度的不均一性。例如，对马里亚纳海槽玄武岩的研究发现，地幔楔在 Nd-Hf 同位素组成上呈现两端元混合的特征（Woodhead et al.，2012）。俯冲板片影响下的地幔源区不均一性是弧岩浆作用研究的重点，但是弧后盆地玄武岩的地幔源区也存在俯冲板片的贡献。在弧后盆地位置，俯冲板片主要以释放流体的形式交代上覆地幔，上覆地幔发生降压和含水部分熔融即形成弧后盆地玄武岩。俯冲流体既可能来自俯冲洋壳或者沉积物，而俯冲熔体多数来自俯冲的沉积物（Escrig et al.，2009；Todd et al.，2010；Tian et al.，2011）。俯冲沉积物还可能直接参与到弧后盆地地幔源区的部分熔融（Ishizuka et al.，2003）。对于弧后盆地挥发分（如 CO_2）的研究还表明，很大一部分洋表挥发分会随着板片的俯冲进入到弧后盆地的地幔源区，暗示了板片表面的挥发分在弧岩浆活动的过程中并没有完全逃逸（Nishio et al.，1998；Newman et al.，2000）。另外，近些年的研究表明，深部地幔物质对弧后盆地玄武岩地幔源区的贡献不能忽视，地幔柱 / 热点物质对于弧后盆地的影响可能比前人认识的大，如 Samoa 地幔柱对于劳海盆玄武岩地球化学组成上的影响（Lupton et al.，2012；Lytle et al.，2012；Nebel and Arculus，2015；Tian et al.，2011）。

（2）部分熔融程度

弧后盆地玄武岩地幔源区的部分熔融程度除了受温度和压力的控制外，还需考虑俯冲板片来源流体的影响，含水量越高熔融程度越大（Macpherson et al., 2000; Hochstaedter et al., 2001; Kelley et al., 2006; Langmuir et al., 2006)。Langmuir 等(2006)将弧后扩张中心地幔熔融区分成两个部分，靠近岛弧一侧的为“湿熔融区”，远离岛弧一侧的为“干熔融区”。两个区的范围比例受到俯冲带距离的控制，随着弧后扩张中心的不断发育并逐渐远离岛弧，“干熔融区”范围增大，“湿熔融区”范围缩小。“干熔融区”发生降压熔融，形成贫 H_2O 富 Fe 的岩浆，而“湿熔融区”发生降压和加 H_2O 的熔融，形成富 H_2O 贫 Fe 的岩浆。弧后盆地玄武岩是这两种性质岩浆发生不同程度混合的产物(Langmuir et al., 2006)。对于弧后盆地玄武岩地幔源区部分熔融的控制因素，Klein 和 Langmuir(1987)认为，地幔潜温决定了地幔的平均熔融程度以及岩浆量。Niu 和 O'Hara(2008)则认为，部分熔融程度受控于地幔的易熔和难熔程度，地幔的易熔程度决定了地幔上涌速率，越是易熔的地幔由于密度小上涌的速率更快，从而导致更高程度的部分熔融。因此，对于弧后盆地而言，尽管地幔部分熔融程度主要由地幔潜温决定(Wiens et al., 2006)，但是对于同一个弧后盆地或者更小尺度的同一个扩张中心而言，俯冲流体的加入无疑是影响熔融程度的更为主要的因素，也极有可能是控制洋壳厚度的主要因素(Martinez and Taylor, 2002, 2006; Hirahara et al., 2015)。

(3)地壳混染和熔体－岩石反应

在 Sr-Nd-Pb 同位素组成上，大陆基底弧后盆地玄武岩相比于大洋基底弧后盆地玄武岩表现出更富集的特点。这可能继承于地幔源区，更可能来自大陆地壳混染的影响(Gamble et al., 1996; Hoang and Uto, 2006)。首先，具有低 Nd 同位素组成的大陆基底弧后盆地玄武岩具有更低的 Ce/Pb 和 Nb/U 值。其次，大陆基底弧后盆地玄武岩在微量元素蛛网图上普遍具有明显的 Pb 正异常。因此，大陆基底弧后盆地玄武岩的成因中，不能忽视大陆地壳的混染因素。对于大洋基底的弧后盆地，岩浆在大洋岩石圈深部(岩石圈地幔或者下洋壳)可能发生熔体－岩石相互作用，以这种方式影响弧后盆地玄武岩的地球化学组成(Kamenetsky et al., 1998; Danyushevsky et al., 2003; Lissenberg and Dick, 2008; Gale et al., 2013)。来自两方面的证据支持这个结论。一方面，大洋核杂岩中的橄长岩被认为是岩石圈地幔中纯橄榄岩与玄武质熔体发生熔体－岩石反应后形成的，如帕里西维拉海盆的大洋核杂岩中的橄榄石橄长岩(Sanfilippo et al., 2013)。在这一反应中，橄榄石被消耗，结晶出新的斜长石和透辉石。另一方面，弧后盆地玄武岩橄榄石斑晶熔体包裹体的异常成分特征也支持存在这种反应。在 MgO 含量相似的情况下，部分熔体包裹体具

有偏高的 Al_2O_3 和 CaO 含量以及相对亏损的强不相容元素组成（如低的 La/Sm 值）的特征（Kamenetsky et al.，1998；Danyushevsky et al.，2003）。单纯的分离结晶模式无法解释这种主量元素上的变化特征，而下洋壳深度发生的熔体－岩石反应（消耗辉长岩中的斜长石）可以同时解释升高的 Al_2O_3、CaO 含量和 Sr/Sr* 值（Kamenetsky et al.，1998；Danyushevsky et al.，2003）。因此，熔体－岩石反应可能也是影响大洋基底弧后盆地玄武岩地球化学组成变化的常见机制。

（4）岩浆演化

弧后盆地玄武岩与洋中脊玄武岩（mid-ocean ridge basalts，MORB）一样以拉斑玄武岩为主。温度和压力的变化是影响两者岩浆分异的主要因素，它们都遵循鲍文反应序列，在弧后盆地玄武岩岩浆不含水的情况下具有相似的岩浆演化趋势线，以橄榄石和斜长石的分离结晶为始，随着分离结晶的进行，辉石会加入到结晶矿物组合中，斜长石由基性斜长石向中性斜长石转变（Herzberg，2004；Gale et al.，2013）。在相对富水的条件下，水的存在可以抑制斜长石的分离结晶，促使单斜辉石优先结晶，从而形成具有高 Al_2O_3 低 CaO 特征的玄武岩（Danyushevsky，2001；Bézos et al.，2009）。此外，玄武岩的含水量和氧逸度之间的关系研究还表明，随着弧后盆地玄武岩地幔源区含水量的增加，地幔的氧逸度也会增加，从而改变地幔部分熔融后形成的岩浆氧逸度。岩浆氧逸度的改变可以进一步影响岩浆的分离结晶进程，促使部分副矿物（如铁钛氧化物）优先分离，从而改变玄武岩岩浆中某些特定元素的含量（Arculus，2003；Bézos et al.，2009；Kelley and Cottrell，2009）。

（5）地幔柱影响

部分弧后盆地在演化过程中存在地幔柱的影响，但可能是以间接的方式。以劳海盆西北部的弧后盆地玄武岩为例，其部分样品的 $^3He/^4He$ 值高达 20 Ra（Ra 为大气中相应比值），与劳海盆东南方位的萨摩亚（Samoa）群岛玄武岩具有相似性，暗示了 Samoa 地幔柱对于劳海盆形成演变的影响（Lupton et al.，2012；Lytle et al.，2012；Nebel and Arculus，2015）。地幔柱物质可能通过板片窗或者板片两侧的空隙进入地幔楔（Lytle et al.，2012）。类似的情况也见于东斯科舍（East Scotia）海盆（Fretzdorff et al.，2002；Leat et al.，2004）。地幔柱物质的贡献在日本海、伊豆－小笠原－马里亚纳海槽和马努斯（Manus）等弧后盆地玄武岩的成因中也被提及，但这些地区缺少明确的地球物理证据支持而存在不确定性。目前并不清楚弧后盆地是否与开放大洋一样，存在普遍的地幔柱活动，可能只是因为构造背景的差异，导致地幔柱信号被俯冲过程稀释。因此，需要重新检视地幔柱对弧后盆地岩浆过程的影响。

3. 关键科学问题

弧后盆地玄武岩的研究程度，总体来说明显落后于玄武岩的整体研究程度。有以下关键科学问题尚待解决。

1）弧后盆地玄武岩地幔源区的岩性尚缺乏有效的甄别。源区物质组成的研究还停留在源区是否有再循环地壳物质以及哪类再循环地壳物质（如俯冲沉积物）的水平上，而决定地幔熔融的关键因素——源区岩性还没有涉及。

2）弧后盆地玄武岩地幔源区的熔融环境尚缺乏有效的约束。目前，对温度、压力、氧逸度等物理化学条件的约束方法偏少，缺少交叉检验，可信度低。

3）弧后盆地玄武岩地幔源区流体的种类和含量缺少有效的约束。例如，除了水之外，CO_2 和其他挥发分（如卤族元素）也可能在弧后盆地玄武岩的源区发挥了重要作用，但是目前的研究程度还很低，基于熔体包裹体方面的观察还很少。

4）弧后盆地玄武岩深部背景的多样性认识不足。例如，俯冲板片的角度、俯冲板片的组成（是否有海山？）、板片是否撕裂、是否有地幔柱的影响等，这些需要岩石学、地球化学和地球物理学的多学科交叉。

（二）现代汇聚板块边缘镁铁质岩浆岩

1. 研究现状和趋势

（1）新生代弧镁铁质火成岩特征和岩石组合

新生代大洋俯冲带不同弧（大洋弧、大陆弧和陆缘岛弧）系统中的玄武岩及共生岩石组合如下：①大洋弧，如 IBM 岛弧，以玄武岩为主，伴有少量安山岩－英安岩，有时还有特殊的高镁安山岩类和埃达克岩（徐义刚等，2020）。其中，玄武岩以拉斑玄武岩和钙碱性玄武岩为主，并有极少量的橄榄玄粗岩。②大陆弧，如北美洲西部及巽他弧，产出大量安山质岩浆岩，并伴随有英安岩和流纹岩等，玄武岩以钙碱性为主，不同于洋内弧玄武岩的多样性（拉斑、低钾、高钾钙碱性和橄榄玄粗质）。③陆缘岛弧，如千岛群岛、日本－琉球岛弧，以玄武岩－安山岩等为主，其中玄武岩以钙碱性为主，并有少量的拉斑质和橄榄玄粗质。上述三种弧系统的玄武岩尽管在地球化学特征上存在一些细微的差别，但都显示了大离子亲石元素富集、高场强元素亏损的典型弧岩浆岩特征，主要与俯冲流体交代的地幔楔熔融有关（Zheng，2019；徐义刚等，2020）。

除了上述典型的弧镁铁质岩浆岩外，一些新生代弧环境中也偶尔会出现一些特殊类型的非弧玄武质（或镁铁质）岩石，包括在地球化学成分上与洋岛玄武岩或洋中脊玄武岩类似的玄武岩以及高Nb或富Nb弧玄武岩（王强等，2020；徐义刚等，2020；Zheng et al.，2020）。其中，高Nb或富Nb弧玄武岩常被认为是由俯冲板片熔融产生的埃达克质熔体交代地幔楔橄榄岩熔融而成的，洋岛玄武岩则与地幔上涌导致的降压熔融或板片熔体交代的地幔在金红石不稳定区发生熔融有关，一些洋中脊玄武岩被认为与洋中脊俯冲有关。

（2）新生代碰撞带镁铁质火成岩特征和岩石组合

大陆俯冲期间，由于大陆地壳较干、缺少流体活动或相对冷的地热梯度，一般认为不产生同期岩浆作用（Zheng and Chen，2016）。但是在一些新生代的大陆碰撞带上盘，则产出较多同碰撞、碰撞后的岩浆岩。新生代碰撞带玄武岩以钙碱性和碱性系列玄武岩为主，不同碰撞类型中岩石组合如下。①在弧－陆碰撞带，如班达、堪察加半岛，除包含洋内弧岩石组合外，玄武岩中大量出现橄榄玄粗岩和白榴岩等碱性系列岩石，具有弧岩浆岩的地球化学特征（Elburg et al.，2002）。多数研究认为，大量碱性岩石的出现与大陆沉积物俯冲交代作用有关（Elburg et al.，2004）。②在陆－陆碰撞带，如阿尔卑斯、扎格罗斯、冈底斯造山带，产出钙碱性和碱性岩石，并且随着时间变化，钙碱性系列岩石逐渐向碱性岩石序列过渡，玄武岩为钙碱性和碱性，碱性玄武岩以橄榄玄粗岩、钾镁煌斑岩、白榴岩等钾质－超钾质岩石为主，显示岛弧型微量元素特征，伴随少量洋岛型碱性玄武岩（Seghedi et al.，2004；Guo et al.，2015）。陆－陆碰撞带上盘不同镁铁质岩石的Sr-Nd-Hf-O等同位素组成变化范围较大，被认为代表了地幔源区包含不同来源及比例的富集组分，如俯冲大洋或大陆板片组分、混杂岩等（Guo et al.，2015）。

2. 板块汇聚带交代作用与镁铁质岩浆源区形成

（1）俯冲板片脱水与流体交代作用

在原始地幔归一化微量元素分布图解上，典型的弧玄武岩相对富集大离子亲石元素和轻稀土元素，亏损高场强元素。这些地球化学特征被认为是岛弧玄武岩的地幔源区受到了俯冲板片释放流体的交代（Tatsumi，2005；Zheng，2019）。富水溶液和榴辉岩组合矿物（石榴子石、单斜辉石和金红石）

在 900～1200 ℃和 3.0～5.7 GPa 条件下的元素分配实验表明，石榴子石和辉石不能导致高场强元素与大离子亲石元素之间的明显分异，但含 1.5% 金红石榴辉岩释放的流体，可以导致高场强元素与大离子亲石元素的分异，并导致地幔楔选择性富集大离子亲石元素而亏损高场强元素（Foley et al.，2000）。传统观点认为，随着大洋板片俯冲深度增加，板片释放的流体逐渐减少（Ishikawa and Nakamura，1994）。也有研究表明，俯冲板片在弧前深度（蓝片岩相条件）已丢失了 5.5 wt% 的 H_2O（Schmidt and Poli，1998），且 13% 的弧前地幔会在 20～60 km 的深度发生蛇纹石化（Savov et al.，2007）。这些浅部蛇纹石化地幔可通过拖曳 - 俯冲或俯冲侵蚀作用进入地幔深部，参与幔源岩浆的形成。在大陆碰撞带，一些研究认为俯冲大陆板片释放含水或 CO_2 流体对造山带中钾质 - 超钾质岩浆岩的形成具有重要贡献（Guo et al.，2015）。

（2）俯冲板片熔融与熔体交代作用

俯冲板片除了释放流体外，在一些特殊条件下（如年轻的、热的洋壳），还会发生熔融，产生埃达克质熔体（Defant and Drummond，1990）。Elliott 等（1997）提出，俯冲板片来源熔体会像流体一样改变或交代地幔楔橄榄岩，使其成为弧玄武岩的潜在源区。实验岩石学和弧下地幔橄榄岩包体的研究表明，含水熔体比富水溶液具有更强的携带轻稀土元素和高场强元素的能力，而且这些熔体会同地幔反应形成富角闪石、金云母和高场强元素的地幔源区（Kepezhinskas et al.，1997）。事实上，一些新生代弧（如菲律宾、巴拿马、墨西哥巴哈半岛、堪察加半岛等）出现了富 Nb 或高 Nb 玄武岩，被认为其源区受到了板片熔体的交代（Sajona et al.，1996；Defant and Kepezhinskas，2001）。俯冲洋壳在弧下深度的部分熔融及后续的熔体 - 橄榄岩反应可能是岛弧系统中一个非常普遍的过程。在大陆碰撞带，俯冲大陆板片组分对地幔楔的交代，会形成同碰撞和碰撞后钾质 - 超钾质岩浆岩的地幔源区。

（3）汇聚带镁铁质岩浆产生的温压条件和动力学机制

俯冲板片释放的流体交代地幔以及弧下地幔中混杂岩的熔融都可能产生玄武质或镁铁质岩浆（郑永飞等，2016；Zheng et al.，2020）。实验岩石学研究表明，含水的地幔橄榄岩在高压（> 2.5 GPa）或者低压高比例（> 25%）下部分熔融会产生玄武质岩浆（Green et al.，2014）。不同实验得到的橄榄岩固相线温度差别很大，在 3 GPa 压力条件下，蚀变橄榄岩固相线可低至约 800℃（Grove et al.，2009；Till et al.，2012），橄榄岩湿固相线高达 1000～1100℃（Green et al.，2012），这种固相线温度的巨大差异会影响对弧玄武质岩浆产生机制的认识（郑永飞等，2015，2016）。另外，在板片俯冲过程中，俯冲板片发生部分熔融形成熔体会交代地幔楔橄榄岩，形成辉石岩（Tamura and Arai，2006），而在板片－地幔楔界面会发生蚀变洋壳、沉积物、蛇纹石化橄榄岩和地幔楔橄榄岩的机械混合形成混杂岩（Gerya and Yuen，2003；Zheng，2012）。这些辉石岩或混杂岩相比橄榄岩更容易发生部分熔融形成弧岩浆岩（Marschall and Schumacher，2012）。

弧系统玄武质岩浆的产生受弧下深部动力学过程、热演化控制。俯冲大洋板片回撤引起的地幔角流和弧下软流圈与岩石圈的相互作用在弧岩浆的产生中发挥了至关重要的作用（Hoernle et al.，2008）。另外，板片撕裂、板片断离、岩石圈拆沉、洋中脊俯冲、无震海岭或大洋高原俯冲、俯冲侵蚀等过程也对弧岩浆岩的形成发挥作用（王强等，2020）。

在一些新生代的碰撞带（如特提斯域、班达碰撞带），则产出有较多同碰撞、碰撞后的岩浆岩，其源区被认为可能受到了俯冲大陆沉积物或不同来源及比例的富集组分交代，或者源区是大陆俯冲产生的混杂岩组成（Guo et al.，2015；Ma et al.，2017）。这些同碰撞、碰撞后的岩浆岩形成的动力学机制包括大陆俯冲、岩石圈减薄或拆沉、混杂岩的熔融等。

3. 存在的主要问题

（1）汇聚带交代作用与岩浆源区形成

俯冲板片释放流体或熔融产生的熔体交代弧下地幔，形成弧玄武岩的地幔源区。实际上，俯冲板片的成分非常复杂。在大洋俯冲带，除了玄武质洋壳外，还包括其上覆的大洋沉积物和下伏的大洋岩石圈地幔；而在大陆碰撞带，除了俯冲大陆地壳之外，还包括其上覆的沉积物和下伏的大陆岩石圈地幔。此外，增生楔沉积物或俯冲上盘的物质也可能通过俯冲、拖曳或俯冲底侵、俯冲侵蚀等过程进入地幔深部。而且，上述物质本身可能含水（如沉积物），经过绿泥石化、蛇纹石化或角闪石化，这些富水岩石在俯冲过程中可携带大量水进入到地幔。另外，在火山弧之下的板片深度，板片来源的俯冲组分也可能是超临界流体、含 CO_2 流体或碳酸盐熔体。因此，由于流体或熔体来源和成分的复杂性，其对弧下地幔的交代作用比人们想象的要复杂得多，而且不同性质的熔体与地幔橄榄岩反应的产物受熔体成分、熔体与地幔橄榄岩比例、反应的温压条件等多种因素控制（徐义刚等，2020）。

（2）汇聚带镁铁质岩浆的产生与演化

产生玄武质岩浆的控制因素一般包括降压、加水和升温等过程（徐义刚等，2020）。一般认为，俯冲带地幔楔橄榄岩的熔融与俯冲大洋板片的挥发分组分的加入有关。精确限定地幔橄榄岩的湿固相线温度对揭示弧玄武质岩浆的产生机制非常关键。但目前不同实验得到的橄榄岩湿固相线温度差别很大（低至 800 ℃，高达 1000～1100 ℃）。因此，不同温压下地幔橄榄岩湿固相线温度的精确确定是一个亟待解决的重要科学问题。另外，研究显示地壳内的岩浆储库大部分时间以晶粥体形式存在，而非传统认为的富熔体相岩浆房（Cooper and Kent，2014；Cashman et al.，2017）。因此，对于喷出到地表的玄武岩，其成分的演化既要研究其地幔源区，也要深入探究喷发前岩浆储库的演化过程。

（3）汇聚带镁铁质岩浆作用与物质循环

汇聚带特别是大洋俯冲带是地球物质循环的重要场所，常常被称为俯冲工厂（Tatsumi，2005）。例如，在大洋或大陆岩石圈俯冲过程中，大洋或大陆板片的沉积物、洋壳或陆壳以及地幔岩石圈通过俯冲作用和俯冲上盘一些物质通过拖曳、俯冲侵蚀等过程进入到地幔中，这些物质释放的流体或熔融产生的熔体交代地幔楔橄榄岩或其本身同地幔混合形成汇聚带镁铁质岩浆岩的源区，然后镁铁质岩浆作用再将进入到汇聚带的地幔组分带回到地表。但是，如何有效识别汇聚带镁铁质岩浆岩中不同循环组分的信息，特别是估算俯冲带物质通量，仍然是当前国际地质学研究的一个难点。

（4）弧玄武岩产生动力学机制与板块构造启始时间

大量的研究将太古宙的岩浆岩同现代弧岩浆岩进行对比，提出板块构造活动（即洋壳俯冲）在太古宙已经出现，其具体启动时间争议巨大（Smithies et al.，2004；Martin et al.，2014；Hastie et al.，2015）。板块构造启动时间争论的一个核心问题是，具有类似现代俯冲机制的板块构造体制是何时出现的？在新生代俯冲带，大洋板片的俯冲不仅形成了地幔楔结构，还产生了起源于地幔楔的弧玄武岩，或熔体与地幔楔强烈相互作用后形成的高 Nb 或富 Nb 玄武岩、高镁安山岩（赞岐岩）、埃达克岩等（王强等，2020；徐义刚等，2020）。大量的研究将太古宙的玄武岩同现代的弧环境玄武岩进行对比来探讨太古宙是否存在板块构造体制。但是，到目前为止，岩石学、地球化学资料还没有得到构造、沉积和高压－超高压变质等其他证据的支持。因此，弧玄武岩产生的动力学机制与板块构造启动之间的关联仍是亟待解决的一个热点问题。

（三）古汇聚板块边缘镁铁质岩浆岩

在大洋俯冲带之上形成的大洋弧或者大陆弧，一般会随着板块构造的演化成为大陆内部的缝合带，因此也是古汇聚板块边缘的标志带。虽然其中的镁铁质岩浆岩出露在现今板块的内部，但是它们已经成为研究古俯冲带地壳物质再循环及其壳幔相互作用的理想对象（Zheng et al.，2020）。在古大洋和大陆地壳俯冲过程中，俯冲的大陆地壳以及牵引陆壳俯冲的古洋壳物质均有可能再循环进入地幔并发生壳幔相互作用，形成富化、富集的地幔源区；同时，古老造山带富集地幔的部分熔融又可以形成镁铁质岩浆岩。因此，古汇聚板块边缘出露的镁铁质岩浆岩，记录了古板块汇聚、碰撞和演化的完整过程，可以记录有关古俯冲带地幔富集机制和地幔深部动力学过程重要的信息。系统研究古汇聚板块边缘镁铁质岩浆岩的形成时代、岩石类型、地球化学特征、源区物质来源及其形成的地球动力学背景，对于认识古俯冲带壳幔相互作用和古老造山带的构造演化具有十分重要的意义。

1. 研究现状和趋势

（1）与俯冲古洋壳再循环有关的镁铁质岩浆岩

在古大洋俯冲过程中，俯冲玄武质洋壳及其上覆沉积物会随着温度和压力的升高，发生变质脱水作用。俯冲板片在弧下深度脱水形成的流体相对富集流体活动性不相容元素（图 3-7），如大离子亲石元素和轻稀土元素，亏损流体不活动性元素（如高场强元素）。这些流体交代上覆地幔楔形成蚀变橄榄岩（蛇纹石化和绿泥石化橄榄岩），该地幔源区发生部分熔融最终形成古大洋弧和古大陆弧镁铁质岩浆岩（Zheng，2019）。

随着洋壳的进一步俯冲，其在后弧深度（>200 km）部分熔融产生的长英质熔体（金红石发生分解）在微量元素上通常表现为富集大离子亲石元素和轻稀土元素、不亏损高场强元素；在放射成因同位素特征上，由玄武质洋壳部分熔融形成的熔体表现为相对亏损，而由沉积物部分熔融产生的熔体表现为相对富集。这些熔体会与橄榄岩发生反应，形成富辉石的橄榄岩、碳酸盐化橄榄岩，或者形成贫橄榄石的辉石岩和角闪石岩。这些超镁铁质的地幔交代岩可以作为一些洋岛玄武岩和大陆玄武岩的地幔源区（Sobolev et al.，2005；Pilet et al.，2008；Zheng et al.，2020a）。因此，理解古大洋俯冲过程中壳幔相互作用的性质是研究镁铁质岩浆作用的关键。

俯冲隧道中所卷入的地壳和地幔物质的性质和比例直接影响到镁铁质岩浆作用地幔源区的岩石学和地球化学性质，从而进一步影响所产生的镁铁质岩浆岩的性质。例如，秦岭－桐柏造山带古生代岛弧型镁铁质火成岩具有岛弧玄武岩类似的微量元素分布特征和亏损的放射性成因同位素组成，记录了俯冲古洋壳在弧下深度析出流体对地幔楔的交代作用（Zheng et al.，

2020b）；造山带中还发育中新生代洋岛型镁铁质岩浆岩，其具有洋岛玄武岩类似的微量元素特征以及亏损的放射性成因同位素组成，记录了先前俯冲古特提斯洋壳在后弧深度来源的熔体交代作用（Dai et al.，2014，2017）。

（2）与俯冲陆壳再循环有关的镁铁质岩浆岩

随着古大洋板片的消失，可以发生大陆地壳的碰撞和大陆岩石圈的俯冲，最终形成碰撞造山带，典型代表为阿尔卑斯－喜马拉雅造山带和大别－苏鲁造山带。碰撞造山带岩浆作用通常具有以下特征（Liegeois，1998；Bonin，2004；Zheng and Zhao，2017）：①缺乏同俯冲岩浆岩，发育大量碰撞后岩浆岩；②碰撞后岩浆岩主要是花岗质岩、含有少量镁铁质岩（主要是钾质和高钾钙碱性）；③岩浆作用通常与沿着主要剪切带的大型水平运动有关。同俯冲岩浆作用的缺失被认为源于大陆地壳较干缺少流体活动或相对冷的地热梯度（Zheng and Chen，2016）。

碰撞造山带镁铁质岩的地球化学特征大致分为两类：一类具有洋岛型微量元素分布和亏损的放射性成因同位素组成，记录了先前俯冲的古洋壳物质再循环；另一类具有岛弧型微量元素和富集的放射性成因同位素，来自大陆地壳熔体交代的地幔源区（Fang et al.，2020；Zheng et al.，2020）。具体来看，大陆俯冲带岛弧型镁铁质岩浆岩在微量元素组成上表现为富集大离子亲石元素和轻稀土元素、亏损高场强元素，在放射成因同位素组成上具有较高的全岩初始 $^{87}Sr/^{86}Sr$ 同位素比值和低的全岩 $\varepsilon_{Nd}(t)$ 值以及低的锆石 $\varepsilon_{Hf}(t)$ 值。特别地，造山带镁铁质岩浆岩中发现有残留的继承锆石核，对应于俯冲陆壳的原岩年龄和变质年龄，为俯冲陆壳循环进入地幔提供了直接的年代学证据（Dai et al.，2011）。这些岛弧型镁铁质火成岩的岩石地球化学特征，表明俯冲陆壳部分熔融形成的长英质熔体交代了岩石圈地幔橄榄岩，形成了富化富集的大陆碰撞造山带岩石圈地幔源区。例如，大别－苏鲁造山带中出露的碰撞后岛弧型镁铁质岩浆岩，是俯冲华南陆壳物质循环进入地幔，并最终通过镁铁质岩浆岩所记录的直接证据（Zhao et al.，2013）。此外，冈底斯造山带中出露的部分碰撞后镁铁质岩浆岩也记录了俯冲地壳物质的再循环（Chung et al.，2005）。

2. 存在问题

相对于现代汇聚板块边缘，人们对古汇聚板块镁铁质岩浆岩的研究相对薄弱。虽然人们已经在古汇聚板块边缘俯冲物质循环和壳幔相互作用、古老造山带地幔性质等方面进行了不少研究，但还存在许多问题亟待进一步研究。

（1）古汇聚板块边缘镁铁质岩浆岩与板片流体交代之间的关系

古板块深俯冲过程中随着温度和压力的升高，由于板片中不同地壳物质及其组成矿物在成分和稳定性上存在差别，俯冲地壳会发生不一致脱水和熔融，产生不同成分和性质的流体。由于这些板片流体在成分和性质上的复杂性，它们与地幔楔橄榄岩反应会形成不同成分的镁铁质-超镁铁质交代岩。这些交代岩在受热后发生不一致部分熔融，所产生的镁铁质熔体在上升过程中一方面会发生结晶分异，另一方面会与地幔橄榄岩发生反应，导致最终侵位到地表的火成岩表现出不同的成分。俯冲地壳成分、板片流体产生条件、地壳交代作用程度、交代岩部分熔融程度、镁铁质熔体结晶分异和同化混染程度等因素都会影响俯冲带镁铁质岩浆岩的成分。此外，超临界流体具有比富水溶液和含水熔体强得多的元素迁移能力（Zheng et al.，2011），特别是能够迁移高场强元素（Zheng，2019），这对俯冲带镁铁质岩浆岩中地壳物质的贡献具有重要意义。然而，有关俯冲带流体，特别是超临界流体与镁铁质岩浆岩成因联系的工作相对较少，它们之间的关系也不清楚。

（2）古汇聚板块边缘镁铁质岩浆岩时空分布规律及其成因

古大洋俯冲过程中弧岩浆岩以玄武岩为主，伴随少量安山岩、英安岩。大陆弧岩浆作用以安山质为主体，由大洋向内陆一侧呈现玄武岩（主要为钙碱性）、安山岩、英安岩到流纹岩变化，地壳厚度随之逐渐增大（Gill，2010）。它们一般都具有典型的岛弧型微量元素分布特征，但大陆弧相对于大洋弧玄武岩更富集不相容元素和放射成因同位素（Gómez-Tuena et al.，2013）。随着大洋板块俯冲以及深度的增加，形成的玄武岩一般会呈现由拉斑、钙碱性到碱性逐渐过渡，在微量元素组成上甚至表现出洋岛型玄武岩的特点（Schmidt and Jagoutz，2017）。为什么在大洋弧出现的火山岩是以玄武岩为主、在大陆弧出现的火山岩是以安山岩为主？是俯冲带流体成分差别导致地幔交代岩成分差异引起的，还是分异结晶和地壳混染的结果？导致这些岩石类型时空分布和地球化学性质差异的具体原因是什么还不明确，值得进一步系统研究。

（3）板块汇聚过程不同阶段的板片流体活动、交代作用与岩浆作用之间的关系

古老造山带的形成经历了洋壳俯冲过程中的变质脱水和部分熔融，到随后洋盆闭合、弧陆碰撞、大陆地壳俯冲过程中的变质脱水和部分熔融（郑永飞等，2015）。不同俯冲板片来源的不同性质流体会对上覆地幔楔进行交代，甚至叠加交代（Zheng，2012，2019）。那么，不同俯冲阶段、不同空间位置

流体交代作用与镁铁质岩浆成分之间有什么对应关系？不同性质流体叠加交代形成的地幔交代岩部分熔融所产生的镁铁质岩浆岩又会表现出什么地球化学特点？这些问题在今后研究中需要整体考虑，并进一步识别和区分。

（4）如何确定并定量估算镁铁质岩浆岩源区中地壳和地幔物质的比例

古板片俯冲过程中，大洋板片的沉积物、火成岩洋壳、陆壳基底和盖层，俯冲地幔岩石圈，以及俯冲上盘物质通过拖曳、俯冲侵蚀等过程进入到地幔中，这些物质释放的流体交代上覆地幔楔橄榄岩形成富集的地幔源区，源区部分熔融又将地幔和地壳组分迁移到镁铁质岩浆岩中（Zheng，2019；Zheng et al.，2020）。如何有效识别俯冲带镁铁质岩浆岩中不同再循环组分的信息，特别是定量估算俯冲地壳和地幔物质的比例，是当前俯冲带研究的一个研究难点，同时也是将来重要的研究方向。

（四）造山带长英质岩浆岩

长英质岩浆岩是指主要由石英、长石等浅色矿物组成的中酸性侵入岩和对应的火山岩，它们可以形成于不同的构造背景，其中汇聚板块边缘是产出长英质岩浆岩最为重要的场所（图 3-9）。在大洋俯冲形成的增生造山带（大洋弧、大陆弧）、大陆俯冲 / 碰撞形成的碰撞造山带和碰撞之后的张裂造山带均广泛分布长英质岩浆岩，揭示它们的岩石学成因机制对认识俯冲带壳幔物质循环、大陆地壳生长和地球化学分异以及造山带构造演化等方面均具有重要的科学意义（Barbarin，1999；Brown et al.，2011；Jagoutz and Klein，2018；Zheng and Gao，2021）。

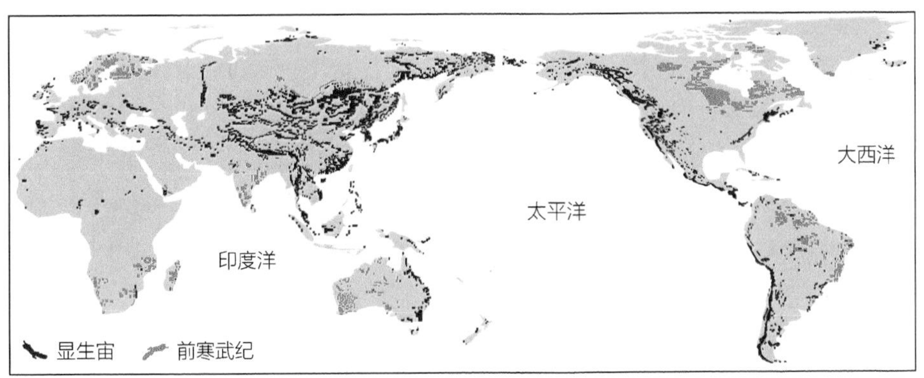

图 3-9 地球上大陆长英质岩浆岩地质分布示意图（修改自 Jagoutz and Klein，2018）

1. 研究现状和趋势

（1）大洋弧长英质岩浆岩

大洋弧一般主要由玄武质岩石组成，但在一些古大洋弧和现代成熟大洋弧也广泛出露长英质岩浆岩，如英云闪长岩、花岗闪长岩、花岗岩等（Jagoutz and Kelemen，2015；Kay et al.，2019）。另外，地震波速度剖面揭示现代活动岛弧普遍含有的中酸性成分的中地壳，被认为是花岗质岩体，如IBM 岛弧（Suzuki et al.，2015）。这些长英质岩浆岩一般具有岛弧型的微量元素分布特征和亏损的放射成因同位素组成，因此被解释为是幔源弧玄武质岩浆结晶分异的产物（Jagoutz and Kelemen，2015；Kay et al.，2019）。但对IBM 岛弧 Tanzawa 英云闪长岩的锆石氧同位素组成研究表明，其源自古老辉长质洋壳的部分熔融（Suzuki et al.，2015）。再者，一些大洋弧出露有埃达克岩，被认为是俯冲洋壳部分熔融的产物（Defant and Drummond，1990）。大洋弧花岗质岩石主体为英云闪长质和奥长花岗质，比平均大陆上地壳的花岗闪长质成分更偏基性，且它们具有比平均大陆上地壳低的不相容元素含量（Saito and Tani，2017）。因此需要额外的过程，如弧－陆或者弧－弧碰撞等过程将大洋弧地壳转换为成熟的大陆地壳，如 IBM 与本州岛碰撞带的花岗岩成分就接近大陆上地壳，这是因为这些弧地壳岩石经历了再造过程，且在此过程中还有成熟地壳物质（如沉积物）等的贡献（Saito and Tani，2017）。

（2）大陆弧长英质岩浆岩

长英质岩浆岩是大陆弧重要的岩石类型，如北美科迪勒拉和中国冈底斯造山带大量分布的中生代花岗质岩基（Ducea et al.，2015a，2015b；Zhu et al.，2019；Castro，2020），其厚度可达 20 余千米。这些长英质岩浆岩通常具有岛弧型的微量元素分布特征和相对亏损或相对富集的放射成因同位素组成。对其岩石学成因的解释主要有两种观点：一是幔源弧玄武质岩浆结晶分异的产物（Lee and Bachmann，2014）；二是玄武质岩浆结晶分异和地壳同化混染，由此提出了熔融－同化－储存－均一（melting-assimilation-storage-hybridzation，MASH）模型和地壳热带（crustal hot zone）的概念（Hildreth and Moorbath，1988；Annen et al.，2006）。无论是上述哪种解释，大陆弧长英质岩浆岩的形成均导致大陆地壳的显著生长。但也有研究表明，一些大陆弧花岗质岩浆岩是新生镁铁质下地壳部分熔融的产物（Petford and Atherton，1996；Tang Y W et al.，2020b），因此代表了大陆地壳成分分异。以上关于大陆弧长英质岩浆岩成因的解释均指示它们形成于高温条件下（900 ℃）。然而，目前对于科迪勒拉长英质岩浆有了较为清晰的认识：该岩浆是在固相线温度（650～750℃）条件下以晶粥体的形式滞留于地壳中数千至几百万年（Cashman et al.，2017；Jackson et al.，2018）。因此，岩石学观察与提出的解释模型之间存在矛盾。

为此，Collins 等（2020）提出了一个新的镁铁质岩浆底垫和硅质岩水化（mafic underplating and silicic hydration，MUSH）模型来解释大陆弧长英质岩浆岩的成因。该模型认为，在莫霍面附近，地幔楔来源的含水玄武质岩浆发生结晶分异，并最终固化形成镁铁质底侵岩石；玄武质岩浆冷却和结晶分异引起水出溶并上升，导致上覆先前形成的镁铁质底侵岩石和其他地壳岩石发生水致熔融，产生大量形成科迪勒拉岩基的低钾初始岩浆；这些花岗质岩浆在绝热上升过程中通过分离残留矿物相，进一步发生分异并重结晶成硅质粥体；这种含水的粥体间歇地被热的、镁铁质的岩浆注入，从而发生岩浆混合并促进岩浆上升和潜在的喷发。该模型既不需要流体缺乏的高温部分熔融，也不需要玄武质岩浆的分离结晶，就能产生科迪勒拉岩基成分多样性的特征，主要强调俯冲带的水在形成花岗岩和大陆地壳中的作用。上述所有模型均认为大陆弧长英质岩浆岩形成于壳内环境，与此不同，Castro 等（2010）认为大陆弧长英质岩浆岩是俯冲混杂岩（洋中脊玄武岩 + 沉积物）在上覆大陆岩石圈之下的地幔深度部分熔融形成的。

（3）大陆碰撞带长英质岩浆岩

在大陆碰撞过程中，可以形成一定量的长英质岩浆岩，它们主要是年轻或古老陆壳物质再造的产物。例如，青藏高原南部同碰撞壳源岩浆岩的锆石 Hf 同位素成分的空间变异很好地反映了新生陆壳和古老陆壳物质再造过程的发生（Zhu et al.，2011）；中国东部苏鲁造山带晚三叠世同折返花岗岩是俯冲华南陆壳物质部分熔融的产物（Zhao Z F et al.，2012，2017a）。弧陆碰撞造山带常常发育增生杂岩，其中含有俯冲带沉积物部分熔融所形成的同碰撞 S 型花岗岩（Collins and Richards，2008）。取决于形成沉积物的化学风化物源，这些 S 型花岗岩可以具有富集的或亏损的放射成因同位素组成（Wu et al.，2006；Zheng Y F et al.，2007，2008）。此外，Niu 等（2013）提出藏南同碰撞林子宗火山岩是俯冲新特提斯洋壳部分熔融的产物，因此认为大陆碰撞造山带也是大陆地壳生长的重要场所。关于同碰撞长英质岩浆岩形成的地球动力学机制则有不同观点：俯冲板片回卷、板片断离、俯冲陆壳近等温 / 升温降压折返等（Zhao Z F et al.，2012，2017a；Zhu et al.，2019）。

碰撞后构造过程是相对于同碰撞构造来定义的，它与原生造山带的构造具有继承和发展的关系（许文良等，2020）。虽然碰撞后构造过程与原生造山带过程隶属不同的地球动力学体制，但是它在空间上继承了先前造山带的位置，是在老造山带构造的基础上发展起来的（郑永飞等，2015；Zheng et al.，2019a）。先前造山带在碰撞后阶段的构造活化可以形成大量的长英质岩浆岩（Liegeois，1998），如加里东造山带、海西造山带、喜马拉雅造山带、大别－苏鲁造山带（Chung et al.，2005；Moyen et al.，2017；Zhao Z F et al.，2017b；Castro，2020）。碰撞后长英质岩浆岩既有S型花岗岩也有I型花岗岩（Moyen et al.，2017；Zhao Z F et al.，2017b；Wu et al.，2020），它们通常具有岛弧型微量元素特征和富集的放射成因同位素组成，因此是古老地壳物质再造的产物。大别－苏鲁造山带早白垩世碰撞后花岗岩普遍含有新元古代和三叠纪残留锆石，部分样品亏损 ^{18}O，表明它们是俯冲华南陆壳再造的产物（Zhao Z F et al.，2017b）。但Moyen等（2017）认为，海西造山带的一些高钾钙碱性花岗岩是幔源玄武质岩浆结晶分异/地壳同化混染的产物，因此代表了大陆地壳生长。此外，藏南冈底斯岩基中的晚新生代花岗岩普遍具有相对亏损的放射成因同位素组成（Chung et al.，2003；Hou et al.，2004），被认为是中生代冈底斯弧新生地壳部分熔融的产物。一些碰撞后花岗岩具有埃达克质地球化学特征（Chung et al.，2003；He et al.，2011），被认为是碰撞加厚地壳部分熔融的产物。碰撞后相继出现的岩石圈抬升、垮塌、拆沉、张裂是碰撞后岩浆作用发生的主要地球动力学机制（Zheng and Zhao，2017）。

2. 存在问题

尽管前人已经对不同类型造山带长英质岩浆岩进行了大量研究，在源区物质来源、部分熔融条件、岩浆演化过程、与大陆地壳生长和地球化学分异之间的关系、形成的地球动力学背景等方面取得了长足进展，但还存在许多问题亟待进一步研究。

（1）物质来源

造山带长英质岩浆岩岩石类型多样，对应的源区物质来源也多种多样，既有古老/新生地壳物质，也可能有幔源物质加入。但地幔超基性岩石部分熔融产生的是玄武质熔体（Wyllie，1984），需要通过结晶分异、地壳同化混染、岩浆混合等岩浆演化过程才能形成长英质岩浆岩；而地壳岩石部分熔融则可以直接产生长英质熔体（Clemens et al.，2019；Zheng and Gao，2021）。文献中对造山带长英质岩浆岩地球化学数据及其成因的解释还存在一些误区，如把亏损的放射成因同位素组成和与正常地幔类似的氧同位素组成解释为地幔物质的贡献，实际上新生地壳物质也可以具有这些特点。因此，人们需要结合野外地质、岩石共生组合、岩石学、矿物学、地球化学等方面对造山带长英质岩浆岩的源区性质进行综合判断。

（2）岩石类型和成分多样性的原因

造山带长英质岩浆岩的岩石类型和地球化学成分取决于诸多因素：源岩的矿物组成和化学成分、部分熔融时的物理化学条件（包括温度、压力和挥发分）、岩浆演化。针对这些影响因素和过程，学者们已经提出了许多机制来解释长英质岩浆岩的成分变化，如源区不均一、不平衡部分熔融、围岩同化混染、岩浆混合、残留体不混合、转熔矿物携带、结晶分异等（Clemens and Stevens，2012；Bonin et al.，2020；Zheng and Gao，2021）。但究竟是哪种或哪些因素在起作用？不同因素对不同种类成分的影响程度有多大？如何识别和区分这些因素和过程？这些都是有待解决的科学问题。

（3）造山带中长英质岩浆储库与岩浆演化过程

造山带广泛出露大型长英质花岗岩基，如北美科迪勒拉岩基、藏南冈底斯岩基。传统认识的熔体相岩浆房在地壳的岩浆系统中只是一种瞬时状态，岩浆大部分时间都是以晶粥体形式储存（Cashman et al.，2017；Jackson et al.，2018），造山带中一些大型花岗岩基也是通过多批次、小体积的岩浆脉冲持续叠加堆积而成的（Coleman et al.，2004）；对岩浆储库的生长和演化过程研究，具有非常重要的理论和实际意义，但仍然存在许多未解决的重要科学问题，如晶粥体中不同相（包括晶体、熔体、挥发分）的分离和相互作用、岩浆过程的时间尺度、构造背景对岩浆储库动力学演化的制约等。未来，需要开展的重要研究内容包括：恢复造山带中大型岩浆储库和复式侵入体形成的精细过程，认识浅部岩浆储库中晶体－熔体和熔体－挥发分的分离过程与反应机制，以及熔体分异过程与岩浆岩矿物 / 全岩成分之间的关系等。

（4）造山带长英质岩浆岩形成的地球动力学背景

造山带长英质岩浆岩的形成是不同造山阶段壳幔岩石部分熔融和岩浆演化的自然结果，与造山过程中的大洋地壳俯冲、大陆地壳俯冲 / 碰撞和碰撞后体制下的张裂造山作用等密切相关（郑永飞等，2015；Zheng and Zhao，2017；Zheng and Gao，2021）。对于大洋俯冲阶段形成的同俯冲大洋弧和大陆弧长英质岩浆岩而言，其产生的动力学机制无疑是：俯冲大洋板片脱水 / 部分熔融释放的流体交代上覆地幔楔，诱导地幔楔发生部分熔融产生岛弧型镁铁质岩浆，镁铁质岩浆上升进入上覆弧地壳，发生结晶分异 / 同化混染或引起弧地壳岩石部分熔融，从而形成长英质岩浆岩。但关于大陆造山带同碰撞和碰撞后长英质岩浆岩形成的地球动力学背景和热量来源则存在较大争议，如板片回卷、板片断离（Davies and von Blanckenburg，1995）、加厚地壳放射成因热积累（England and Thompson，1984）、加厚岩石圈地幔底部对流移除（Houseman et al.，1981）、碰撞加厚岩石圈根部拆沉（Bird，1979）、张裂造山作用（Zheng and Zhao，2017）等。因此，综合考虑岩浆产生机制和岩浆作用过程、造山带构造演化历史，建立变质－深熔－岩浆作用之间的联系，可能是未来将俯冲带长英质岩石与造山过程关联起来的努力方向。

（5）造山带长英质岩浆岩成因与大陆地壳生长和演化

长英质岩浆岩广泛出露于造山带，但关于它们的成因机制及其与大陆地壳生长和演化之间的关系还存在巨大争议：幔源镁铁质岩浆结晶（和/或同化混染）分异（Hildreth and Moorbath，1988; Annen et al.，2006; Kemp et al.，2007; Lee and Bachmann，2014; Moyen et al.，2017; Castro，2020），代表显著的大陆地壳生长；新生/古老地壳再造（Petford and Atherton，1996; Zhao Z F et al.，2017b；Tang Y W et al.，2020; Zheng and Gao，2021），代表大陆地壳成分分异。对造山带不同阶段的长英质岩浆岩进行研究，揭示它们的岩石学成因机制及其与俯冲带壳幔物质循环、大陆地壳生长和成分分异以及造山带构造演化等方面之间的关系，是将来该领域的重要研究方向。

四、俯冲带成矿作用

伴随着板块构造理论的产生，与之对应的俯冲带及其相关成矿研究也逐渐深入。经过近半个世纪的发展，俯冲带成矿作用的研究已经成为全球成矿理论研究体系中最为重要的组成部分，并持续成为矿床学研究中热点最为集中的领域。目前俯冲带成矿作用的研究成果和热点领域主要体现在以下几个方面，即俯冲带成矿专属性、俯冲带热液成矿规律、俯冲带关键金属资源。这些领域中尚未解决的众多科学问题又进一步决定了俯冲带成矿作用在未来的发展趋势和研究方向。

（一）俯冲带成矿专属性

1. 研究现状与发展趋势

成矿专属性是指一定的成矿作用及其产物（矿床）与一定的地质作用及其产物（地质体）的专属关系，一般用于描述成矿作用与岩浆作用之间的关系，但也可以指构造背景和地体属性对矿床的控制。俯冲带不仅是壳幔作用的集中场所，也是大型矿床的重要产地。俯冲带构造、岩浆和流体等条件有利于多种矿床的形成，如斑岩型矿床、火山块状硫化物矿床（volcanic massive sulfide deposits，VMS）、铬铁矿矿床等。俯冲带表现出显著的成矿专属性，大洋俯冲带和大陆俯冲带的成矿作用具有明显差别。

在大洋俯冲背景下，在大陆弧形成的主要是斑岩 Cu-Au 矿床和高硫型的浅成低温热液 Au 矿，在弧后盆地发育的是火山块状硫化物矿床。然而在大陆俯冲带，主要是陆壳岩石重熔形成的 S 型花岗岩或者是 I 型与 S 型混合的花岗岩，伴生的成矿作用主要是斑岩 Mo 矿和碰撞型的 Sn 矿。板片断离、拆沉和撕裂等深部过程引发的幔源钾质－超钾质岩浆活动和新生下地壳的重熔而形成的埃达克质岩广泛分布，主要形成斑岩 Cu-Mo 矿和岩浆－热液相关的 Au-Sb 矿床（侯增谦等，2006；王瑞等，2020）。未来，俯冲带成矿专属性的发展趋势体现在以下几个方面。

（1）俯冲带成岩成矿专属性

大洋俯冲带的成岩成矿专属性虽然相对比较简单，如在大陆弧主要产出中酸性钙碱性岩浆岩与对应的斑岩 Cu-Mo-Au 矿床。但具体到不同大洋俯冲带则又表现出一定的复杂性，如南美洲安第斯成矿带主要是由洋－陆俯冲产生大量中酸性的花岗闪长岩类和斑岩 Cu-Mo 矿床，而在洋－洋俯冲的西南太平洋成矿带，则主要表现为更偏中基性的花岗闪长岩类和斑岩 Cu-Au 矿床。即使同在新生代安第斯成矿带，马里昆加（Maricunga）成矿带以产出富 Au 的斑岩成矿系统著称，与其西侧的巨型斑岩 Cu-Mo 成矿带形成鲜明对比。然而，这两套成矿系统的成矿母岩并没有明显差别。相比于大洋俯冲带，大陆俯冲带岩浆岩、构造和成矿都表现为多样性的特征。大陆俯冲带的前身是大洋俯冲带，因而带内既有早期大洋俯冲形成的岩浆岩，也有大陆俯冲碰撞过程中形成的岩浆岩。矿床也表现为多期次叠加的特征，以特提斯成矿域为例，带内既有俯冲成因的斑岩 Cu-Au 矿床，也有大量大陆碰撞背景下形成的斑岩 Cu-Mo 矿床和造山带型 Au 带。

（2）区域成矿和局部巨型成矿作用的深部动力学控制

地壳浅表的成矿作用往往受控于深部的过程，如俯冲板片的性质（洋壳或者陆壳）、成熟度、角度和极性，板片回撤、撕裂和断离等都会控制成矿过程。这些深部动力学过程导致成矿作用在特定时间和特定的构造边界发育，也就是区域成矿。例如，特提斯成矿域最大规模的成矿事件发生在中新世（～15 Ma）；全球最大的 10 多个斑岩型铜矿都分布于中安第斯成矿带（智利和秘鲁）。对区域成矿的系统研究除了可以丰富成矿理论，还可以从大尺度指导找矿勘查。除了区域成矿，在某个构造带内，也常有一个或两个巨型的矿床在局部发育，成为“难以超越”的“巨无霸”矿床，如蒙古国南戈壁省晚古生代斑岩型成矿带中的奥尤陶勒盖（Oyu Tolgoi）金铜矿（铜储量为3110万t、黄金储量为1328t、白银储量为7600t）。

（3）成矿金属组合和分布规律

俯冲带内金属通常表现出一定的组合规律，如洋-洋俯冲通常发育斑岩Cu-Au矿床，洋-陆俯冲通常发育斑岩Cu-Mo矿床，而在安第斯陆缘弧远端拉张环境下又形成Cu-Au斑岩成矿系统。在北美大陆内部也可以形成陆内环境下复合的Cu-Au-Mo斑岩成矿系统（如Bingham），但在华南陆内成矿作用过程中则形成以斑岩Cu-Mo为主的斑岩型矿床（如德兴）。大陆碰撞带表现出更为复杂的金属组合和分布。以拉萨地体为例，从南到北依次为造山型Au矿、冈底斯带斑岩Cu-Mo矿床、中拉萨地体斑岩Mo矿和北拉萨地体Pb-Zn矿。

2. 关键科学问题

1）如何在未来用大数据、岩石学、地球化学、矿床学来系统研究对比大洋与大陆俯冲带成岩成矿的特殊性和专属性？

2）如何深入理解区域内大规模金属富集和局部巨型矿床成因的深部动力学控制机制？

3）成矿分布规律除了受岩浆来源和地壳属性的控制，还受到什么影响？俯冲带构造对于金属在壳幔作用中的分配和富集有何控制作用？

（二）俯冲带热液成矿规律

1. 研究现状与发展趋势

大洋俯冲带之上的斑岩型矿床紧邻缝合带，与大陆弧岩浆作用关系密切，成因比较清楚（图3-10）。一般经历了洋壳玄武岩和海底沉积物的俯冲脱水脱硫，在地幔楔形成富化富集的区域，然后部分熔融形成安山质弧岩浆，通过结晶分异形成中酸性岩浆，由此在浅部发生热液流体出溶而成矿（Richards，2003；Zheng et al.，2019b；陈华勇和吴超，2020）。整个斑岩成矿过程具体包括：①俯冲板片的脱水熔融并交代地幔楔；②地幔楔中的石榴辉石岩发生部分熔融或者直接形成安山质岩浆，或者先形成玄武质岩浆上升至下地壳底部并经历熔融-同化-储存-均一后再形成安山质岩浆；③安山质岩浆自下地壳底部上升至上地壳，在这个过程中发生结晶分异和挥发分出溶，形成成矿热液流体；④成矿流体汇集和金属沉淀等众多关键地质过程。

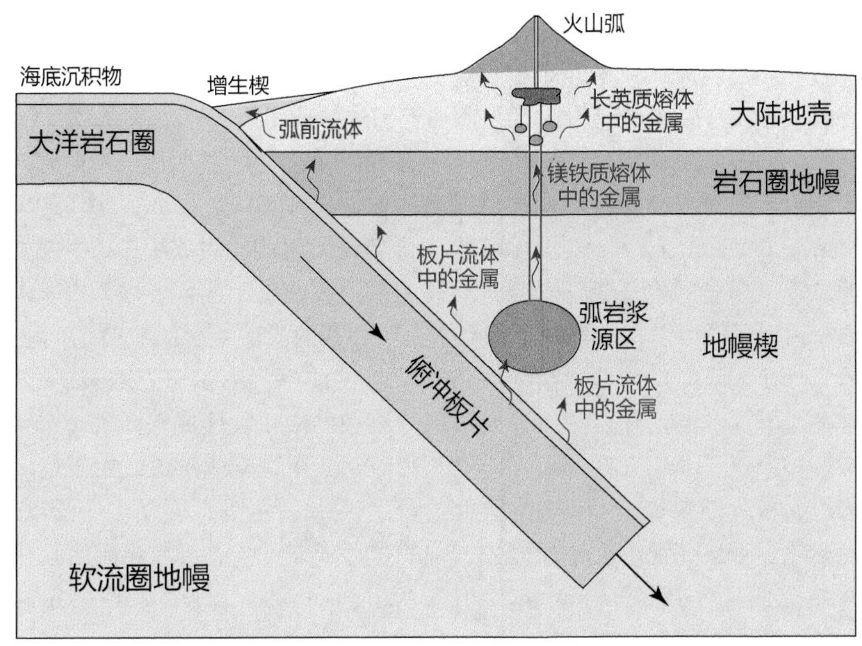

图 3-10　大洋俯冲带斑岩型矿床形成过程示意图（修改自 Zheng et al.，2019b）

与俯冲成矿相关的岩浆主要是钙碱性的玄武 - 安山 - 流纹质岩浆。在一些构造背景下，如洋脊俯冲、平板俯冲和成熟的大陆弧也发育高 Sr/Y 和 La/Yb 的埃达克质属性的岩石（Richards and Kerrich，2007；Richards et al.，2012；Sun et al.，2012；Wang et al.，2015；王瑞等，2020）。

对大陆碰撞带上盘形成的岩浆热液矿床，由于原来的大洋俯冲带已经消失，成矿作用虽然发生在大陆碰撞带，但是属于碰撞过程本身成矿还是大陆弧岩石圈在碰撞后阶段再活化成矿，是个值得研究的问题（Zheng et al.，2019b）。大洋俯冲成矿的一个重要机制是大洋板片俯冲和脱水造就了富水、高硫和高氧逸度环境，使得弧下深度金属以硫酸盐相迁移而带入到浅部成矿系统（Richards，2003；Cooke et al.，2014）。由于碰撞环境下并没有洋壳俯冲、沉积物脱水交代地幔楔等一系列深部过程，传统的基于大洋俯冲带的斑岩型矿床理论并不能合理解释大陆碰撞环境下斑岩型矿床的形成过程。为此，已有研究提出了陆陆碰撞环境下斑岩型铜矿的成因模式，该模式的核心内容包含以下三点：①大陆碰撞导致新生下地壳熔融；②埃达克质岩浆上侵形成大的岩浆房；③岩浆房流体出溶形成斑岩铜矿（侯增谦等,2004；侯增谦，2010；陈衍景和李诺，2009；陈衍景，2013）。这个新生下地壳主要是大陆俯冲改造的下地壳 + 部分岩石圈地幔。

由于大型斑岩铜矿的形成需要大量的岩浆水、金属、S 和 Cl 等成矿元素，早期俯冲洋壳阶段的成矿元素积累对于后期碰撞成矿至关重要，在大陆碰撞之前总是存在洋壳俯冲作用，这为活动大陆边缘成矿提供了先决条件（Wang R et al.，2018；Zheng et al.，2019b）。从拉萨地体南部冈底斯造山带内俯冲和碰撞型矿床的时空分布可以明显地看出这种相关性（Hou et al.，2015；Wang R et al.，2017；Zheng et al.，2019b）。大洋俯冲阶段的斑岩成矿作用主要发生在近弧带，那里的岩浆较富水且具有高的氧逸度，易于成矿；远弧岩浆由于低的氧逸度环境可以将来自深部的幔源金属和其他成矿元素如 S 和 Cl 储备在下地壳。在大陆碰撞阶段，活动大陆边缘之下的岩石圈地幔发生再活化，地幔楔在弧下深度的富化富集组分发生部分熔融再活化，使成矿元素和水在镁铁质熔体中进一步富集（Zheng et al.，2019b）。在后碰撞阶段，由于造山带的垮塌、板片断离或岩石圈地幔的减薄等作用，新生下地壳发生广泛重熔，从而活化了下地壳深度镁铁质侵入岩中的这些成矿元素，导致出现大规模的岩浆热液成矿作用（Hou et al.，2017；Wang R et al.，2017，2018；Zheng et al.，2019b；王瑞等，2020）。

板块俯冲促进了壳幔相互作用和热液流体活动，进而活化了地球深部的金属，为各类矿床的形成创造了优越的条件。俯冲带内普遍发育与中酸性岩浆有关的 W、Sn、Mo、Cu、Fe、Pb、Zn、Hg、Sb、Ag、Au 等热液矿床，其中最主要的矿床类型是斑岩型矿床和火山块状硫化物矿床。斑岩型矿床及其相关矿床（如夕卡岩型和浅成低温型矿床）主要分布在大陆边缘弧，而火山块状硫化物矿床主要分布于弧后盆地或洋中脊环境中。近年来的研究表明，在弧后盆地闭合过程中还可以形成以铁-铜-金复合成矿为特征的铁氧化物铜金（即 IOCG 型）矿床。针对俯冲带成矿的研究以斑岩型矿床最为集中，并一直作为成矿理论研究的热点。这是因为斑岩型矿床具有大吨位、低品位、大规模热液蚀变和易于开采的特征。这类矿床提供了全世界 75% 的 Cu、90% 的 Mo、25% 的 Au 和很多稀有金属（Sillitoe，2010）。中国对于 Cu 的自给率普遍低于 30%，对于 Cu 的需求逐年在增高，因而对于斑岩型矿床的研究显得尤为重要。以斑岩型矿床为例，全球大型矿床都普遍集中于大洋俯冲带，如全球三大成矿域中的环太平洋成矿域和古亚洲洋成矿域都是在大洋俯冲的背景下形成的。针对大洋俯冲带的成矿作用的研究已有近百年的历史，人们积累了丰富的经验。近些年来，在全球三大成矿域之一的特提斯成矿域，发现了一系列的大中小型斑岩和浅成低温热液矿床。定年技术的提升拓展了人们对这些矿床形成年龄的认识，然而人们对此类矿床成矿规律的研究明显不足。

未来，俯冲带热液成矿（特别是斑岩型矿床）过程与规律发展趋势主要体现在以下几个方面。

（1）物质来源和运移（金属预富集机制）

陈华勇和吴超（2020）通过综合对比认为在斑岩型矿床形成过程中存在一种连续预富集过程：在俯冲带 MASH 区域底部延伸到上地壳的空间岩浆处于一种类似“晶粥”的环境（Magee et al.，2018），不断发生着矿物的结晶和岩浆滞留与上升，而在岩浆滞留过程中可能伴随硫化物的连续预富集。这种极端的预富集模型可能解释大型斑岩铜矿成矿事件往往是岩浆弧演化晚期产物的地质事实，即晚期岩浆弧内积累了丰富的堆晶硫化物因而更具成矿潜力。普通弧岩浆与成矿岩浆在原始物质组成上有无差别，预富集过程对 Au 和 Cu 分异的影响，成矿物质在地壳内富集过程的识别与评估，预富集硫化物的“活化”机制和精细过程等关键问题需在今后的研究中进一步厘清。Chen 等（2020）综合对比了大陆弧和大洋弧堆晶的 Cu-Ag 含量，发现地壳厚度对于弧岩浆硫化物饱和早晚有显著的控制。在加厚的大陆弧，硫化物饱和较早，弧堆晶具有高的 Cu 含量和 Cu/Ag 值，而在相对薄的大洋弧，硫化物饱和较晚，演化的弧岩浆相对富 Cu。硫化物饱和早晚会影响弧堆晶中金属元素的组成，这对于后碰撞环境下斑岩型矿床的形成有重要影响。

（2）构造控矿（挤压－伸展转换和成矿）

南美安第斯斑岩成矿带是全球最为瞩目的巨型成矿带，构造背景为东太平洋俯冲带之上的大陆弧背景。该成矿带内的斑岩型矿床主要分为两期，早期为古新世—始新世，主要是一些小型的斑岩型矿床；晚期为始新世—渐新世，产出巨型的斑岩 Cu-Mo 矿床，被认为和法拉隆板片的平板俯冲有密切的关系。在西南太平洋俯冲带之上的大陆内部也发育大量的巨型斑岩 Cu-Au 和浅成低温热液金矿床，它们的成因一直是个有争议的问题。有研究提出，这些陆内热液的矿床形成与无地震的洋脊、海山链和洋底高原的俯冲有关。但是，这种构造转换是否能够引发平板俯冲、地壳加厚抬升和剥蚀？埃达克岩的形成是否有利于斑岩型矿床和浅成低温热液矿床的形成和出露？

北智利古新世—渐新世和美国西南白垩纪—古新世的斑岩 Cu-Mo 矿床可能有着类似的大洋俯冲带之上的构造背景。由于海山或者洋底高原具有正浮力，它们发生俯冲时会改变俯冲的角度，形成低角度俯冲。Cooke 等（2005）通过对世界主要斑岩铜矿带成矿背景的综合研究，发现大洋板片的低角度俯冲非常有利于挤压背景的形成。但是，在挤压背景下发育的是阿尔卑斯型蓝片岩相–榴辉岩相变质作用（Zheng，2021c），一般对应于板块俯冲的早期阶段。已有的研究表明，斑岩型矿床常形成于构造机制转换阶段（Solomon，1990；Sillitoe，1997；Kerrich et al.，2000；Richards，2003；Cooke et al.，2005），如由挤压向伸展转换阶段（Richards，2003），俯冲角度变化过程（James and Sacks，1999），非常有利于斑岩型矿床的形成。的确在板块俯冲的晚期阶段，板片俯冲的角度在重力作用下变陡，其上盘演化为拉张背景，弧后地区的岩石圈会发生减薄，先前形成的大陆弧安山岩会发生重熔形成花岗质岩浆，由此结晶分异产生岩浆流体最终形成斑岩型矿床（Zheng et al.，2019b）。

（3）热液蚀变矿化过程（热液流体出溶与元素富集机制）

一般认为，斑岩成矿的钾化阶段最初是自岩浆分异的高温单相流体，随着岩浆上升而分异出高密度的卤水相和低密度的气相（Rusk et al.，2004）；绢英岩化阶段的挥发分是岩浆直接分异出的低温液相（Sillitoe，2010）；青磐岩化阶段流体则是由岩浆卤水和侵入体加热的循环地下水混合形成的（Cooke et al.，2014）。一系列研究表明，斑岩成矿岩浆与普通岛弧岩浆的演化路径一致且两者成矿物质含量并无显著差异（Richards，2003，2015；Zhang and Audétat，2017；Du and Audétat，2020），指示岩浆流体出溶和萃取熔体内成矿物质效率也可能是影响斑岩能否成矿的关键因素。

2. 关键科学问题

1）对新生下地壳的岩石组合进行系统的岩相学和地球化学研究，可以进一步厘定 Cu-Au 等成矿元素在不同构造背景下的深部地壳中的赋存状态，进而探讨成矿元素的预富集和再活化机制。

2）冈底斯造山带巨型斑岩成矿带的形成应该得益于从挤压向伸展转化，但其具体受何种动力学机制控制仍不清楚。

3）目前，对斑岩成矿环境岩浆–热液富硫体系内影响成矿元素分配和赋存的主要机制，以及该过程对斑岩型矿床形成的具体贡献尚未完全厘清，限制了对斑岩系统成矿元素富集过程的认识。例如，有利于斑岩成矿的流体出溶条件是什么？岩浆–热液阶段成矿元素的分配行为、主控因素及机制是怎样的？该阶段成矿元素的主要赋存形式，以及这些主控因素是如何影响元素的分配和赋存行为的？这些成矿元素如何在斑岩成矿的各个热液阶段运移和沉淀，并最终造成蚀变矿化分带等一系列未解决的问题都指示进行精细的热液成矿机制研究，特别是加强模拟斑岩热液成矿过程的实验工作势在必行。

（三）俯冲带关键金属资源

1. 研究现状与发展趋势

2016年11月，国务院批复通过的《全国矿产资源规划（2016—2020年）》首次将24种矿产列为战略性矿产。2017年，美国地质调查局列出35种关键矿产资源，为保障美国安全和经济繁荣，时任美国总统特朗普签署了《确保关键矿产安全和可靠供应的联邦战略》特别行政令。欧盟2018年发布的《关键原材料和循环经济》报告中所列的关键金属为27种。在当今国际环境下，世界各国普遍提升了对战略性关键金属的重视，虽然每个国家的关键矿产清单不尽相同，但是总体上包括三类金属，即稀有金属、稀土金属和稀散金属。17种稀土元素包括元素周期表中IIIB族中的镧系元素以及钇和钪，中国稀土资源丰富，产量居世界第一位。在空间分布上，具有北方轻稀土富集、南方重稀土富集的特征。稀散金属含镓、铟、铊、锗、硒、碲及铼7种元素。其由于独特的物理化学性质和成矿规律，常被作为副产金属综合回收。中国主要的稀散金属资源十分丰富，但是国内对该金属的利用率不高，因此大量出口到美国和日本。稀散金属主要以“稀”“伴”“细”的特征伴生于其他矿床，相应的独立矿床十分少见。目前，对稀散金属的地球化学性质和行为的研究还较为薄弱，在元素超常富集机理的问题上仍存在较大争议（温汉捷等，2019）。中国拥有丰富的稀土和稀散矿产资源，但是部分稀有金属资源现状并不乐观。稀有金属元素主要包括锂、铍、铷、铯、铌、钽、锆、铪、锶等，具有独特的物理化学性质，被称为“白色石油”、“能源金属”或“高能金属”，在军工、电子、石油加工、冶金、机械、能源、轻工和农业等领域都得到了广泛的应用，是新一轮科技革命的必备矿产资源。

花岗岩型和伟晶岩型矿床在稀有金属矿床中具有重要地位。前人已经对稀有金属矿床的成因类型及特征、花岗岩体地球化学特征及成因（袁忠信等，2003；London，2009；Ero and Ekwueme，2009；Lai et al.，2012）、稀有金属矿物学（Dill，2010；Sheard et al.，2012；王汝成等，2019）、成矿元素富集机制（Zhu et al.，2001；Linnen et al.，2012）、岩浆－热液演化与成矿关系（Audetat et al.，2008；Thomas et al.，2008，2011；Pekov and Kononkova，2010）、成岩成矿年代学（Camacho et al.，2012）等方面进行了研究，取得了初步的进展。

俯冲作用可能对某些稀散金属元素富集成矿有积极作用。以Re为例，Golden等（2013）统计了全球135个地区422件辉钼矿中Re的含量，发现辉钼矿中 Re 含量随着成矿年龄变小具有逐渐升高的趋势，尤其是从 300 Ma 开始辉钼矿中 Re 含量增幅更加明显，反映了陆壳风化带入的 Re 增多。同时，海底黑色页岩中 Re 含量随着沉积体系年龄减小而逐渐升高，与辉钼矿中 Re 含量逐渐升高的趋势相同，对应大气氧含量升高的地质过程（Sheen et al.，2018）。那么这样的趋势就为 Re 富集成矿提供了一个可能，即俯冲洋壳可以将上覆黑色岩系带入俯冲带中，黑色岩系中赋存的 Re 随着板片脱水交代进入上覆地幔楔，从而对产生的熔融体中 Re 含量具有重要的影响，而且越年轻的地壳沉积物，Re 富集成矿的可能性越大（温汉捷等，2019）。世界上分布有大量的斑岩型矿床，Re 常伴生在斑岩型矿床的辉钼矿中，而且斑岩型矿床 Mo 的品位与辉钼矿的 Re 含量呈反相关关系，辉钼矿中 Re 含量最高的为斑岩型 Cu-Au 矿床，其次依次为斑岩型 Cu 矿床、斑岩型 Mo-Cu 矿床，最低的为斑岩型 W-Mo 矿床（Sinclair et al.，2009；Millensifer et al.，2014）。这样的含量差异可能与不同类型矿床的岩浆氧逸度、岩浆热液演化过程中的物理化学条件有关，但是具体控制作用需要进一步研究。铟（In）矿床广泛分布于活动大陆板块边缘以及地热梯度急剧变化的造山带附近（Schwarz-Schampera and Herzig，2002），与汇聚板块边缘温压结构和动力体制变化密切相关。从全球范围来看，与板块俯冲作用有关的富 In 矿床主要分布在西太平洋板块边界（如东亚和东北亚地区）、玻利维亚和纳斯卡－南美板块边界、秘鲁的北美西板块边缘以及中欧的海西和阿尔卑斯造山带。这些富 In 的矿床普遍与 A 型或 S 型花岗岩有关。由于 In 元素在岩浆分异过程中属于不相容元素，所以与高分异花岗岩也有关系，至于能否形成有价值的工业矿床，可能受到花岗岩源区的金属含量以及演化过程的制约，有待进一步研究。

针对目前研究的薄弱区域，关键金属未来的研究方向具体包括：关键金属元素地球化学性质；重大地质事件、圈层物质循环与成矿物质基础；关键金属元素超常富集的苛刻条件；关键金属矿床成矿规律；关键金属元素赋存状态；关键金属元素高效清洁利用等主要方向（翟明国等，2019）。根据国家自然科学基金委员会发布的“战略性关键金属超常富集成矿动力学重大研究计划 2019 年度项目指南”，未来的研究会侧重于花岗岩－伟晶岩型矿床、碱性岩－碳酸盐型矿床和风化－沉积型矿床，实现三大科学目标：①揭示关键金属元素超常富集成矿的苛刻条件，建立关键金属超常富集成矿理论，实现成矿理论突破；②揭示关键金属成矿规律，确定关键金属元素矿床新类型，实现指导找矿突破；③查明微观尺度关键金属元素赋存状态，攻克关键金属强化分离理论瓶颈，实现分离理论突破。

2. 关键科学问题

1）俯冲作用是造山作用的开端，该时期形成的是以白云母为主、稀有金属矿化程度很低的伟晶岩，其后稀有金属矿化才逐渐增加。俯冲作用是否为稀有金属的聚集提供了预富集？如果是，具体的机制是什么？

2）Re、In 等稀有稀散金属元素的富集与俯冲带岩浆氧逸度、岩浆热液物理化学条件有何关系，其具体控制作用需要进一步研究。

3）元素地球化学行为对关键金属富集的控制，包括其成矿的专属性、元素共生分异、分配系数研究，物理化学条件与化学动力学机制等。

4）深部地幔岩浆作用和浅部地壳高分异岩浆作用对关键金属超常富集的控制机制。

第二节　国内板块俯冲带研究的优势与薄弱环节

一、国内本学科的优势领域及分析

近 40 年来，特别是进入 21 世纪以来，中国在板块俯冲带领域取得了很多具有国际影响力的重要成果，极大提高了国际地位。这些优势方向主要包括大陆俯冲带和古大洋俯冲带，采用的方法包括岩石地球化学、同位素年代学、同位素地球化学、地球物理学等。中国具有类型丰富、出露完好的碰撞造山带，如喜马拉雅造山带、天山造山带、阿尔金造山带、祁连 - 柴北缘造山带、秦岭 - 桐柏造山带、红安 - 大别 - 苏鲁造山带等，在这些大陆碰撞带形成之前是原特提斯洋、古特提斯洋、新特提斯洋俯冲，古大洋俯冲带先后转变成大陆俯冲带。所有这些古大洋和大陆俯冲带为人们研究板块俯冲带提供了良好的天然实验室。

在地球物理方面，地球物理手段在研究俯冲带的几何形态、物理性质、形变和动力学演化中起着关键作用，而针对俯冲带大地震的地球物理研究也是认识地震发震机理和防震减灾的重点。中国大陆内部地震频发，具备了坚实的地震研究队伍，在地震分布、地震破裂过程、地震成像等研究方向上具备了国际前沿水平。这些以地震学为基础的地球物理研究已很好地拓展到针对俯冲带的研究中。包括中国地震台网在内的中国密集地震台网为西太平洋俯冲带的地震成像提供了得天独厚的条件。在大尺度上，从基于传统体波走时成像，到考虑有限频走时的成像，以及最新的基于全波形的伴随成像的模型，这些研究结果提供了涵盖东亚及其邻近区域下方直到下地幔顶部的地震波速结构，刻画了西太平洋俯冲的不同形态，对研究俯冲板片的动力学起到重要作用。在区域尺度上，以中国东北下方停滞在转换带的俯冲板片结构为例，中国学者进行了大量研究。通过层析成像获得了中国东北下方俯冲板片的形态并讨论了其和板内火山的关系；通过波形模拟三重震相获得更精细的俯冲板片结构特征；利用接收函数、转换波以及 ScS 多次波获得的上地幔不连续面的起伏特征，对研究俯冲板片在转换带的空间分布和是否存在板片“撕裂”起到关键作用，同时也是厘清和中国东北板内火山相关的热上涌起源的关键。在更小尺度上，中国学者也进行了大量的工作，如获得伊豆－小笠原俯冲板片精细结构及其地震精定位，提供了该区域 680 km 的深源地震的发生背景的关键认识。对日本本州岛下方双地震带的精细结构成像研究，为中源深度地震的发震机理及其对应的矿物脱水过程提供了约束。进一步，对到达下地幔的俯冲板片的成像提供了认识地幔动力学演化过程和板块重构的关键数据。不仅在结构成像上，中国在俯冲带大地震研究中也占有重要的一席。以 2011 年日本东北大地震为例，中国在结合大地测量数据的有限断层反演、反投影法确定震源破裂特征的研究都处于世界前沿。基于俯冲带大地震的震后形变的研究结果，为俯冲大洋软流圈流变特征提供了重要约束。同时，中国在世界上首次针对地球“第四极”马里亚纳海沟南段开展海洋探测，通过结合海洋地质和地球物理，将是对俯冲带研究的重要推动。地球动力学是认识俯冲带演化历史和过程的重要手段，尽管该方向在中国还是处于起步状态，但已具备了优秀的人才队伍和计算资源的保证，针对俯冲的起始及其对应的喷发火成岩特征、俯冲板片演化及其对应的地表构造演化特征已开展了大量的工作。

地球化学手段在研究俯冲带原岩性质、变质、交代和岩浆活动时间确定、变质过程中流体和元素活动、壳幔相互作用、俯冲地壳物质的再造和再循环等方面起到关键作用。地球化学研究需要合适的可供研究的样品以及可以获得大量准确地球化学数据的分析方法和仪器。基于这两个因素，中国俯冲带地球化学研究具有三个显著优势。①中国出露了数量众多的俯冲带，有古俯冲带也有现代俯冲带，时间跨度大，类型多样，为俯冲带随时间和空间演化的研究提供了理想的天然实验室。例如，西南天山是已知世界上出露有洋壳超高压榴辉岩的两个大洋俯冲带之一，红安造山带和柴北缘造山带同时出露有洋壳和陆壳俯冲形成的榴辉岩，记录了大洋俯冲向大陆俯冲的构造转换。②中国众多高校和研究所拥有数量众多的地球化学分析实验室/仪器，如国内拥有世界上数量最多的离子探针仪器，可以稳定高效地产出高质量的数据。③中国有大量的高水平人才进行地球化学研究，除了应用地球化学数据解决俯冲带问题外，还积极进行分析方法建设和新方法开发，从实验和理论上研究元素分配和同位素分馏机理。例如，各种金属稳定同位素分析方法的开发，各种矿物微区 U-Pb 定年和同位素分析方法的开发，氧同位素分馏系数的理论计算，矿物间金属同位素分馏系数的理论计算和实验标定等。

近年来，利用这些俯冲带地球化学研究的地质优势和分析技术上的进步，我国在俯冲带岩石地球化学、同位素地球化学和同位素年代学方面取得了显著进展。在岩石地球化学方面，如利用大别－苏鲁造山带超高压变质岩独特的氧同位素负异常特征，确定了大别－苏鲁造山带俯冲地壳具有华南陆块属性。通过将大别－苏鲁造山带碰撞后中生代岩浆岩与华南陆块鉴定性年代学和地球化学特征的对比，明确这些岩浆岩的源区为俯冲大陆地壳自身及其交代的华北克拉通岩石圈地幔，建立了从大陆地壳的俯冲和折返到碰撞造山带构造垮塌的构造过程，将陆内造山作用引入板块构造理论体系中。在同位素年代学方面，如开展了大量锆石 U-Pb 定年分析，结合岩石学研究，建立了不同俯冲带超高压变质连续而完整的 *P-T-t* 演化轨迹，确定了陆壳深俯冲过程的时间序列及其在地幔深部的短暂居留，认识到高压－超高压变质岩石的折返过程经历了复杂的多板片、多期次和不同动力学机制的构造过程，建立了大洋俯冲到弧陆碰撞和陆陆碰撞及碰撞后再造的时间序列。在国际上率先开发了锆石 U-Pb 定年、微量元素和 Lu-Hf 同位素联机分析技术，并将其应用到俯冲带变质岩中，建立了区分不同成因锆石的地球化学判别指标，为俯冲带变质和交代过程中时间的确定奠定了很好的基础。利用新开发的矿物 U-Pb 定年方法，解决了一些俯冲带变质年龄争议或确定特殊岩石的变质年龄。例如，通过金红石 U-Pb 定年明确了西南天山榴辉岩相变质作用时间；通过钙钛矿 U-Pb 定年确定了西南天山异剥钙榴岩的超高压变质时代。在同位素地球化学方面，如通过榴辉岩氧同位素异常的保存，提出大陆地壳的深俯冲和折返以"快进"和"快出"为特点；利用建立的矿物氢同位素和含水量分析技术，发现深俯冲陆壳折返过程中存在可识别的流体活动；通过多种同位素体系联合分析，识别出大陆俯冲带地幔楔受到多期俯冲洋壳和陆壳来源流体的交代作用。将近年来发展起来的金属稳定同位素应用到俯冲带研究中，促进了对俯冲带流体活动和壳幔相互作用的认识，将其与传统稳定同位素结合，发现大陆俯冲带存在反向的地幔楔来源流体对俯冲板片的交代作用。

在矿床学上，中国独特的地质条件，特别是俯冲带相关地质作用的广泛发育，为中国矿床学的原始创新研究提供了得天独厚的条件。也正是因为如此，随着国家重大项目的支持和分析测试手段的提高，近年来俯冲带相关矿床学研究也取得了质的飞跃，在国际矿床学界的影响力也正在迅速提高。俯冲带矿床学研究在中国具有两个显著优势。①研究区域十分辽阔，中国一半以上的国土由造山带组成，而近些年取得重要找矿突破的区域多位于这些造山带中，为成矿机制研究提供了大量良好的实例。②不同造山带中俯冲带形成的时间跨度大、形式多样，为多种成矿类型的阶段性集中产出提供了良好的地质条件。例如，中亚造山带自早古生代一直发育到中生代，由早期增生造山作用（俯冲＋地体拼贴）为主过渡到后期拉张作用为主（后俯冲阶段），成矿类型也由斑岩型和 VMS 型等俯冲作用直接相关的矿床转变为后俯冲阶段的钼矿和钨矿床等。青藏高原的特提斯成矿带则经历了自中生代俯冲为主到新生代碰撞（后俯冲）为主的构造背景转变，其主要代表性成矿类型——斑岩铜矿，也从藏北侏罗－白垩纪俯冲形成的斑岩铜矿（中小型为主）过渡到后俯冲阶段的冈底斯斑岩铜矿带（多产出大型矿床）。近年来，利用这些俯冲带成矿研究的独特地质优势和其他有利因素，中国俯冲带相关矿床学研究进展非常显著。例如，青藏高原“大陆碰撞（后俯冲）环境斑岩铜矿成矿模式”建立，得到了国际矿床学界的公认，极大地拓展了对斑岩铜矿已有的认识。新疆北部古弧盆体系的成矿综合研究也提出了在增生造山过程中“幕式成矿”的特色模式，并进一步建立了“古斑岩成矿系统叠加成矿模型”和“弧相关盆地闭合铁铜金成矿模型”等新的成矿模式，丰富和完善了俯冲带相关矿床的成矿理论体系。作为全球最为典型的俯冲带相关成矿类型——斑岩铜矿的研究也在中国俯冲带成矿研究中得到了最充分的发展，除了传统的洋－陆俯冲环境产生的经典斑岩铜矿模型完善（如中亚造山带古生代斑岩铜矿），中国学者还提出了全新的“大陆碰撞（后俯冲）环境斑岩铜矿模式”（如青藏高原冈底斯斑岩铜矿带）以及在华南和东北地区较为特色的“陆内环境（俯冲带后部）斑岩铜矿模式”。这些成果显著促进了全球斑岩铜矿成矿机制研究的发展，增强了中国在俯冲带相关矿床学研究中的国际地位和影响力。

实验分析技术的改进和分析方法的创新，是推动地球科学研究取得进展的重要驱动力。近年来，国内许多地球科学研究单位在实验室建设和各种分析手段方面取得了重大突破，尤其在同位素定年、矿物微区元素和同位素、非传统稳定同位素等分析方法方面和国际同行之间的差距越来越小，实现了从追赶到并行，甚至在部分方面已经成为技术引领者。在地球化学研究方面，随着原位分析技术的快速进步，矿物和熔体包裹体的元素和同位素微区分析手段在俯冲带过程示踪中的优势越来越明显。例如，根据俯冲带岩浆岩中不同时期结晶的矿物及其熔体包裹体可以记录不同阶段的岩浆成分，从而反映源区的性质和岩浆演化过程。随着高灵敏度多接收等离子体质谱仪的广泛应用，高精度的非传统稳定同位素的分析方法得到了快速发展，使得非传统稳定同位素地球化学成为研究俯冲带过程的新工具。中国多个科研院所购置了 MC-ICP-MS，建立了 Fe、Cu、Mg、Zn 等分析方法，并且在俯冲带领域的研究中得到很好的应用。例如，俯冲带镁铁质岩浆岩及其地幔包体，以及碳酸盐化榴辉岩的研究表明，地幔可能具有不均一的 Mg 同位素组成，轻 Mg 同位素特征结合其他元素和同位素特征可能指示了俯冲洋壳物质（特别是碳酸盐岩）的交代作用。此外，对于俯冲带成矿方面的研究，针对矿床学研究样品多期次、多组分、多来源的特点，需要突破传统全分析手段难以提供高时空分辨率信息的瓶颈问题，发展复杂基体条件下的微区原位分析新技术和新方法，包括离子探针原位分析技术、激光微区原位分析技术等。例如，通过测定单矿物 / 单个包裹体微米尺度元素含量和同位素组成，可以精细刻画元素在岩浆和 / 或热液体系中的分配行为和富集过程。俯冲带的结构和过程十分复杂，仅使用单一的地球物理或地球化学方法难以揭示复杂的地质过程。因此，在常规分析手段基础上，结合一些新发展起来的分析手段，将能更为有效地研究俯冲带过程。

二、薄弱环节的定性分析

尽管中国在板块俯冲带领域取得了长足进展，但从国际上来看，在很多方面仍然比较薄弱，特别是实验岩石学、地球动力学、高精度同位素定年等方面。例如，在实验岩石学方面，国际上从 20 世纪上半叶就开始了大量基础研究，包括岩石相变和相图、不同体系岩石的固相线、熔体与矿物之间的微量元素分配系数和稳定同位素分馏系数，以及超临界流体出现的 P-T 条件及黏滞度、电导率、元素迁移行为等物理化学特征。特别是在极高压力（如核幔边界、下地幔、地幔过渡带等）条件下的矿物相变、矿物物理性质及化学行为方面，总体上国内研究仍非常缺乏，也缺乏相应的研究基础设施。

在地球动力学方面，国际上对俯冲带结构、过程和产物的数值模拟也取得了很多重要的认识，如俯冲带的几何结构、温压结构和地质结构。国内研究在喜马拉雅造山带和青藏高原的壳幔结构方面取得了突出进展，但很多其他典型的造山带结构并不清楚。在俯冲带动力学演化方面，目前仍缺乏对各种控制元素特别是流体的影响的研究。

在高精度同位素定年方面，国际上很早就开展了通过化学溶蚀（CA）-同位素稀释（ID）-热电离质谱（TIMS）U-Pb 方法对锆石进行定年，并广泛应用在高精度地层定年、花岗岩岩基演化的时间尺度、造山带部分熔融的持续时间等。国内在这个方面仍然刚刚起步。针对俯冲带变质过程定年，国际上也开发了对石榴子石的微区 Sm-Nd 和 Lu-Hf 同位素等时线定年等先进技术。这些可以很好地揭示俯冲带变质持续时间和俯冲碰撞过程中的热事件等。目前，国内相关实验室刚刚开始建设，研究力量还很薄弱。

在俯冲带成矿研究领域，俯冲带成矿研究与俯冲带其他领域（变质、交代、熔融、分异等）研究缺乏结合。长期以来，围绕俯冲带各种成矿类型的研究虽然如火如荼，但主要都是从矿床学的各种研究方法和角度，如成矿物质来源、成矿流体演化、矿质沉淀机制等方面进行探讨。在成矿研究的同时，俯冲带其他领域，如俯冲带岩石成因、化学动力学、起始机制及其判别、地球物理结构解析等方面都取得了飞速进展。总体来说，目前俯冲带成矿作用的研究还鲜有与这些领域进展紧密结合并产生重要原始创新成果的实例。学科交叉融合程度也明显不够。矿床本质上是具有经济价值的矿物集合体，矿床的形成受一系列其他地质因素的控制，这决定了矿床学必然与矿物学、岩石学、地球化学等其他地质分支学科有着极其紧密的联系。最近十年以来，以大数据分析和数值模拟为代表的新一轮科技手段革新正对传统学科的理论创新起到显著的影响和推动作用。目前，大数据分析和机器学习等也开始应用到地学领域中，并取得了一些重要的理论创新成果（Liu et al., 2019）。然而，大数据在矿床学中的具体应用和相关研究还相对较少，也缺乏利用大数据分析有效推动成矿理论创新的实例。此外，材料学科的测试新手段、新方法对成矿元素的赋存状态和富集机制研究有很好的推动作用，但目前矿床学研究与材料学科的深度融合才刚刚开始。

除以上几个具体的领域外，国内板块俯冲带研究还存在学科交叉不足的问题。板块俯冲带研究仍需要地质学、地球化学和地球物理学等不同学科背景的科学家通力合作，同时在与环境和灾害、生命演化、碳循环、超大陆形成和裂解以及宜居地球的形成等当前的地球科学研究热点结合方面也较弱。大数据分析和机器学习方兴未艾，已经在众多学科和领域中深度应用和广泛发展，而与地球科学的结合刚刚开始，在板块俯冲带研究方面还有很大的发展空间。

此外，在俯冲带研究方面目前仍然面临着重复性和跟踪式研究过多，而原始创新性成果偏少、重大成果更少的局面。在重大仪器设备方面，仍以购置国外仪器为主，自主生产研发仪器极少，面临着“卡脖子”的问题。在国际合作交流方面，前沿性大科学计划仍然主要由国际学术界发起，国内主导的国际合作交流项目仍然较少。

第三节　学科建设与人才队伍情况

一、总体经费投入与平台建设情况

中国在板块俯冲带研究方面投入了大量研究经费，包括从国家自然科学基金委员会、科学技术部、中国科学院、自然资源部等部委提供的资助。例如，科学技术部从20世纪90年代起至21世纪20年代陆续支持了4个973计划项目，分别为“大陆深俯冲作用”（首席科学家从柏林研究员）、“大陆板块会聚边界的地幔动力学与现代地壳作用”（首席科学家许志勤院士）、“深俯冲地壳的化学变化与差异折返”（首席科学家郑永飞院士）、“大陆俯冲带壳幔相互作用”（首席科学家郑永飞院士）。进入21世纪以来，国家自然科学基金委员会先后启动了“Pangea超大陆重建”和“大陆地壳演化与早期板块构造”等重大项目，以及“华北克拉通破坏”、“特提斯地球动力系统”和“西太平洋地球系统多圈层相互作用”等重大研究计划，同时正在执行的国家专项还有“第二次青藏高原综合科学考察研究”、中国科学院A类战略性先导科技专项“泛第三极环境变化与绿色丝绸之路建设”等。

在平台建设方面，国家自然科学基金委员会基础科学中心、国家重点实验室和中国科学院等部委的重点实验室获得了持续资助，在实验室平台建设上取得了长足进展。目前，已经拥有世界一流的实验室和仪器设备，包括元素、常规和非常规同位素地球化学成分精确分析，高精度的同位素年代学分析，高温高压实验，数值模拟，地球物理探测等。可以说，在硬件方面，在很多方面已经与国际先进实验室比肩。在分析方法和分析技术领域也有了很多突破，逐步与国际接轨，在一些领域已经走在世界前沿。

二、人才队伍情况

总体上，中国俯冲带研究建立了以中国科学院院士领衔、中青年优秀人才为骨干的人员队伍。特别是近 20 年来，中国的一些大学和科研机构形成了相对稳定的从事板块俯冲带相关研究的人才队伍。值得注意的是，近 30 年来俯冲带研究方面培养了诸多高端人才，1991～2021 年，在中国科学院新增选的地学部院士中，与板块俯冲带研究有关的院士占 29%。特别是进入 21 世纪以来，新当选院士中从事俯冲带相关学科领域研究的包括俯冲带岩浆作用、俯冲带元素地球化学行为、俯冲－碰撞成矿作用、俯冲带地球物理结构、俯冲带变质作用和构造演化等。近 20 年来，与俯冲带研究相关的国家杰出青年科学基金获得者约占地球科学部杰出青年科学基金获得者总数的 1/6，在专业上包括岩石学、地球化学、矿床学、地球物理、构造地质学等研究方向。其中，中国科学技术大学以郑永飞院士为首的研究团队在几轮 973 计划和国家自然科学基金项目等的连续资助下，以大陆俯冲带壳幔相互作用和汇聚板块边缘化学动力学为主题，不仅培养和稳定了一支具有国际知名度的学术队伍，并建立了相应的实验平台，而且使中国在大陆深俯冲领域具有重要的国际影响力，在一些研究领域已经进入国际一流甚至国际领先行列。西北大学以张国伟院士为首的研究团队在中国中央造山带的构造演化等方面取得了系列重大进展。中国科学院地质与地球物理研究所朱日祥院士、吴福元院士和张宏福院士等对华北克拉通破坏的研究受到国际学术界的关注。中国地质大学（北京）以李曙光院士为首的团队开拓了俯冲带碳循环的金属稳定同位素示踪研究方向，深化了俯冲带化学地球动力学研究。

此外，针对喜马拉雅－青藏高原造山带，各个科研院所、高校都有科研人员在开展相关的研究工作，建立了多学科交叉、规模宏大、实力强劲的研究队伍。这支队伍主要聚焦青藏高原，而在境外（如伊朗、土耳其、地中海沿岸－阿尔卑斯等）开展工作的仍然不多。另外，与古俯冲带和新生代的青藏高原相比，在新生代和现今大洋俯冲带开展研究的队伍仍比较缺乏，力量比较薄弱。此外，地球动力学数值模拟是俯冲带过程和机制研究的重要手段，目前初步建立了相关研究团队，如中国科学院大学李忠海研究团队和中国科学技术大学冷伟研究团队等，但总体研究力量仍然相对薄弱。

第四节　重要举措与存在的问题

目前在板块俯冲带领域，资助模式、学科交叉、人才培养、创新环境建设等方面还面临不少问题。为了解决这些问题，提出以下举措。

1）持续稳定资金支持。俯冲带学科的基础研究和国家战略相关研究需要持续稳定的资金支持，这对学科发展、人才培养和重大成果产出极为重要。除了自主选择课题进行探索研究外，考虑建立类似国家自然科学基金委员会基础科学中心的资助模式，在国内遴选一两个俯冲带科学研究方面的顶尖团队，进行持续稳定支持，对推动本学科发展具有重要意义。

2）促进学科交叉。板块俯冲带是典型的交叉学科，俯冲带结构、过程和产物的研究需要地球物理学、地球化学、地质学的有机结合，需要观测、实验、模拟的深度融合，需要结合各个学科的优势共同攻关。因此，这种多学科强交叉的研究需要结合重大项目开展，需要建设综合性的研究平台等来促进板块俯冲带的学科交叉，促进高水平成果的产出。

3）拓展研究广度。板块俯冲带是板块构造的核心问题，其时间范围很广，各种时代的古俯冲带广泛发育，在中国大陆多有发育，研究比较深入。另外，俯冲带的空间范围也很广，现今地球上正在进行的大洋和大陆俯冲带数量众多，形式多样，但是由于偏离中国大陆，因此国内科学家开展的研究较少。俯冲带跨时空尺度的综合对比具有重要意义。

4）促进人才培养。在板块俯冲带领域，通过稳定资助、平台建设，引进和培养具有国际视野的优秀青年人才，通过国际重大项目合作，涌现引领学科发展的战略科学家。建立年龄结构合理、领域分布均衡又各有特色的人才队伍。在高水平科研平台建设的同时，推动高水平技术人才的培养。

5）营造良好的创新环境。引导科学家建立良好的学术道德，合理的科研评价体系，注重科研本身的创新性和学术价值。

第四章 关键科学问题、发展思路与发展方向

第一节 未来学科发展的关键科学问题与挑战

一、关键科学问题之一：俯冲带结构

（一）科学挑战 1：俯冲带几何结构

1. 问题的提出

俯冲带几何结构是俯冲带的基本特征，其核心是俯冲板块在岩石圈深度及其之下到地幔内部的几何形态。俯冲板块在浅部和深部的几何形态既是板块运动驱动力的重要影响因素，又对造山带的构造样式、岩浆活动和动力学过程起着决定作用。因此，对俯冲板片在不同深度几何形态的研究是认识俯冲带动力学的基础。一方面，不同的俯冲板片深部形态是其与上覆板块和周围地幔发生相互作用的结果，为深入认识俯冲过程对俯冲带地幔形变、构造样式、岩浆分布以及浅表地质特征等的控制作用提供了深部约束。另一方面，俯冲板片形态随深度的变化反映了地幔结构和性质的分层性，已经成为揭示地幔物理属性、化学组成及其分层性成因的基本依据。

根据板片倾角的大小，可将俯冲带几何结构分成三种：①低角度（＜30°），对应于缓俯冲（图 4-1A）；②中角度（40°～50°），对应于正常俯冲（图 4-1B）；③高角度（＞70°），对应于陡俯冲（图 4-1C）。在现代汇聚板块边缘，大多数大洋俯冲带属于正常俯冲和陡俯冲，只有大约 10% 属于缓

俯冲（其中大多位于东太平洋俯冲带）。

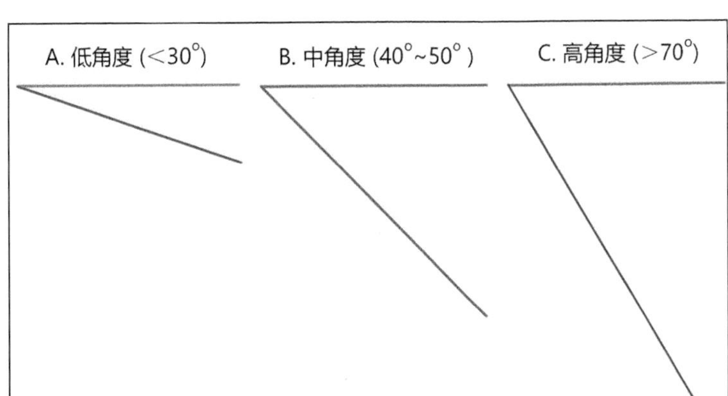

图 4-1　板块俯冲角度示意图

不论是大洋板块俯冲还是大陆板块俯冲，在板块向下俯冲的过程中，在动力学、运动学和几何因素等的共同作用下，俯冲板片本身的形态会出现不同的变化（Li Z H et al.，2019）。例如，俯冲板片会出现水平或者垂直方向上的撕裂，在极端情况下板片会发生碎片化。俯冲板片可能会停滞在地幔过渡带或者穿过地幔过渡带进入下地幔直达核幔边界。基于俯冲板片的几何结构，利用地球动力学模拟，可以用来约束俯冲板片本身及其周围介质的黏滞结构、温压结构、板片年龄、俯冲海沟的后撤和前进速率，进而为俯冲过程和板块运动的重构提供更好的约束（Li Z H et al.，2019）。

板片俯冲，尤其是大洋板片俯冲，经常伴随着地震的发生。地震不仅发生在俯冲板片和上覆板块的界面，而且还发生在俯冲板片的内部。在很多大洋俯冲带，俯冲板片都存在双地震带现象，即上层地震发生在俯冲板片的地壳中，下层地震发生在俯冲板片的地幔中（Brudzinski and Chen，2003）。俯冲带地震在深度上的分布呈现双峰特征，第一个峰值出现在 0～30 km 深度，并且呈指数衰减到 300 km 深度，第二个峰值出现在 550～600 km 深度，最深可以达到约 700 km（Frohlich，2006）。因此，可以应用俯冲板片内地震的分布来勾画俯冲板片的形态（Hayes et al.，2012）。但是，并不是所有的俯冲板片从浅到深都有地震发生，因此不能只基于地震的分布来确定俯冲板片的形态。

相对于大洋板块俯冲来说，由于大陆地壳相较于地幔密度低，人们过去认为大陆岩石圈不可能俯冲进入地幔。自从首次在意大利西阿尔卑斯

（Chopin，1984）和挪威西片麻岩区（Smith，1984）变质岩中发现柯石英后，人们开始认为大陆岩石圈能够发生深俯冲（Coleman and Wang，1995；Chopin，2003；Liou et al.，2009）。大陆俯冲地球动力学过程的探讨既需要大尺度（整个俯冲带尺度）信息又需要小尺度（几千米尺度）的信息，同时也要兼顾俯冲、变质以及折返过程的强烈空间各向异性（Jolivet et al.，2003；Guillot et al.，2009）。与俯冲的大洋岩石圈相比，俯冲的大陆岩石圈不一定都伴随地震的发生。

为了确定俯冲带的几何结构，除了利用俯冲带中地震的分布外，各种地球物理成像方法被用来研究俯冲带的结构，其中地震成像是最常用和最有效的方法。确定俯冲板片的几何结构也有利于更好地理解与俯冲带相关的地震和海啸灾害。

2. 研究现状

几乎所有的火山弧之下都发育了贝尼奥夫地震带，与地震层析成像揭示的高速异常分布一致，为追踪大洋俯冲板片构造变形和应力状态提供了重要信息。地震学观测揭示，俯冲大洋岩石圈既可以滞留在地幔过渡带，也可以穿过地幔过渡带抵达下地幔乃至核幔边界。大洋俯冲板片的倾角大多在40°～60°，但是也存在缓俯冲、陡俯冲，或者浅部缓俯冲、深部陡俯冲等复杂的几何形态（图 4-2），而且沿俯冲带的走向板片形态可发生突然变化。多波束、多参数联合地震反演技术以及多种地球物理方法高密度和高精度观测联合约束的快速发展，特别是地震层析成像技术的发展，推动了对俯冲带精细几何结构等方面的认识。近年来，地震学结构成像分辨率和俯冲带地震定位精度的大幅度提高，为构建完整的全球俯冲带板片深部形态图像提供了关键信息。这些图像揭示，板片形态具有显著的横向差异性和三维结构多样化特征。

对于全球地震活跃的俯冲大洋板片，Hayes 等（2012）主要是基于全球 EHB 地震目录[①] 和全球地震震源机制目录，通过垂直于海沟的一系列二维剖面对俯冲带的形态进行拟合。在大洋俯冲带的浅部，尤其是在海沟附近，高分辨海洋地形和主动源地震成像结果都可以用来约束俯冲板片的形态。基于

① EHB 地震目录是 Robert Engdahl、Rob van der Hilst、Raymond Buland 三个人每人名字中的一个字母提出来命名的一种地震定位方法，基于国际地震中心（International Seismological Centre，ISC）发布的地震目录的震相数据，利用改进的定位方法和地球速度模型，通过识别和利用深部震相和后续震相，对事件进行重新定位后得到的目录。

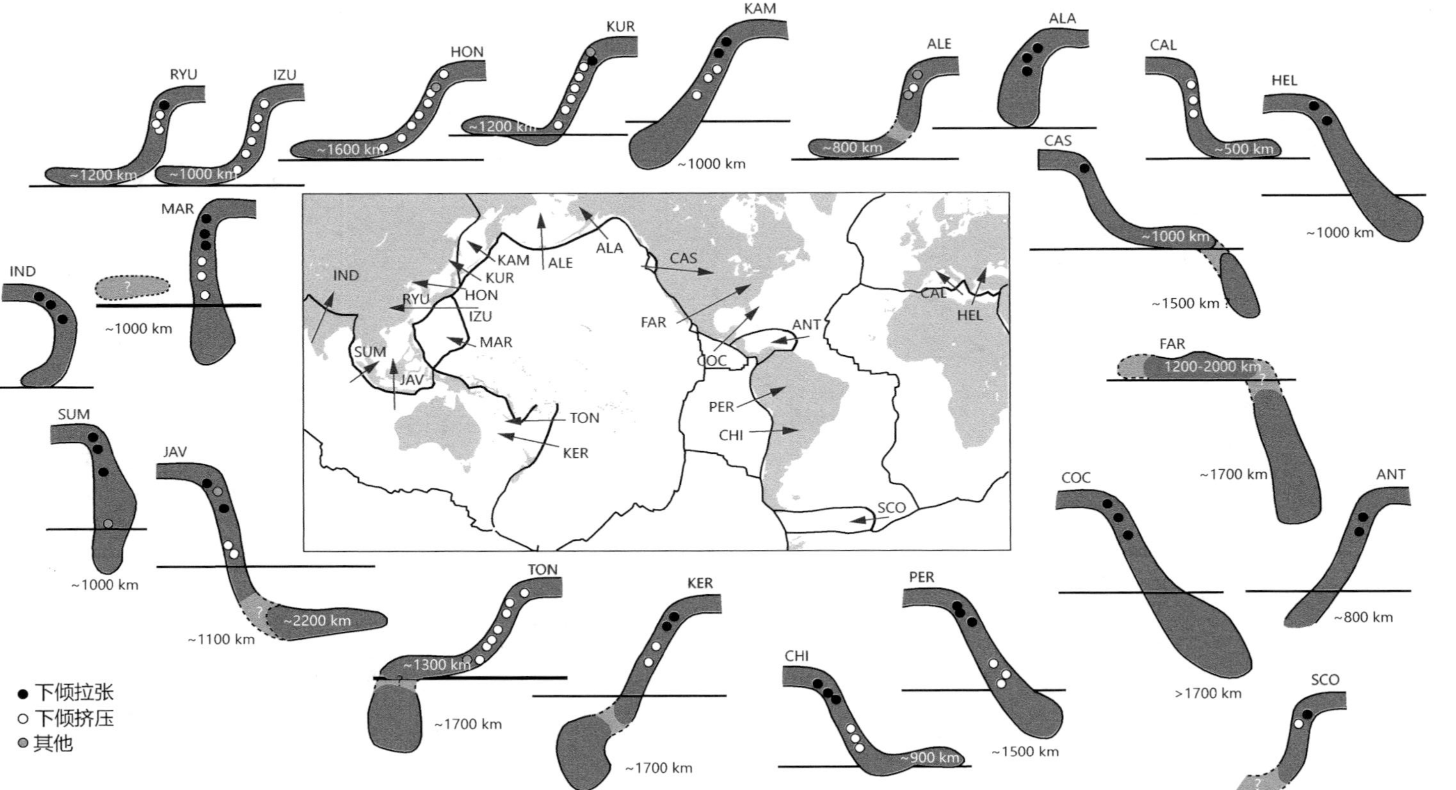

图 4-2　全球主要俯冲带地幔过渡带附近板片结构形态（引自 Goes et al.，2017）

黑线代表地幔过渡带底部 660 km 间断面，字母表示各俯冲带名称 RYU- 琉球；IZU- 伊豆；HON- 本州；KUR- 千岛群岛；KAM- 勘察加半岛；ALE- 阿留申群岛；ALA- 阿拉斯加；CAL- 卡拉布里亚；HEL- 希腊；IND- 印度；MAR- 马里亚纳；CAS- 卡斯卡迪亚；FAR- 法拉隆；SUM- 苏门答腊；JAV- 爪哇；COC- 科科斯；ANT- 安地列斯；TON- 汤加；KER- 克马德克；CHI- 智利；PER- 秘鲁；SCO- 南斯科舍这样的策略，Hayes 等（2012）给出了覆盖大约全球 85% 大洋俯冲带的俯冲板片形态模型 Slab1.0。但是 Slab1.0 模型存在一些明显的缺点，包括俯冲板片的形态受限于地震的分布，即在没有地震约束的深度，获得的俯冲板片形态存在较大的误差；俯冲板片的形态是由一系列二维曲线插值得到的，不是一个真正的三维形态；不能提供俯冲板片的厚度信息。

与周围介质相比，俯冲板片相对较冷，在地震波速上通常呈现相对高速异常，因此地震成像可以用来确定俯冲带的结构以及俯冲板片的形态。其中，一个代表性的工作是 Fukao 和 Obayashi（2013）针对全球俯冲带的 P 波速度成像。他们利用超过 1000 万个地震走时数据，并且考虑了地震射线的有限频效应，在前期速度模型的基础上，获得了新的全球速度模型。对于环太平洋俯冲带，俯冲板片存在四种不同的类型：①板片在 660 km 间断面滞留；②板片穿过 660 km 间断面；③板片在 660～1000 km 滞留；④板片俯冲到下地幔。

在 Slab1.0 的基础上，Hayes 等（2018）采用直接的三维建模策略，利用更多的数据来约束俯冲板片的几何形态，包括更多的主动震源成像结果、接收函数、局部和区域地震重定位结果，以及天然地震成像结果，建立了 Slab 2.0 俯冲带模型。与已有的全球俯冲带模型相比，Slab2.0 俯冲带模型给出了更高分辨率的俯冲板片形态。例如，对于巴布亚新几内亚俯冲板片的几何结构，Slab2.0 俯冲带模型更好地刻画了反转的“U”形结构。

大陆俯冲带的几何学主要涉及折返超高压变质岩区之下的上地幔结构，对其约束通常来自远震体波层析成像（Piromallo and Morelli，2003）。地震层析成像的结果通常结合其他地球物理探测的资料进行补充、分析，如接收函数的分析旨在探测大陆板块下方莫霍面的位置及其在深度上的延续性（Li et al.，2003）。对于西阿尔卑斯造山带，Zhao 等（2016）推测其下有陡倾的大陆板片连续体以及与之相连的大洋板片，板片深度超过 300 km。此类地震观测分析的分辨率主要依赖于走时数据的质量、数量以及地震台阵的密度。在分辨率较差的情况下，很难判断一个不连续的速度结构体是反映了活动板块边界地球动力学性质，还是地震观测分辨率低导致的结果。

3. 面临的挑战

基于俯冲板片内部和界面处所发生的地震分布和地震成像结果来制约俯冲带几何结构，首先依赖于地震定位和成像的精度。对于全球大洋俯冲板片来说，大部分俯冲板片位于海底，而在海底布设海底地震仪还存在挑战，因此俯冲板片上方的区域地震台站覆盖较为稀疏，导致与俯冲带相关的地震定位和速度成像精度较低。例如，地震定位的误差，通常在 5 km 以上。对于俯冲板片成像，通常也存在较大误差，如 Fukao 等（2009）对伊豆－小笠原－马里亚纳俯冲带太平洋俯冲板片进行了地震层析成像研究，结果显示俯冲板片几乎填满了整个地幔过渡带。Ribe 等（2007）的俯冲板片成像结果显示，位于爪哇岛（Java）下方 660 km 间断面下的板片厚度高达 400 km，而位于中美洲下方的科科斯群岛（Cocos Islands）俯冲板片的厚度高达 460 km。如果俯冲板片厚度的估计可靠，意味着俯冲板片在穿过 660 km 间断面后发生了挠曲（slab buckling），这可以用来约束俯冲板片本身以及周围地幔的黏滞性和温度分布。但是，如此厚的俯冲板片也可能是俯冲带成像分辨率较低所产生的假象。

对于大陆俯冲带来说，地震成像可用来描述大陆俯冲带的内部结构及其周围地幔的变形特征。在世界范围内，自前寒武纪（Jahn et al.，2001）到新近纪（Baldwin et al.，2004）的超高压变质岩露头均有发现，但大陆板片俯冲到地幔中的直接地震证据依然十分稀少（Schneider et al.，2013；Zhao et al.，2015；Zhang et al.，2021）。地震层析成像为喜马拉雅碰撞带西部的板片几何形态提供了良好的约束（Negredo et al.，2007）。在约 250 km 深度可观测到正在减薄的高速异常体，沿 N-S 向延伸，横贯兴都库什地区，传统上将之解释为正在断裂的板片（Koulakov and Sobolev，2006）。Zhang 等（2021）通过近震双差层析成像方法，反演获得了缅甸中部地壳和上地幔顶部高分辨率的三维 V_P、V_S 和 V_P/V_S 波速比模型，成像结果清晰揭示了喜马拉雅造山带东部印缅山脉和中央盆地下方 100 km 深度范围内存在印度大陆岩石圈，该岩石圈以 25° 倾角东倾俯冲，在北纬 22° 的大陆地壳厚度约为 30 km，向北逐渐增厚，暗示缅甸下方印度板块可能存在从北边的大陆岩石圈向更南边的大洋岩石圈的过渡。另外，即使超高压变质岩并未折返到地表，地震探测也可为大陆俯冲提供证据。例如，地震观测结果表明，减薄的伊比利亚地壳俯冲到比利牛斯山下方约 70 km 深度（Chevrot et al.，2015）。

4. 主要的研究方向和应对策略

为提高俯冲带地震的定位和获得足够的俯冲带高分辨率图像，设计大型、密集的海底地震仪观测点至关重要（Brisbourne et al.，2012；Zhao et al.，2016）。在现有全球和区域地震观测系统的基础上，为了提高俯冲带成像的精度，发展高精度的地震定位和成像算法是关键。在俯冲带地震定位方面，传统的基于地震绝对到时的 EHB 地震目录的地震定位精度仅能达到 10 km 左右，因此很难精确刻画俯冲板片的上界面。然而基于地震对相对到时的地震定位算法，如使用全球台站的远震双差定位，可以获得类似于区域地震的定位精度（Pesicek et al.，2014）。

对于俯冲带的地震成像，因为在俯冲带缺乏区域地震台网，所以俯冲带的成像精度较低。例如，对于大陆俯冲带，远震体波成像给出的一级几何学限制十分重要，但通常不能与折返超高压带保留的变形结构对比。对于俯冲带成像，在提高地震定位的基础上，可以利用多尺度双差成像算法提高俯冲带成像的精度（Pesicek et al.，2014）。与传统的走时成像相比，双差地震成像可以对震源区附近的结构进行精细成像。例如，对于日本东北部下方的太平洋俯冲板片，基于双差地震成像算法和俯冲板片中存在的双层地震带，给出了俯冲板片的高分辨率速度结构（Zhang et al.，2004）。对于伊豆－小笠原俯冲带，利用远震双差地震成像算法，基于全球台站接收到的地震走时数据，Zhang 等（2019）给出了俯冲板片的高分辨率形态变化，发现俯冲板片存在撕裂和反转。

除了利用地震直达波的到时信息外，其他类型的震相到时也可以用于更好地约束俯冲带的成像结果（Zhao et al.，2004）。因此，发展提取深度震相的算法，如基于波形自相关提取深度震相的算法（Fang and van der Hilst，2019），至为重要。全波形成像方法利用了地震波形中更多的信息，有潜力给出更精细的俯冲带几何结构（Chen et al.，2015）。在地震数据之外，还可以利用其他类型的地球物理数据对俯冲带进行成像。例如，Syracuse 等（2016）基于地震和重力联合成像算法，利用地震体波走时、面波频散和重力数据确定了哥伦比亚下方的俯冲板片形态。对于西阿尔卑斯造山带，远震体波和面波联合反演成像（Lyu et al.，2017）以及接收函数分析（Zhao et al.，2015）显示，该区域下方存在从地表超高压变质带延伸到上地幔的连续高速异常体，指示了连续俯冲的欧洲陆壳和牵引陆壳俯冲的特提斯大洋板片。

地震实验表明，新生代之前发生的超高压俯冲，其完整俯冲结构很难较好地保留，只有随后发生折返、后期构造变形和/或同化作用，才可以探测到少量残留物。一旦俯冲板片与周边地幔再次热平衡，远震体波成像将不能探测到这些结构。古老大陆俯冲带问题需要其他方法来分析。探讨大陆俯冲板片及周边上地幔的复杂变形或许是一种解决方法。这种变形在地震上常反映为地震波各向异性（Margheriti et al.，2003；Barruol et al.，2011；Salimbeni et al.，2013）。地震波各向异性主要来自岩石的塑性变形导致的橄榄石、角闪石等矿物的晶格优选定向（Savage，1999）。实验表明，在无水、低应力条件下，橄榄石的 a 轴趋向于平行于地幔流动方向（Zhang and Karato，1995），导致橄榄岩的 P 波和 S 波快波的偏振方向平行于地幔流动方向。这建立了地震波各向异性与上地幔塑性流动的关系。

数值模拟表明，俯冲过程中，俯冲隧道内向下的俯冲流与向上的折返流共同控制岩石变形，当俯冲停止、上覆载荷移走触发折返时，岩石变形不再明显（Malusà et al.，2015）。对于大陆俯冲带，一些地区的超高压变质岩并未表现出同折返变形特征（Lenze and Stöckhert，2007）。由于不清楚折返前俯冲隧道的内部结构，因此获取俯冲隧道内部结构及相关变形岩石的高分辨率图像是未来地震探测的主要方向。这既需要合理设计地震阵列密度以更好地覆盖板块边界缝合带，也需要加强对超高压变质岩的物理化学特征与古老俯冲通道内岩石的联系（郑永飞等，2013）。综合分析地球内部的速度、密度、热流和电导率来制约深部岩石的组分和结构，从而降低数据解释的不确定性。例如，电导率受流体强烈影响，大地电磁测深获得的电导率异常可制约板块边界区的流体含量（Wannamaker et al.，2009；Le Pape et al.，2012）。因此，综合不同地球物理方法的观测实验，如地震学、重力和大地电磁（Liu et al.，2016），将为进一步研究大陆俯冲过程中复杂的地质过程带来曙光。

（二）科学挑战 2：俯冲带温压结构

1. 问题的提出

俯冲带的温压结构主要由俯冲板片年龄、板片汇聚速率、板片俯冲角度、上覆板块厚度、剪切加热速率以及地幔楔的性质等参数决定（图 4-3）。一般来说，板片年龄越大，汇聚速率越大，俯冲带温度越低；板片年龄越小，汇聚速率越小，俯冲带温度越高。俯冲带的温压结构控制了俯冲岩石中含水矿物脱水的位置和条件，影响了弧岩浆作用的有无及其发生的位置（郑永飞等，2016；Zheng，2019）。

近 30 年来，通过应用解析模型和数值模型以及地质学和地球物理学观测等多学科交叉的方法，已经对俯冲带温压结构取得了很多重要的认识。尤其是随着地球动力学数值模型的计算速度和网格精度的提升，学界对地幔楔的黏性结构和流场的认识进一步加深，与矿物学和岩石学的结合也更加紧密，同时也带来了很多新的问题和挑战（冷伟和毛伟，2015）。另外，超高压变质矿物和含水矿物在地幔深度的发现指示，它们的形成和保存要求相当低的俯冲带地热梯度，而这个特殊条件只有通过快速的板块俯冲才能实现（郑永飞

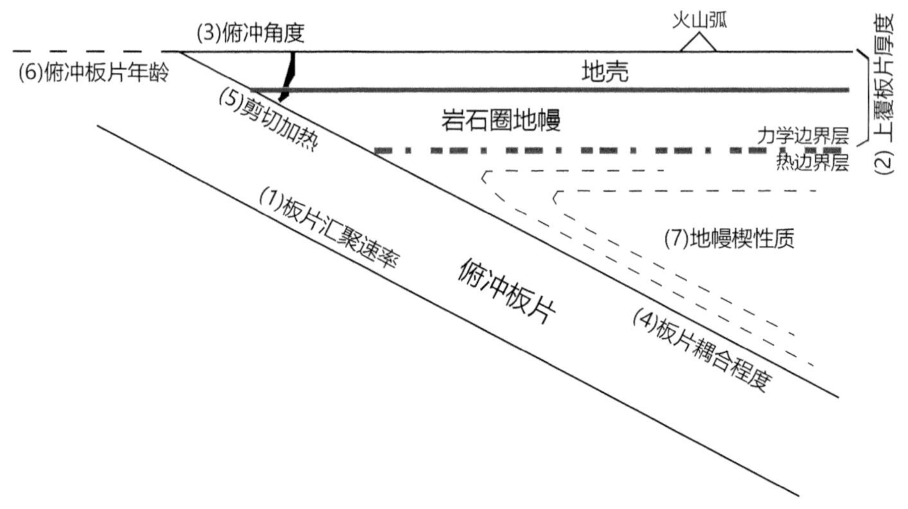

图 4-3　俯冲带温压结构控制因素（修改自 Zheng，2019）

等，2016；Zheng，2019）。因此，对于不同的俯冲带以及同一俯冲带在其演化的不同阶段，板块汇聚速率、上覆板块厚度、俯冲角度、板块耦合程度等都是影响俯冲带温压结构的因素（图 4-3）。

2. 主要研究方法与发展方向

前人很早就应用解析方法发现，俯冲板片的垂向速度和年龄是俯冲带温压结构的两个主控因素（McKenzie，1969）。板块开始俯冲后，其自身逐渐被周围地幔加热，而周围地幔则不断冷却，该过程持续直至二者达到稳态的温度分布（Molnar and England，1990，1995）。通过设定俯冲板片与上覆板块之间边界层的厚度和黏度参数，解析模型能对该边界层的温度分布进行约束（Davies，1999；England and Wilkins，2004）。俯冲板片发生地震的最大深度与其温压结构密切相关（Molnar et al.，1979；Kirby et al.，1991；Hacker et al.，2003）。这很可能是由于地震的发生由岩石的屈服强度控制，而屈服强度受温度的直接影响（Jarrard，1986）。然而，解析模型受方法本身限制，无法在模型中设定更多的地幔参数，因此数值模型是研究俯冲板块温压结构的主要方法。

典型的俯冲板块数值模型中关键参数主要包括流进和流出的边界条件，俯冲板块本身的速度和初始温度场，地幔楔中由于板块俯冲诱发的流场，上覆板块的结构和流变参数，以及可能影响温压结构的剪切生热和黏性生热等。Kincaid 和 Sacks（1997）的模型考虑了板块自身的负浮力对俯冲的驱动作用，并且对俯冲板块和周围地幔使用了不同的化学组分来加以区别。他们发现，对于俯冲速度较快和年龄较老的板块，俯冲板片表面温度较低，这和之前解析模型得到的结果一致。Peacock 和 Wang（1999）在模型中给定了板块的俯冲速度和角度并限定地幔楔的黏度为常数，结果显示日本西南部的俯冲板片比东北部的年龄更轻并且速度更慢，所以西南部俯冲洋壳的温度要高 300～500℃，导致西南部俯冲板片在浅部弧前深度已经显著脱水，流体难以进入弧下地幔深度，因此不能形成在东北部观察到的深源地震和岛弧火山作用。

当地幔楔的黏度由常数改变为随温度和压力变化时，地幔楔中的流场在拐角处产生汇聚作用，使地幔楔和俯冲板片表面的温度都有显著的提升（van Keken et al.，2002）。在数值模型中，角流拐点标志着俯冲板块与上覆板块的解耦位置，需要对其进行专门处理。不同的解耦深度以及上覆岩石圈厚度对弧前区域的热－力学和岩石学过程具有重要的影响（Wada et al.，2008；Liu M Q et al.,2017）。使用地表观测热流作为主要约束条件，Wada 和 Wang(2009）发现对典型的俯冲带解耦点的深度大多位于 80 km 左右。Syracuse 等（2010）综合了前人的研究结果，对全球 56 个主要俯冲带的温压结构进行数值拟合，模型中包括各俯冲带板块俯冲的速度、俯冲板块年龄、上覆板块结构以及沉积层厚度等约束条件，并对俯冲板块与上覆板块解耦点的不同情况进行了讨论。模型的结果显示，对同一俯冲带，不同的解耦点设置对俯冲板片的莫霍面温度、在岛弧下的板片表面温度以及地幔楔中的最大温度都影响不大。

板块俯冲过程中伴随着矿物岩石的相变和脱水作用，这些作用会对其温压结构产生较为显著的影响，可以将数值模型结合矿物相变模型和流体输送模型进行研究（van Keken et al.，2011）。除了影响温压结构，这些相变和流体作用也会对俯冲板片的密度分布和动力学效应产生影响（Wilson et al.，2014）。特别是，在脱水和熔融的综合作用下，俯冲板片上表面可能由于瑞利－泰勒不稳定性而产生温度低于周围地幔楔的上升冷柱（Gerya and Yuen，2003；Hasenclever et al.，2011）。由于矿物相变模型和流体输送模型中的不确定参数较多，还需要更多的工作以确定各种相变和流体过程对俯冲板片温压结构的影响。

3. 面临的挑战

前人研究发现，对俯冲板片表面温度起主要控制作用的是板块汇聚速率、俯冲板块年龄和俯冲角度（图 4-3）。板块的垂直俯冲速率越小，俯冲板片表面温度越高；垂直俯冲速率越大，则俯冲板片表面温度越低。但是，针对某一个俯冲带，需要根据其结构和地质观测分析影响该俯冲带温压结构的其他因素。例如，Peacock（2003）发现剪切热对俯冲板块温度的影响不大，这与弧前区域观测到的热流结果一致。当考虑俯冲板块的三维几何效应时，在相同的地表热流情况下俯冲板片表面温度可能会有100℃左右的差异（Morishige and van Keken，2014）。

洋壳俯冲带变质岩记录的变质温压条件是恢复俯冲带温压结构的重要手段，进而对理解板块构造演化乃至地球的动力学模型都有指示意义。然而，现代大洋俯冲带的温压结构与自然界洋壳俯冲带变质岩记录的结果之间存在一定差异。Penniston-Dorland 等（2015）通过世界各地古洋壳俯冲带折返的蓝片岩和榴辉岩峰期 *P-T* 条件的总结，发现它们所记录的峰期高压变质温度大部分高于计算模拟的结果（>200℃），由此认为模型低估了俯冲带温度。Kohn 等（2018）通过计算提出，汇聚板片之间的剪切热可大大提高俯冲板片的温度，由此可以吻合古俯冲带折返的高压变质岩记录。但 van Keken 等（2018）计算后得到，虽然增加合理的剪切热可以导致俯冲洋壳表面温度提高，但升温幅度不会超过 50℃。van Keken 等（2019）进一步提出，现今看到的蓝片岩和榴辉岩并不能代表实际俯冲带的一般特点，因为这些岩石的折返具有选择性，在暖俯冲带更有利于蓝片岩和榴辉岩的折返，结果在较高地热梯度下叠加变质。另外，对古俯冲带高压变质岩峰期 *P-T* 条件的估算也有很多不确定性，高压岩石在折返过程中受到改造可能造成其峰期压力被低估，或者温度被高估。

对俯冲板片温压结构的数值模拟研究面临的一个关键问题是如何通过观测验证数值模拟得到的温压结构。地表热流提供了对俯冲板片的浅部温压结构的直接约束（Wada and Wang，2009），数值模拟研究应结合地震层析成像给出的板块精细结构和岩石学研究给出的岩石的温度–压力路径曲线，以更好地理解俯冲板片的深部温压结构。Syracuse 等（2010）对现代大洋俯冲带温压结构的定量概括显示，俯冲洋壳表面地热梯度主要为 5～10 ℃/km（图 4-4），对应于阿尔卑斯型变质相系的形成（Zheng and Chen，2017，2021）。在俯冲带浅部的弧前深度，俯冲板片表面岩石经历升温升压过程，但在快速俯冲到约 80 km（2.5 GPa）时，板片表面的温度可能尚不到 400℃，地热梯度很低（<4.8℃/km）。在 80 km 深处，由于俯冲板片与地幔楔之间的力学耦合，板片表面岩石受到快速加热，温度可升至 700～900℃，对应的地热梯度从 4～5℃/km 升至 8.4～10.8℃/km；软流圈顶部的弧下深度，俯冲板片表面以

升压为主，地热梯度有所降低。

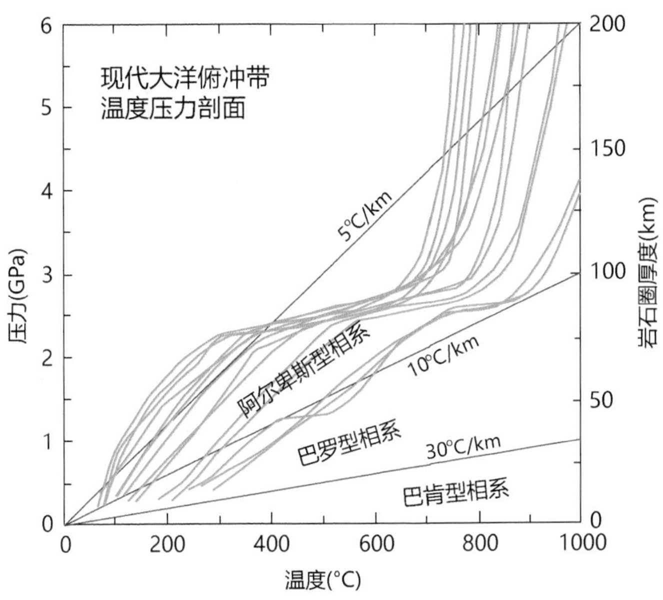

图 4-4　现代大洋俯冲带温压结构（修改自 Syracuse et al.，2010）
其温度压力条件主要对应于阿尔卑斯型变质相系的形成（Zheng and Chen，2017）

此外，俯冲板片的温度从俯冲开始到抵达稳态的过程中，在 1 亿年的时间尺度上表面温度可能会下降 100℃左右，稳态模型得到的温度场可能低估了实际俯冲板片的表面温度（Hall，2012）。因此，针对某一个俯冲带，研究其在俯冲过程中温压结构随时间的演化，并通过地质－地球物理观测进行约束和验证是未来俯冲带温压结构研究的重要方向。前人对俯冲带温压结构的数值模拟研究和地表观测约束主要针对大洋俯冲带（Wada and King，2015）。随着对大陆俯冲带认识的不断深入（Zheng and Chen，2016），对俯冲带温压结构的研究扩展到大陆俯冲带与大洋俯冲带温压结构的对比。大陆俯冲带和大洋俯冲带在温压结构的形成和演化上具有相似性。无论是大洋俯冲带还是大陆俯冲带，其演化可以分成早晚两个阶段：①早期阶段，板片具有低的角度，地热梯度较低，同步升温升压形成阿尔卑斯型变质相系；②晚期阶段，板片具有高的角度，地热梯度较高，同步升温降压形成巴肯型变质相系（Zheng and Chen，2017，2021）。由于大陆俯冲板块的性质和含水量对板块温压结构的影响（Zheng and Chen，2016；郑永飞等，2016），在大陆俯冲带区域所产生的高压－超高压变质岩的温压特征可以为恢复大陆俯冲带的温压结构提供很好的约束。因此，对大陆俯冲带的研究为理解俯冲带温压结构的变化规律提供了新的天然实验室。

4. 应对策略与预期目标

针对以上问题与挑战，需要对俯冲带温压结构进行深入研究。第一，应大力发展通过各种观测手段对俯冲板块的温压结构进行直接约束的方法。例如，发展更好的岩石学和地球化学温压计，通过地表观测的元素含量及比值得到俯冲板块在特定深度的温压分布。第二，应结合俯冲带的具体情况，如剪切生热效率、动态演化过程等，精细分析特定俯冲带的温压结构。第三，应补足对大洋俯冲带和大陆俯冲带温压结构随时间演化研究相对不足的短板。通过这些手段，有望在较短时间内取得对俯冲带温压结构研究的重要进展，并得到符合地表观测和深部约束的各个俯冲带的温压结构分布，为俯冲带研究的其他领域提供准确的参考温压场。

（三）科学挑战3：俯冲带地质结构

1. 问题的提出

俯冲带地质结构包括俯冲板片以及板块界面的物质组成、变质作用和构造变形（Stern，2002；Zheng and Chen，2016）。由于后期构造作用的叠加改造，对俯冲带地质结构的识别主要依据现代大洋俯冲带（环太平洋）和新生代大陆俯冲带（喜马拉雅－冈底斯造山带和阿尔卑斯造山带）的研究，以及对全球超高压变质带的对比研究。由于俯冲板片的物质组成、密度、黏度、俯冲角度、俯冲速率和地热梯度不同，俯冲带地质结构具有显著的差异。对俯冲带地质结构的研究主要以汇聚板块边界出露的构造增生楔、蛇绿岩、变质岩、岩浆岩和矿床等为对象，通过这些地质体的物质组成、时空分布、构造和变形特征来恢复板块俯冲前后的地质学变化。

2. 面临的挑战

俯冲板块界面是俯冲带地质结构的重要组成部分，不仅控制了俯冲板片与地幔楔的相互作用，也是高压－超高压变质岩折返和流体活动的通道（Hermann et al.，2006；Vannucchi et al.，2012；Zheng，2012；Raimbourg et al.，2018）。面临的关键科学问题如下。①如何通过俯冲板块界面的物质组成和流变结构来约束俯冲带的构造环境和时空演化？②如何追踪俯冲带流体活动与变质变形作用和岩浆作用之间的时空耦合关系？③俯冲带岩石相变和流体活动如何影响俯冲板块界面的构造变形和超高压变质岩的折返？

此外，对俯冲带地质结构的研究也是理解俯冲过程产物——大陆弧安山岩和大型斑岩铜矿以及火山、地震和海啸的基础（Elliott et al.，1997；Gaetani and Grove，2003；Richards，2003）。由于大陆地壳的平均成分为安山质，板块汇聚过程如何影响陆壳的生长与再造是限定板块构造起始时间的关键。因此，对俯冲带地质结构的研究对理解大陆形成与演化、圈层相互作用以及地震、火山和矿产资源的分布具有重要意义。

3. 重点研究方向

（1）俯冲带的应力状态与矿物相变的关系

虽然 410 km 和 660 km 间断面的起伏以及地幔过渡带的增厚常用于追踪俯冲板片（Agius et al.，2017；Yu Y et al.，2017），由于地震分辨率不够，高温高压实验难以直接测量地幔过渡带和下地幔矿物的流变性质，人们对滞留板片和穿越地幔过渡带的俯冲板片的物质组成、含水量、流变学结构和应力状态的认识还非常有限，这影响了对俯冲带海沟后撤历史与板片几何形态相关性的探索。此外，与 410 km 和 660 km 间断面相比，由瓦兹利石向林伍德石相变导致的 520 km 间断面并不是全球的波速不连续面，表明受板块俯冲的影响，地幔过渡带物质组成极不均一。将矿物物理的高温高压实验与理论计算、地震学观测与动力学模拟相结合，有望推进对板块俯冲过程如何影响地幔物质组成和动力学演化这一科学问题的认识。

（2）大陆俯冲带的超深俯冲与折返机制

在前方俯冲大洋岩石圈的拖曳力和后方洋中脊的推力下，在板块汇聚边缘的大陆岩石圈板块会继续俯冲到另一个大陆之下，形成超高压变质岩石，参与深部的物质循环并影响上覆岩石圈板块的构造变形和岩浆活动。全球已发现了 30 余条超高压变质带，大多形成于显生宙（Liou et al.，2009，2014；Zheng，2012）。因此，大陆深俯冲是全球板块构造的重要一环，对地幔不均一性的形成具有重要作用。超高压变质带以长英质和泥质片麻岩为主要组成，含少量榴辉岩和橄榄岩。在表壳岩中发现的超硅石榴子石、斯石英假象等变质矿物以及在石榴橄榄岩中发现的高压矿物出溶体表明大陆深俯冲可达 300 km（Dobrzhinetskaya et al.，1996；Scambelluri et al.，2008；Liu et al.，2018）。目前确定的最古老的超高压变质带的年龄是新元古代（Caby，1994；Genade de Araujo et al.，2014），而更古老的超高压变质作用信息在古元古代和太古宙的高压麻粒岩地体中还有待发现（Chopin，2003）。

实验表明，当俯冲深度超过 250 km 后，大陆上地壳岩石可转变为硬玉斯石英岩，密度显著升高，相对于地幔橄榄岩不再表现为低密度，并可以在重力作用下沉入地幔过渡带。因此，250 km 被称为超高压变质岩的“不折返深度”（Wu et al.，2009）。对新生代皮特凯恩群岛（Pitcairn Islands）洋岛玄武岩的地球化学研究表明，太古宙的大陆物质可以俯冲至地幔过渡带并长期保存（Wang X J et al.,2018）。但是，近年来对超高压变质岩的详细研究揭示，陆壳岩石可深俯冲到斯石英稳定域的地幔深度（～300 km）并快速折返（Liu L et al.，2007，2018）。对于相对高密度的大洋岩石圈，“不折返深度”约为100 km。值得注意的是，近年来在西藏罗布莎蛇绿岩、乌拉尔 Ray-Iz 蛇绿岩中发现金刚石等超高压、超还原矿物，表明上地幔底部甚至地幔过渡带的物质可以通过地幔柱作用加入到大洋岩石圈（杨经绥等，2013；Dilek and Yang，2018），也可能通过俯冲隧道折返回到浅部（Yu Y et al.，2017）。因此，蛇绿岩保留了大洋俯冲带参与深部物质循环的重要信息。

因此，关键科学问题有两个。①低密度的大陆地壳如何到达地幔过渡带并长期保存？②为什么有些超高压变质岩的折返深度大于 200 km？对上述问题的回答将有助于确认大陆超深俯冲的普遍性和大洋俯冲带的物质循环，从而更好地解释地幔中俯冲带的地球化学信息、板块汇聚边界的构造演化与壳幔相互作用。

（3）俯冲带的构造活动与流体迁移

俯冲板片与上覆板块之间的相对运动通过板块界面大型逆断层的滑动来调节，常导致大地震。俯冲板块界面指俯冲板片与上盘的构造边界，包括顶板滑脱层、底板滑脱层以及这两个构造边界中间的部分（Vannucchi et al.，2012）。沿俯冲板块界面，流体在 15～40 km 不同的破裂域之间迁移，不仅控制了俯冲带大地震的复发周期，而且导致地幔楔蛇纹石化（Lay et al.，2012）。俯冲板块界面与俯冲隧道的区别在于：在俯冲隧道这个有限的空间内部存在俯冲物质的连续回流（图 4-5），而俯冲板块界面复杂的几何学形貌控制了隧道流能否出现（Gerya et al.，2002；Ring et al.，2020）。

大洋板块俯冲时，以洋底沉积物为主的大洋板块地层碎片、弧前沉积物、岛弧岩石以及沿板块界面折返的高压－超高压变质岩，通过一系列倾向岛弧的叠瓦状逆断层混杂堆积在一起，形成构造混杂岩（图 4-6）。这些叠瓦状逆冲断层向下倾角变缓，收敛于顶板滑脱层。在板块俯冲过程中，俯冲板块界面的力学耦合随时间和空间的变化控制了板块界面的几何学、界面内岩石的相对运动（折返、埋藏、拆离、底垫）以及俯冲板片和上覆板块及大地幔楔的相互作用（底垫、基底侵蚀）（Agard et al.，2018）。俯冲板块边界的流变学性质控制了俯冲带的初始演化和底板滑脱层的发育（Agard et al.，2016）。

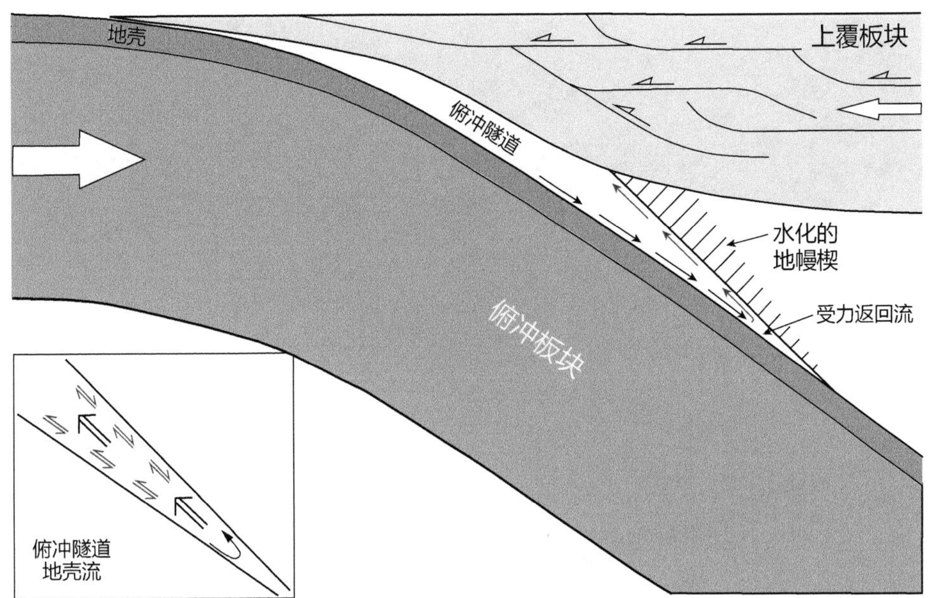

图 4-5　俯冲隧道内部物质流动示意图（修改自 Ring et al.，2020）
地壳岩石在俯冲隧道内的行为参见郑永飞等（2013）

近年来发现贝尼奥夫地震带具有双层结构：上层地震带位于俯冲洋壳中，可能是玄武岩和辉长岩在蓝片岩相－榴辉岩相变质过程中的脱水致裂（Zhang et al.，2004）；下层地震带位于俯冲的岩石圈地幔中，可能与蛇纹石或者其他含水矿物的脱水有关（Peacock，2001；Hacker et al.，2003）。俯冲带不仅有正常的地震，还存在慢地震。在大洋岩石圈很年轻的热俯冲带，地热梯度较高，断层在地幔楔角附近出现间歇性加速运动的现象，并伴随低频地震，能量在几周内缓慢消耗，称为幕式震颤与慢滑移（Rogers and Dragert，2003），这可能是地幔楔角附近的高压流体造成的（Gao and Wang，2017）。

4. 应对策略

由于俯冲带的构造变形常常伴随着变质作用、流体活动和部分熔融，因此对俯冲带地质结构的研究必须将构造地质学、岩石学、地球化学、地球物理观测、高温高压实验与动力学模拟相结合，从而追踪俯冲带不同深度的矿物相变、元素迁移、岩石变形以及流体和熔体活动。

5. 预期目标

通过对高压－超高压变质带和蛇绿岩带的地质填图和精细的构造地质学、岩石学、地球化学、同位素年代学研究，有望建立洋－洋俯冲、洋－陆俯冲、

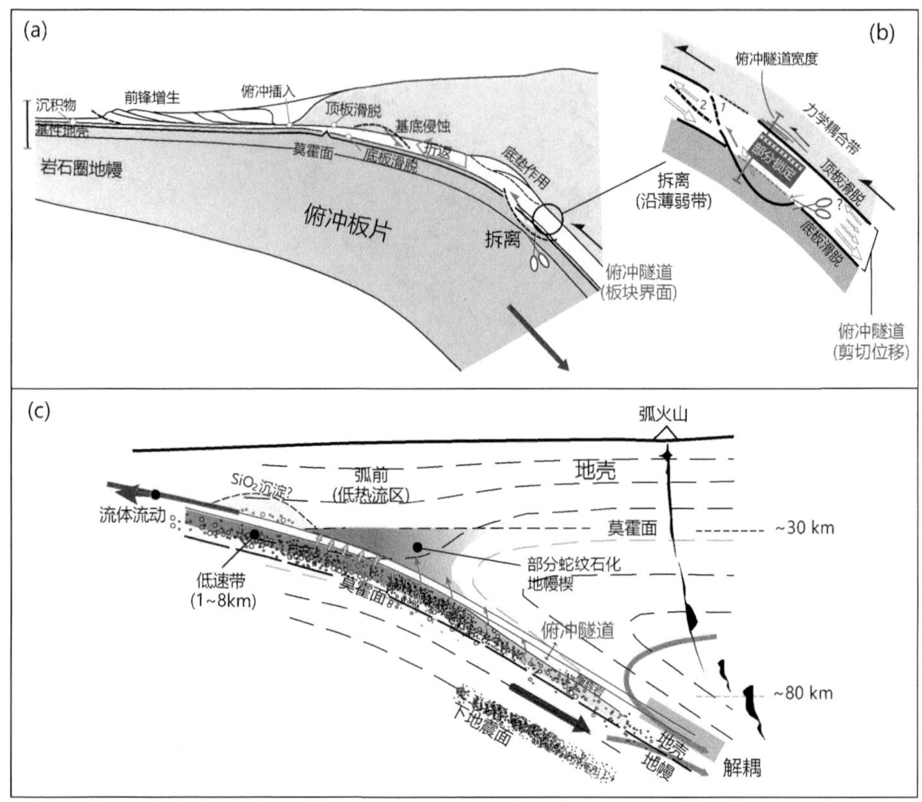

图 4-6　大洋俯冲带的结构与板块界面（修改自 Agard et al.，2018）

板块界面的厚度被放大以显示顶板滑脱层、底板滑脱层和物质折返。（a）顶板滑脱层的向下迁移使俯冲隧道内的物质或者俯冲板片的物质增生到上覆板块底部，称为底垫作用，而顶板滑脱层的向上迁移使上覆板块底部的物质进入板块界面，称为基底侵蚀。（b）俯冲板片内薄弱带的应变集中导致底板滑脱层向下迁移，从而把板片内的岩石加入到板块界面，导致拆离作用。（c）俯冲带中利用地球物理方法识别的主要地质过程和特征，包括地震活动、流体活动和岛弧形成等

弧－陆俯冲、陆－陆碰撞等构造环境下俯冲带地质结构的三维模型，并通过动力学模拟验证。将俯冲带高精度地震观测与岩石物理实验相结合，可以更好地约束俯冲带的流体活动及其迁移规律。

二、关键科学问题之二：俯冲带过程

（一）科学挑战 4：俯冲带变质脱水和部分熔融

1. 问题的提出

俯冲板片携带存在岩石孔隙和含水矿物中的水向地球深部俯冲。随着

俯冲过程中温度和压力的升高，俯冲板片发生显著的变质脱水反应，释放出富水溶液、硅酸盐熔体以及超临界流体等多种类型的板片流体（Bebout，2014；郑永飞等，2016）。发生深俯冲的板片在折返过程中同样存在变质脱水和部分熔融作用（Zheng，2009，Zheng et al.，2011）。板片变质脱水和部分熔融是俯冲带物质循环过程的关键一环，也是引起壳幔相互作用的重要机制。它不仅改变俯冲板片的化学组成，也导致地幔不均一性与弧岩浆作用。俯冲板片变质脱水和部分熔融也可以引发俯冲带多种地球物理效应，如中深源地震、地震波速降低和能量衰减及电导率异常等。

人们通过对俯冲地壳岩石的研究，结合高温高压实验、相平衡模拟、理论计算以及现代地球化学测试技术，在俯冲带变质脱水和部分熔融识别、发生条件和过程、产生流体性质和来源及其效应等方面取得了很多重要进展。但作为俯冲带物质和能量传输的主要机制和载体，俯冲带变质脱水和部分熔融也面临着诸多的关键科学问题与挑战。

2. 面临的挑战和重要研究方向

（1）俯冲带变质脱水和部分熔融的控制因素

俯冲板片中含水矿物分解是板块俯冲带主要的流体来源，也是控制俯冲带许多重大地质事件的内因（Zheng and Hermann，2014）。含水矿物随温度压力变化在俯冲带不同深度发生脱水和部分熔融作用（Schmidt and Poli，2014）。目前，学界已经认识到含水矿物的稳定性和板片在俯冲过程中脱水行为受俯冲带温压结构控制（图 4-7），并建立了一些俯冲板片流体释放随深度变化的模型（如 Hacker，2008；van Keken et al.，2011；Schmidt and Poli，2014；郑永飞等，2016）。此外，尽管名义上无水矿物含水量较低，它们却是俯冲地壳在弧下和后弧深度重要的水储库，在深俯冲地壳折返过程中能提供大量流体来源（Zheng，2009；Zheng and Hermann，2014）。然而，对名义上无水矿物中水如何影响俯冲板片岩石的含水性进而影响其变质脱水和部分熔融还不十分清楚。

未来需要重点开展研究解决如下问题：俯冲板片不同类型岩石中含水矿物稳定的温压条件？俯冲板片中岩石组成和含水矿物成分如何影响含水矿物的稳定性？俯冲带温压结构受哪些因素控制？俯冲带温压结构如何随时间和空间发生变化？名义上无水矿物在俯冲带不同条件下含水性如何？名义上无水矿物含水量变化受哪些因素控制？这些问题的解决需要综合折返高压－超高压岩石中含水矿物的稳定性研究，名义上无水矿物含水量测定，变质岩准

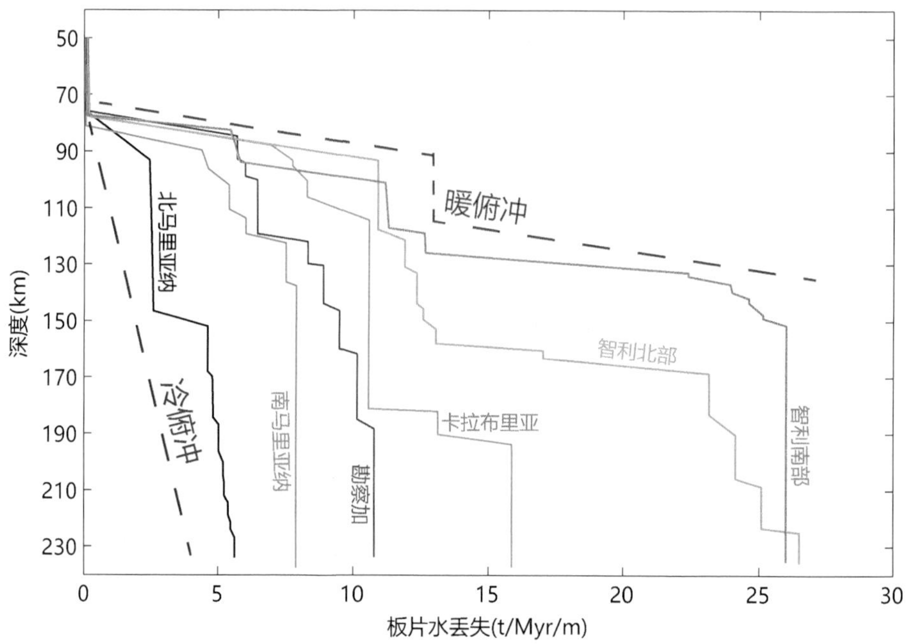

图 4-7　板片脱水深度与俯冲带温压结构之间的联系（修改自 van Keken et al.，2011）

确的 *P-T-t* 轨迹恢复，高温高压实验确定含水矿物稳定性，相平衡模拟研究定量确定全岩组成、温度、压力等对各种岩石体系变质脱水和部分熔融的控制作用。

（2）部分熔融发生的条件和机制

在热的大洋俯冲带，当洋壳岩石的俯冲 *P-T* 路径穿过其湿固相线时，可发生水化熔融，在更高的温压条件下可以发生脱水熔融（Schmidt and Poli，2014）。学者已经认识到大陆俯冲带超高压变质岩在大陆碰撞的晚期阶段会发生深熔作用（Zheng et al.，2011，Hermann and Rubatto，2014；Chen Y X et al.，2017）。深熔作用可以通过脱水熔融反应或者水化熔融反应这两种机制发生，且不同类型岩石发生脱水熔融和水化熔融的条件也存在很大差异（Zheng and Chen，2017，2021）。对于部分熔融发生的条件和机制，已有的研究主要是根据高温高压实验结果和岩石 *P-T* 轨迹来推测，相对缺乏实际的岩石学制约。此外，高温高压实验研究发现，俯冲洋壳在特定条件下甚至可以形成碳酸盐熔体（Poli，2015）。在大陆俯冲带超高压变质岩中也发现了碳酸盐熔体存在的可能证据（Korsakov and Hermann，2006）。碳酸盐熔体具有不同于含水熔体的元素溶解和迁移能力，其在俯冲板片－地幔楔界面出现，对于俯冲带元素迁移和壳幔相互作用具有重要意义。

未来需要重点开展研究解决如下问题：如何建立部分熔融与岩石 *P-T* 轨迹之间的准确对应关系？俯冲板片不同类型岩石脱水熔融和水化熔融分别在什么条件下发生？不同含水矿物的脱水熔融在俯冲带发生的条件？俯冲板片发生水化熔融的流体来源和如何迁移？导致俯冲板片发生脱水熔融的机制是什么？

鉴于高压–超高压变质岩在折返过程中普遍经历了较高地热梯度下的变质叠加改造，需要针对不同体系的天然样品开展系统的现代岩石学研究，结合高温高压实验、相平衡模拟以及地球化学分析，才能准确限定俯冲板片脱水熔融和水化熔融发生的条件，并确定它们在俯冲带发生的具体机制。

（3）俯冲带不同深度变质脱水和部分熔融产生的流体种类

俯冲板片在浅部弧前深度主要脱出富水溶液，在弧下和后弧深度，板片地壳有可能发生部分熔融形成含水熔体，甚至形成超临界流体（Kessel et al.，2005；Hermann et al.，2006；Zheng et al.，2011；Ni et al.，2017；Zheng，2019）。超临界流体具有溶解迁移元素能力强的特点，使其成为俯冲带元素迁移的重要载体。然而，超临界流体一旦与其他介质反应或离开它稳定的温压条件，就会发生相分离或相转换从而失去超临界流体的特性（Kawamoto et al.，2012），结果导致其识别存在很大的困难。虽然天然样品观察指示了俯冲带超临界流体作用产物的存在（Ferrando et al.，2005；Zhang et al.，2008；Zheng et al.，2011），但是能确切指示超临界地质流体存在的地球化学证据还很缺乏（Zheng，2019），还没有建立超临界地质流体存在的鉴定性地球化学识别指标（Ni et al.，2017）。对某些硅酸盐–水体系超临界流体的形成条件还存在很大争议，如对基性岩–水体系的二次临界端点不同的高温高压实验给出的结果存在显著差异，对天然样品的研究也给出不一样的结果（Ni et al.，2017）。

未来需要重点开展研究解决如下问题：如何从岩石学和地球化学角度识别俯冲带超临界流体的出现？超临界流体产生的条件及其地质效应？在俯冲板片–地幔楔界面是否会有碳酸盐熔体产生？碳酸盐熔体是如何产生的？如何识别不同阶段不同性质流体之间的演化？

鉴于超临界流体和碳酸盐熔体形成的高温高压条件及其高的反应活性，需要以俯冲带记录这些流体活动的超高压地质样品为基础，结合高温高压实验获取的流体–固体状态方程、高温高压状态下原位物理化学性质测试分析、计算模拟等研究手段对不同深度流体的形成和性质进行综合限定。

（4）变质脱水和部分熔融对元素迁移的影响

通过高温高压实验、折返的高压/超高压变质岩及其中脉体（特别是其中包含的流体包裹体）的研究以及理论计算，目前对俯冲带变质脱水和部分熔融形成的流体对元素迁移的影响已建立起框架性认识（Schmidt and Poli，2014；Zheng and Hermann，2014；Frezzotti and Ferrando，2015；Zheng，2019），但是对流体的地球化学组成还缺乏定量制约。大多数研究主要关注富水溶液活动（Manning，2004），但是对于俯冲地壳高压－超高压部分熔融过程中元素的活动性，只有少量研究对淡色脉体和岩体进行全岩成分研究（Zheng and Hermann，2014），将矿物成分和全岩成分分析结合起来的研究更少。已有研究认识到，矿物的溶解和保存对熔体的组成起到很重要的控制作用，但并不清楚不同成因矿物（变质矿物、转熔矿物、深熔矿物）对熔体组成的控制作用，它们在深熔作用过程中的行为以及如何控制元素和同位素分异也有待深化研究（Zheng and Hermann，2014）。已有的研究主要关注流体中主要组分——水和微量元素，而对其他挥发分（如碳、氮、硫、卤素等）在流体中的组成及其效应关注较少。

氧化还原性（通常以氧逸度表示）和酸碱度（通常以pH表示）是流体的两种重要的状态参量，不仅控制着流体中变价元素的赋存状态和溶解能力，也极大地制约着流体－岩石反应过程中的相平衡关系和同位素分馏行为。然而，目前对俯冲板片不同类型岩石在俯冲带释放流体的氧化还原性存在很大争议（Debret et al.，2015；Frezzotti and Ferrando，2015；Evans et al.，2017；Li et al.，2020）。俯冲板片中的流体活动往往是幕式的，可能具有不同的源区、形成机制和时代，特别是俯冲大陆地壳岩石普遍经历了折返和后期退变质流体的改造。因此，关于板片流体氧化还原性认识的争议可能是其具有较强的时空不均一性导致的。由于流体中很难保存元素存在具体形式的信息，通过天然样品或高温高压实验评估流体酸碱度的研究十分稀少，也只有少数研究对流体酸度进行了理论模拟。

未来需要重点开展的研究包括：①不同成因矿物的识别及其对熔体组成的控制作用；②不同类型流体在不同温度－压力－成分－氧逸度－酸碱度条件下对微量元素的溶解和迁移能力；③挥发分相关矿物相在俯冲带不同岩性、不同深度的存在形式；④挥发分在俯冲带流体中的溶解度及控制因素；⑤俯冲带不同时空条件下产生流体的氧逸度和酸碱度。

（5）大洋俯冲带和大陆俯冲带板片变质脱水和部分熔融异同点

前人研究认为，大陆俯冲带由于俯冲板片相对缺水，缺乏流体活动（Rumble et al.，2003；Zheng et al.，2003）。然而，超高压条件下稳定含水矿物的发现、名义上无水矿物中微量水的存在、超高压变质岩中高压－超高压脉体以及超高压矿物中流体包裹体的发现，指示大陆俯冲带同样存在显著的流体活动（Zheng，2009；Hermann and Rubatto，2014）。根据陆壳和洋壳中含水矿物的比较，推测在弧下深度，大陆俯冲带具有与大洋俯冲带类似当量的流体活动（郑永飞等，2016）。虽然大洋俯冲带和大陆俯冲带在俯冲地壳物质组成上存在一定的差异（Zheng and Chen，2016），但是由于两者在俯冲带温压结构上的相似性，它们在俯冲带流体活动和壳幔相互作用上也存在一定的相似性。俯冲到弧下深度经受超高压变质作用的洋壳岩石极少能折返到地表，对大洋俯冲带深部流体活动以及元素分异的具体过程和机制还缺乏直接的变质岩石学证据，对俯冲大洋板片组分进入地幔楔的方式、反应过程和机制主要还是依据弧火山岩的地球化学研究。因此，有必要对少数折返的洋壳超高压变质岩和陆壳超高压变质岩进行对比研究，围绕大洋和大陆俯冲带温压结构和物质组成上的差异，揭示大洋俯冲带和大陆俯冲带在弧下深度变质脱水和部分熔融上的异同点。

未来需要重点开展的研究包括：①大洋俯冲带和大陆俯冲带温压结构异同点及其影响因素；②沉积岩和镁铁质岩在不同温压结构条件下变质脱水和部分熔融行为；③水化橄榄岩在大洋俯冲板片变质脱水和部分熔融过程中的作用；④长英质和镁铁质岩石变质脱水和部分熔融行为差异。

3. 应对策略和预期目标

俯冲带变质脱水和部分熔融涉及俯冲板片中各类岩石在俯冲和折返过程中的流体行为。其根本目标是确定俯冲板片中各种类型流体产生的条件和机制，查明它们产生的物理化学效应。推进俯冲带变质脱水和部分熔融研究，需要从以下几个方面着手。

1）经历弧下深度变质作用而后折返的超高压岩石提供了俯冲带深部变质脱水和部分熔融的直接记录。然而，这些岩石在折返过程中通常比较低压力下的岩石受到更强烈的叠加改造作用，需要对这些岩石进行以微区微量分析为主的现代岩石学分析，厘清不同成因和不同期次生长矿物，才能将俯冲带变质脱水和部分熔融过程与岩石 *P-T-t* 轨迹准确联系起来，从而揭示变质脱水和部分熔融发生的条件和机制。

2）对高压–超高压变质岩中流体活动直接记录的脉体和流体包裹体进行系统岩石学和地球化学研究。要结合流体包裹体的均一化实验，为俯冲带不同深度变质脱水和部分熔融产生流体种类（特别是超临界流体和碳酸盐熔体）提供准确限定，为变质脱水和部分熔融产生流体组成提供定量制约。

3）开展俯冲带条件下各种类型岩石和矿物的溶解和熔融实验。要注重进行硅酸盐熔体和超临界流体产生温压区间的高温高压实验，在实验设计上不仅聚焦于简单的系统（如单种矿物的溶解度），还要聚焦于组成上与天然俯冲带岩石相似的复杂系统。通过这些高温高压实验研究，为含水矿物的稳定性提供实验制约，确定名义上无水矿物含水量的控制因素，为相平衡模拟提供基础的热力学参数，探讨组成、温度、压力等因素对变质脱水和部分熔融的影响，了解各种条件下流体的状态参量及其中各种元素的存在形式和含量。

4）随着实验热力学参数的不断完善和计算能力的增强，加强对基性岩、沉积岩、花岗质岩石和超基性岩相平衡关系的理论模拟研究。要与实验资料相结合，深入了解这些岩石俯冲变质过程中的流体行为，揭示俯冲板片变质脱水和部分熔融的具体细节和过程；要开展第一性原理和机器学习相结合的多尺度分子模拟研究，获得精细的流体微观结构和物理化学性质计算数据。

5）多学科交叉融合研究俯冲带变质脱水和部分熔融。鉴于俯冲板片折返岩石多阶段复杂演化和流体难以保存的特点，需要有机结合高温高压实验、理论计算、天然样品研究、地球物理探测等多种研究手段，来研究俯冲带变质岩和岩浆岩所记录的脱水和熔融信息。

（二）科学挑战5：俯冲带双向交代作用

1. 问题的提出

俯冲带存在正向交代作用和反向交代作用。正向交代作用是指俯冲板片释放流体交代地幔楔（图4-8），该过程是实现地球内部物质循环和能量交换的关键途径，深刻影响着板块汇聚边缘变质–岩浆–成矿作用以及人类宜居环境。反向交代作用是指水化地幔楔岩石脱水交代俯冲地壳的过程，对俯冲板片物理化学变化和碰撞动力学过程具有重要影响。

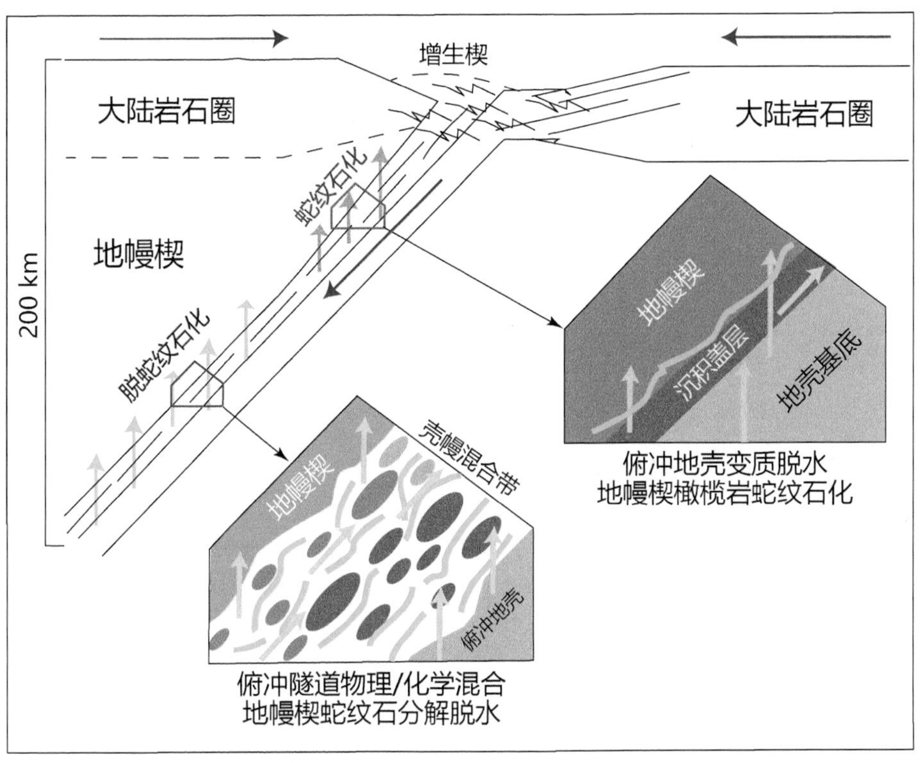

图 4-8　大陆俯冲隧道中板片界面流体正向交代示意图（郑永飞等，2013）
板片界面既有不同物质的物理混合，又有化学反应和流体交代作用

人们通过俯冲地壳和地幔楔岩石的研究，在交代流体的来源和性质以及流体与橄榄岩交代的时间、过程和化学效应等方面均有重要突破性进展（Zheng，2019）。然而，由于俯冲带流体的多来源和复合性、俯冲带构造背景的多样性（洋－洋俯冲、洋－陆俯冲和陆－陆碰撞）以及壳幔物质循环过程的复杂性，目前在俯冲带关键元素循环、俯冲交代作用的时空差异、地幔楔物理化学性质演变、反向交代在壳幔演化中的作用等方面面临新的挑战。

2. 面临的挑战和重点研究方向

（1）俯冲带关键挥发分元素循环

与生命息息相关的地表碳、硫、氮和氢等关键挥发分元素通过俯冲带交代作用能运输到地幔内部，使得地幔物理化学性质发生显著改变，诱发地幔部分熔融，然后通过岩浆作用再释放到地球浅表系统，从而完成俯冲带碳、硫、氮和水的循环（Métrich et al.，1999；Hirschmann，2006；Dasgupta and Hirschmann，2010；Stüeken et al.，2016；Plank and Manning，2019）。然而，在俯冲带交代作用过程中挥发分元素如何从板片迁移到地幔内部？在交代地幔楔内赋存状态如何？在这一宏观过程中各自通量是多少？不同挥发分元素之间如何相互作用？这些都是极具挑战性的科学问题。

未来重点研究方向包括：①关键挥发分元素在俯冲板块和地幔楔中的微观赋存状态；②影响关键挥发分元素通量的控制因素；③不同挥发分元素之间相互作用（协同演化）的过程和机理；④俯冲带关键挥发分元素循环与宏观地质效应的内在联系。

（2）俯冲带交代作用的时空差异

俯冲板片在不同深度经历不同的变质反应和脱水熔融作用（Poli and Schmidt，2002；Zheng et al.，2011），因此俯冲带交代作用在不同深度对地幔楔的改造也相应不同。在现代板块构造体制下，俯冲带在小于 60 km 的弧前深度以富水溶液交代为主，在 80～160km 的弧下深度和大于 200km 的后弧深度以熔体交代为主。俯冲地壳在不同深度释放的流体成分差异极为显著（Zheng，2019）。不同的交代反应会导致地幔楔的化学性质在纵向和横向空间上高度不均一，进而致使地幔楔部分熔融形成的板内岩浆和岛弧岩浆成分迥异（Pilet et al.，2008；Zheng，2019）。此外，板块俯冲从初始到成熟阶段，俯冲带温压结构会发生明显变化，导致俯冲地壳对地幔楔的交代随时间而演变，如何准确限定不同时限（不同俯冲阶段）的交代作用差异仍是难题。

未来的重点研究方向包括：①俯冲带不同深度流体性质和交代作用的差异性；②不同时空尺度俯冲带交代作用的差异性。

（3）地幔楔物理化学性质演变

虽然关于板块构造在地球上何时启动还存在较大争议，但是大多数学者认为板块构造在太古宙就已出现（Brown，2006；Shirey and Richardson，2011；Zheng and Zhao，2020）。由于太古宙的地幔温度比显生宙偏高 200～300℃，俯冲带温压结构的变化会显著影响板块俯冲样式，进而影响地幔楔的物理和化学性质，因此可预见俯冲带交代作用在地球早期与现代板块构造体制下存在明显差别。太古宙—显生宙的岛弧岩浆岩和洋壳玄武岩中 $Fe^{3+}/\sum Fe$ 和 V/Sc 值显示其氧逸度逐渐升高，亦表明地质历史时期俯冲带交代作用导致地幔物理化学性质在不断演变（Stolper and Keller，2018；Stolper and Bucholz，2019）。已有研究表明，大陆起源与地球早期俯冲带交代作用密切相关（Martin et al.，2005；Moyen and Martin，2012），克拉通岩石圈地幔的形成也可能与地球早期俯冲带地幔过程有关（Kelemen et al.，1998；Wittig et al.，2008）。

探究从地球早期到显生宙地幔楔物理化学性质演变这一科学挑战，需聚焦的重点研究方向包括：①不同地质历史时期俯冲带地幔岩浆性质演变；②地幔楔氧逸度演化；③板块俯冲型式转变对地幔楔交代作用的影响。

（4）反向流体交代的识别及其在壳幔演化中的贡献

目前，对地幔楔蛇纹岩流体反向交代俯冲板片以及弧下橄榄岩的认识极为有限。反向交代会极大地影响深俯冲板片和所交代地幔橄榄岩的地球化学组成。弧岩浆岩的一些地球化学特征（如高 $\delta^{11}B$）指示地幔源区中有弧前地幔楔蛇纹岩来源流体的贡献（Tonarini et al.，2011；Spandler and Pirard，2013）。俯冲带中富镁交代岩异常的 Mg-Fe 同位素组成也要求反向流体交代的出现（Chen et al.，2016，2019）。然而，如何从岩石学和地球化学上认识反向交代作用？这种反向交代出现的环境、流体出现的规模和对弧岩浆影响的范围如何？反向交代如何影响深俯冲板片化学分异、碰撞动力学过程？这些都是极不清楚、极具挑战的前沿问题。

因此，为揭示反向流体交代在壳幔演化中的贡献，需重点聚焦于以下几个研究方向：①地幔楔来源流体的岩石学和地球化学识别；②地幔楔来源流体反向交代发生的过程和机制；③大洋 / 大陆俯冲带板片 - 地幔楔界面反向流体交代对弧岩浆和深俯冲地壳地球化学组成的影响。

3. 应对策略与预期目标

俯冲带双向交代作用过程极为复杂，在不同时间和空间尺度下表现出巨大的差异行为。为解决上述多种科学问题的挑战，需从以下几个方面开展深入研究。

1）针对不同类型的造山带同时开展俯冲地壳和地幔楔岩石观察、高温高压实验和计算模拟，明确关键挥发分元素在俯冲板块和地幔楔岩石中的赋存状态，深度挖掘不同挥发分元素之间的相互作用和协同演化规律，厘清挥发分元素间相互作用的机理和过程，从而有助于评估不同元素在交代过程中的行为，探究从微观赋存到宏观地质效应的内在关联。

2）加强典型造山带内部不同深度 / 不同时间尺度的造山带地幔楔橄榄岩、金伯利岩中的橄榄岩包体与岛弧和板内岩浆岩的综合对比研究，结合不同类型地幔交代岩部分熔融产物的实验和理论计算研究，将能深入挖掘俯冲交代作用的时空差异。

3）针对前寒武纪造山带地幔楔橄榄岩与显生宙造山带地幔楔橄榄岩开展综合对比研究，结合高温高压实验和模拟计算，评估不同地质历史时期俯冲带交代作用的差异和演化，这不仅能揭示不同地质历史时期地幔楔物理化学性质演变规律，还对深入理解地球早期演化和板块俯冲型式转变有重要指导意义。

4）针对世界典型的大洋俯冲带和大陆俯冲带造山带橄榄岩、流体交代岩和弧岩浆岩进行精细的岩石学和地球化学分析，准确限定反向交代流体的来源、出现的地质条件及其地球化学特征，探讨反向交代的地球动力学过程。特别是对全球范围内弧岩浆岩指示蛇纹岩来源流体的关键指标（$\delta^{11}B$、As-Sb-Cl 含量等）进行深入分析，制约反向交代对弧前和弧下地幔楔影响的范围、程度及其与俯冲带结构的关系，进而反演反向交代在壳幔系统地球化学演化中的贡献。

（三）科学挑战 6：地幔楔部分熔融

1. 问题的提出

地幔楔是俯冲带的重要组成部分，为俯冲板片之上、上覆板块地壳之下的楔形地幔区域（图 4-9），包含岩石圈地幔楔和软流圈地幔楔两类。其中，软流圈地幔以低速高衰减、对流热传导为特征，而岩石圈地幔在不同的判别标准下具有不同的含义，包括弹性岩石圈、热岩石圈和地震学岩石圈等多种定义。因此，在不同的岩石圈定义条件下，地幔楔的构成可能存在一定差异。弹性岩石圈代表了岩石圈内长期稳定的刚性板片，以 650 ± 100 ℃等温面为界限（Anderson，1994）。热岩石圈以温度和热传导的方式为出发点，以特定温度（如 1300℃）所在的深度代表岩石圈厚度（McKenzie and Bickle，1988）。地震学岩石圈认为，岩石圈地幔代表了高速层，而软流圈地幔是一个力学上的薄弱层，以低地震波速和高衰减为主要特征（Karato，2012）。在俯冲带中，由于深部俯冲板片释放流体会降低上覆地幔的黏度和地震波速。因此，根据地震波、流变学定义的岩石圈－软流圈边界可能发生变化。

在地球化学组成上，岩石圈地幔与软流圈地幔之间可能既有差异性也有相似性。克拉通地区岩石圈地幔代表原始地幔经过大规模镁铁质熔体抽离之后的残留（Lee et al.，2011），通常以主量上难熔、微量和同位素上富集的方辉橄榄岩为主。但是，新生的非克拉通型（包括俯冲带）岩石圈地幔与软流圈地幔之间在岩石地球化学上很难区分。例如，中国东部新生代岩石圈地幔与下伏软流圈地幔之间，可能既有直接转换成因，也有受到俯冲太平洋板片

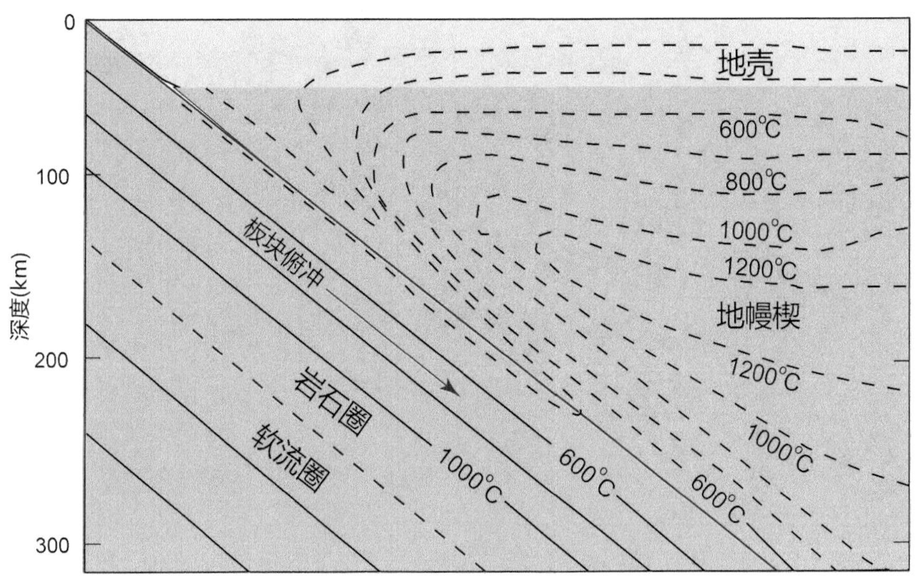

图 4-9 俯冲带早期阶段板片及其上覆地幔楔温度分布示意图

在板块俯冲的早期阶段，板片与地幔楔之间处于耦合状态，板片与地幔楔之间界面的温度最低，对应的地热梯度也最低（<15℃ /km）

衍生流体交代改造的部分（吴福元等，2008；Xu et al.，2017）。软流圈地幔岩石学上主要由二辉橄榄岩组成，包含橄榄石、斜方辉石和单斜辉石（Frisch et al.，2011）；微量元素以亏损不相容元素为特征，被认为是洋中脊玄武岩的地幔源区；一般具有低的 $^{87}Sr/^{86}Sr$ 值以及高的 $^{143}Nd/^{144}Nd$ 和 $^{176}Hf/^{177}Hf$ 值（Hofmann，1997），代表了地幔中最亏损的端元（亏损地幔）。大量研究表明，在软流圈中存在不同尺度的不均一性，可能保存有古老的难熔的地幔（Liu et al.，2008），或洋 / 陆壳俯冲释放的富集组分（Zheng et al.，2020）。

全球地震层析成像显示，俯冲板块进入软流圈地幔之后有两种形态，一种是穿越地幔过渡带进入下地幔，另一种是平躺滞留在地幔过渡带（Huang and Zhao，2006；Li and van der Hilst，2010；Goes et al.，2017）。Zhao 等（2004）将在 400～660 km 的地幔过渡带中滞留的大洋板片之上的上地幔称为大地幔楔。一般将板片俯冲到小于 200km 的弧下深度所形成的地幔楔称为小地幔楔，而将板片俯冲到大于 200km 的后弧深度所形成的地幔楔称为大地幔楔（图 4-10）。无论地幔楔大小，它们都是俯冲带系统重要的组成单元，既有差异又相互关联（徐义刚等，2018；郑永飞等，2018）。

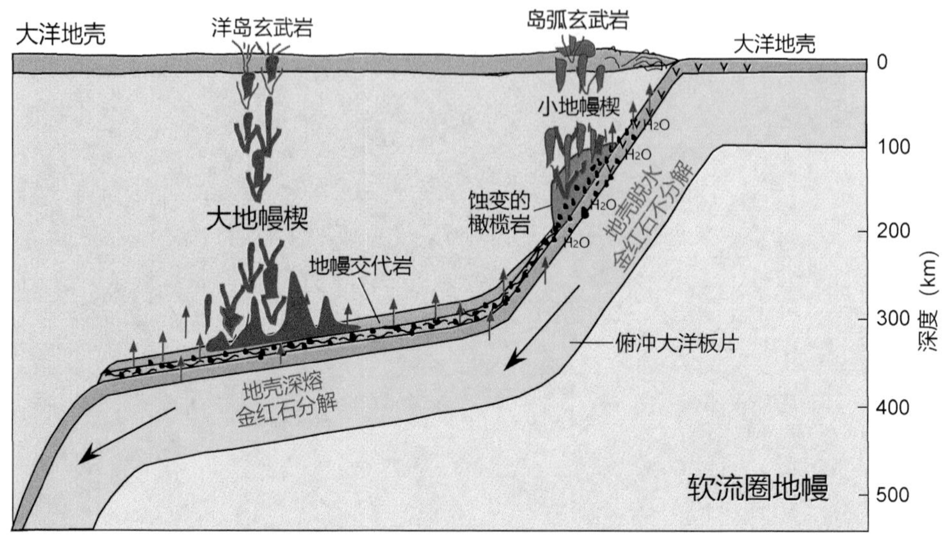

图 4-10 洋－洋俯冲形成不同大小的地幔楔示意图（修改自 Ringwood，1990）

地幔楔随板块俯冲深度增加越来越大，小地幔楔为岛弧玄武岩的地幔源区，大地幔楔为洋岛玄武岩的地幔源区。俯冲洋壳在弧下深度脱水和后弧深度熔融分别形成不同性质的流体，交代不同大小的地幔楔

小地幔楔的形成过程，即板块俯冲早期阿尔卑斯型变质过程至晚期俯冲带成熟过程。地幔楔在形成过程中的地球化学成分会受到两个因素的影响：板片来源流体的交代和地幔楔熔融对易熔元素的提取。俯冲板块的下插、后撤等过程必然导致地幔楔形成过程中发生广泛的降压熔融，加上这一阶段新生地幔楔尚不存在稳定、大尺度的地幔对流，因此伴随着岩浆活动的进行，成熟地幔楔会逐渐变得更加贫瘠（Shervais et al.，2019; Li et al.，2019a）。成熟地幔楔具有比新生地幔楔更高的温度，因此它对俯冲板片的加热效率相对早期俯冲系统要高。晚期俯冲的板片会发生部分熔融，板片熔体交代上覆地幔楔并促使它发生熔融，岩浆抽取会导致地幔楔变得更为贫瘠（Foley et al.，2002; Li et al.，2019a）。降压熔融和流体助熔并存是地幔楔演化过程中的一个重要特征，在弧后拉张阶段地幔以降压熔融为主向流体助熔为主转变，代表了一个成熟俯冲带的建立（Li et al.，2019a）。随着板片俯冲的持续进行，板片会进入到地球更深处（Goes et al.，2017）。

地幔楔熔融是俯冲带深部过程的重要表现形式之一。弧下地幔楔是原始弧岩浆最重要的源区。因此，地幔楔的特征及部分熔融过程的研究一直是当前国际俯冲带科学研究的热点。但是，不同地幔楔类型（如软流圈地幔楔与岩石圈地幔楔、小地幔楔与大地幔楔）及其物理化学条件对弧岩浆成分有何影响？不同性质流体交代地幔楔对地幔楔的物质组成有何影响？不同类型地幔楔的形成过程如何？它们对壳幔物质系统演化有何影响？这些问题一直是俯冲带地幔楔熔融过程研究的前沿与热点。

2. 面临的挑战

（1）软流圈地幔楔、岩石圈地幔楔如何影响弧岩浆的起源

一般认为，弧岩浆的源区深度主要为 80～160 km，在洋 - 洋俯冲带或者大陆岩石圈相对薄的洋 - 陆俯冲带，对应深度主要是俯冲交代的软流圈地幔（图 4-11），即俯冲板片来源的组分和软流圈地幔楔共同决定了大洋弧玄武岩或者大陆弧安山岩的成分（Hawkesworth et al.，1993；Kelemen et al.，2014；Zheng et al.，2020）。在大陆岩石圈相对较厚的区域，该深度可能对应岩石圈地幔（Zheng and Chen，2016；Chapman and Ducea，2019）。

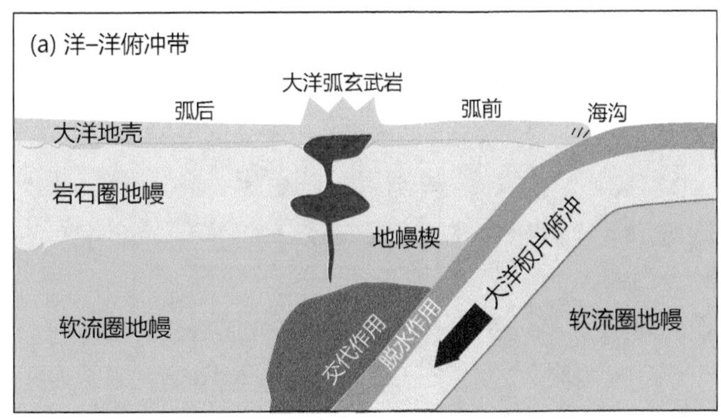

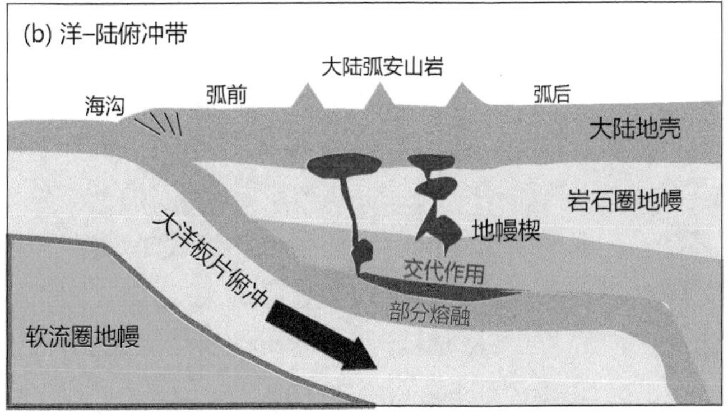

图 4-11　洋 - 洋俯冲带和洋 - 陆俯冲带地幔楔部分熔融示意图
（引自郑永飞等，2016）

大陆弧玄武岩往往比大洋弧玄武岩具有更高的不相容元素含量，有三种

可能的原因:①弧玄武岩的源区加入了富集的古老的岩石圈地幔组分（Pearce，1983）；②弧玄武岩浆在上升过程中受到上覆厚的古老地壳的混染（Pearce and Peate，1995）；③由于大陆弧比大洋弧具有更高的海拔和地表风化剥蚀速率，导致大陆弧的弧前海沟具有更多的沉积物被俯冲板片带入地幔楔（Jagoutz and Kelemen，2015）。

确定弧玄武岩地幔源区是否有上覆古老的岩石圈地幔参与，难点在于岩石圈组成的不均一性、深部俯冲组分改造以及浅部地壳物质的干扰。部分研究者认为，弧岩浆源区识别出的岩石圈地幔组分可以是之前俯冲来源的富集组分或弧岩浆在岩石圈地幔中的残留，因此很难区分是现今俯冲组分的变化还是混入了岩石圈地幔中的古老俯冲组分导致的弧岩浆成分变化。

（2）不同类型地幔岩石部分熔融的物理化学条件

俯冲带地幔楔形成过程中存在俯冲板片脱水和熔融过程，板片释放的流体和挥发分（如 H_2O 和 CO_2）以及流体会交代地幔楔橄榄岩。在不同温压条件下，不同性质和比例的板片流体会与地幔橄榄岩反应生成主微量元素和岩性不同的地幔交代岩（Su et al.，2019；Wang et al.，2020；Zheng et al.，2020），如含水橄榄岩、含 CO_2 橄榄岩、辉石岩、角闪岩、石榴子石岩、异剥橄榄岩、纯橄岩等，其中含水橄榄岩、含 CO_2 橄榄岩、辉石岩在弧下地幔中比例可能较大。

含水橄榄岩部分熔融的温压条件与体系中的含水量变化有关。当含水量为 0.05 wt%～0.5 wt% 时，水主要储存在地幔韭闪石中（压力≤3 GPa），此时发生的是角闪石橄榄岩脱水熔融，固相线温度为 950～1100℃（1.5～3.0 GPa）（Mandler and Grove，2016）。当含水量超过地幔橄榄岩储水能力时（>0.5 wt%），即发生水饱和橄榄岩熔融，但是对其固相线的最低温度仍然存在巨大的分歧。例如，在 3 GPa，有的高温高压实验结果低达 800℃，而有的高达 1000～1100℃（Green et al.，2012；Till et al.，2012）。争议的焦点是：水饱和橄榄岩在不同温度下，首次出现的含水相到底是熔融形成的含水硅酸盐熔体，还是脱水形成的富水溶液，抑或是高温高压条件下形成的超临界流体。因此，未来的研究需要寻找更好地判断橄榄岩熔融的指标，而不是通过传统淬火方法测试玻璃组成（Ni et al.，2017）。

含 CO_2 橄榄岩熔融的温压条件与压力有关。由于低压下 CO_2 在玄武岩的溶解度很低，低压下（<2.0 GPa）CO_2 不会大幅度降低橄榄岩的固相线，其固相线温度为 1100～1250℃（0～2.0 GPa）（Dasgupta，2018）。但是当压力超过 2 GPa 时，地幔橄榄岩会与 CO_2 反应生成白云石橄榄岩（70～100

km）和菱镁矿橄榄岩（＞100 km）（Falloon and Green，1989；Dasgupta and Hirschmann，2006），而这些碳酸盐化橄榄岩的固相线温度在 3 GPa 时为约 1050℃，且其固相线位置不随着体系 CO_2 浓度变化而变化，这一点区别于水的影响（Dasgupta and Hirschmann，2007）。

熔体交代的地幔橄榄岩也会影响弧岩浆源区熔融的温压条件。板片来源的富硅熔体可能与地幔橄榄岩反应先形成硅过剩的石榴辉石岩，这类辉石岩在地幔楔中将优先发生部分熔融（Straub et al.，2011；Zheng，2019）。但目前对石榴辉石岩部分熔融的实验岩石学研究相对不足，已有实验主要关注对洋岛玄武岩和洋中脊玄武岩组成的研究（Hirschmann et al.，2003；Kogiso et al.，2004b），对交代成因辉石岩的研究相对缺乏。不过，根据辉石岩与橄榄岩在矿物组成和化学成分上的差异，可以预计辉石岩的固相线温度主要与其全碱含量呈负相关，与 Mg# 呈正相关，整体比橄榄岩固相线要低 100～300℃。

（3）弧下地幔部分熔融产生熔体成分的控制因素

俯冲带的原始岩浆岩至少有 6 种类型，分别为钙碱性玄武岩、安山岩、拉斑玄武岩、拉斑安山岩、橄榄玄粗岩和低硅岛弧玄武岩（Kelemen et al.，2014；Schmidt and Jagoutz，2017），并且单个弧岩浆带可以在同一位置不同时间，或是在不同位置不同时间发育不同类型的原始岩浆岩（Stern et al.，2003；Durkin et al.，2020）。这种现象表明，弧岩浆岩在时空和地球化学组成上具有明显的不均一性。传统的弧岩浆岩成因模式认为，这种不均一性代表了弧岩浆岩源区交代组分及部分熔融条件的差异（Hawkesworth et al.，1993；Turner et al.，2017；Zheng，2019）：俯冲板片随着俯冲深度的增加，会不断地释放流体、熔体甚至超临界流体交代上覆地幔楔橄榄岩，形成富集的地幔交代体，成为弧岩浆岩的源区（Grove et al.，2012；Mallik et al.，2016；Zheng，2019）。不过，对地幔楔部分熔融形成弧岩浆深度的认识存在变化，早期认为该深度主要在 100 ± 40 km 的范围内，越来越多的研究指示这个深度主要在 120 ± 40 km（即在 80～160 km）。此外，部分熔融程度对熔体成分有着显著的影响。

不同于传统的弧岩浆成因模型，计算地球动力学研究表明，俯冲带变质混杂岩在弧下深度从板片拆离进入地幔楔内部，在地幔楔中与橄榄岩一起发生部分熔融可以产生不同成分的弧岩浆（Gerya and Yuen，2003；Castro et al.，2010；Behn et al.，2011；Marschall and Schumacher，2012；Nielsen and Marschall，2017；Cruz-Uribe et al.，2018）。这种物理混合作用不同于俯冲带流体交代的化学混合作用，所产生的弧岩浆成分取决于混杂岩中不同岩石类型的相对比例；并且假设部分熔融的压力在1.5～2.5 GPa，对应于地幔楔的核部（Castro et al.，2010；Cruz-Uribe et al.，2018）。传统的弧岩浆岩模型更强调俯冲带整体的化学动力学行为，而混杂岩模型虽然着重解释弧岩浆岩的地球化学属性，但是对于主量元素和微量元素以及稳定和放射成因同位素在两个混合端元之间的比例并没有进行一致性检验。此外，混杂岩模型目前的支持主要来自计算地球动力学结果，并没有实验岩石学和地球化学证据。

此外，大量的研究发现，弧岩浆岩存在幕式活动的特点，即岩浆爆发期（持续1000万～3000万年）和岩浆间歇期交替出现，这种现象在大陆弧地区表现得尤为明显。何种因素控制了弧岩浆幕式爆发的特征存在很大争议。前人研究提出的模式包括岩石圈拆沉引发的软流圈上涌、交代上覆岩石圈地幔的俯冲、俯冲板片周期性地释放挥发分等。任何一种弧岩浆岩的动力学模型都应该将弧岩浆岩幕式活动的特征考虑在内。

（4）大地幔楔对板内岩浆作用的影响

小地幔楔、大地幔楔对壳幔系统的演化具有重要影响（图4-10）。小地幔楔在弧岩浆岩的形成中发挥了关键作用，而大地幔楔在全球物质循环、岩石圈地幔改造和板内岩浆作用的触发方面发挥了重要作用（徐义刚等，2018；郑永飞等，2018；Zheng et al.，2020）。以东亚大地幔楔为例，滞留在地幔过渡带的俯冲板片可进一步脱水、脱碳交代过渡带上覆的深部上地幔，对东亚深地动力系统和地幔熔融/交代作用产生了重大影响：再循环洋壳流体或熔体、再循环碳酸盐流体或熔体等物质对上地幔进行改造并产生板内玄武岩（Li et al.，2017；Li et al.，2016a，2016b，2019a，2019b；徐义刚等，2018；Zheng et al.，2020）。需要指出的是，俯冲板片流体或熔体可能是板片在金红石稳定深度发生脱水熔融交代小地幔楔，而在更深的地幔深度发生金红石分解，向深部地幔释放流体和熔体，可以解释为什么同样受到板块俯冲影响，岛弧岩浆和板内岩浆表现出截然不同的微量元素特征，如岛弧岩浆表现出铌的负异常而板内岩浆表现出铌的正异常（Zheng，2019）。

如何通过岩石地球化学研究来识别、还原一个俯冲带在地质历史时期的形成过程是俯冲带科学研究的难点。不同构造背景、不同时代的洋壳在俯冲过程的早期和晚期阶段可能具有不同的岩浆响应。另外，尽管揭示大地幔楔系统的形成与运作机制对于完善板块构造理论具有深远意义，但是目前对于它的形成、演化、消亡以及对壳幔系统演化的影响等方面还知之甚少。

3. 重点研究方向

1）软流圈与岩石圈两种地幔楔的差异及其对弧岩浆起源的影响：软流圈与岩石圈地幔楔组成及形成机制；地幔楔组成和结构（如岩石圈厚度）演化与弧岩浆岩产生。

2）地幔楔中不同类型地幔岩石部分熔融的物理化学条件：石榴辉石岩不同温压条件下的熔融实验；水饱和橄榄岩的固相线的限定；不同含水橄榄岩（含角闪石、蛇纹石和绿泥石等）脱水熔融固相线的限定；不同类型流体的交代过程与机制。

3）弧下地幔部分熔融条件与熔体成分的控制因素：弧下地幔部分熔融条件；熔体成分的控制因素；混杂岩熔融的岩石学约束；弧岩浆幕式活动的控制因素。

4）大地幔楔的形成过程及其与陆内壳幔系统演化的关联：俯冲板片滞留于地幔转换带的原因；板片从俯冲到滞留再到下沉进入下地幔的全过程；俯冲和滞留板片对水和碳循环的影响；再循环物质对上地幔不均一性和地幔熔融作用的影响；大地幔楔形成与板内岩浆作用之间的成因联系。

4. 应对策略

中国有很多新生代以前的古俯冲带或古火山弧带，这是中国独特的优势。中国学者对这些俯冲带进行了深入的研究，取得了许多重要的成果。但是，中国缺少现代活动的大洋俯冲带和火山弧，这导致一些重要的地质现象无法直接观测，很多研究只是根据已经发生的地质现象去推断过去发生的事情。事实上，地球科学最基本的原理就是“将今论古”，一些经典的地幔楔理论和模型主要是基于对现代大洋俯冲带的研究。因此，要确定地幔楔特征及其部分熔融机制，需要采取以下应对策略。

1）加强国际合作，选取现代大洋俯冲带，如环太平洋以及澳大利亚－东南亚之间的印度洋新生代大洋俯冲带，采用地球物理学、地质学、地球化学等多学科交叉手段对典型区域进行解剖。以西太平洋为例，该构造域拥有全球 1/3 左右的俯冲带（Stern and Scholl，2010），板块俯冲作用将大量的沉积物、不同年龄的蚀变洋壳和岩石圈地幔带到了东亚地幔深部。每年通过板块俯冲进入地幔的俯冲沉积物达 2.5km^3，势必对东亚大地幔楔系统产生重大影响。滞留在地幔过渡带的俯冲板片可进一步脱水、脱碳交代过渡带上覆的深部上地幔，对东亚深地动力系统和地幔熔融 / 交代作用产生了重大影响（徐义刚等，2018）。因此，西太平洋俯冲系统及其产生的小地幔楔、大地幔楔系统是将是需要研究的核心。

2）加强古大洋俯冲带与现代大洋俯冲带之间的对比研究，构建古俯冲带的小地幔楔、大地幔楔系统及岩浆形成模型。

3）以现代和古代大洋俯冲带小地幔楔、大地幔楔系统研究成果为基础，采用高温高压实验、物理和数字模拟、大数据等手段，再现地幔楔的形成、演化与熔融的过程。

5. 预期目标

预期目标主要包括如下几点。

1）查明大小地幔楔中软流圈地幔楔与岩石圈地幔楔的特征、形成过程及其对壳幔物质系统演化的影响；

2）揭示地幔楔中不同类型地幔岩部分熔融的物理化学条件及其弧岩浆成分的影响和控制作用；

3）揭示不同性质流体交代地幔楔作用及其对地幔楔物质组成的影响；

4）阐明不同类型混杂岩对地幔楔组分的改变及其产生弧岩浆岩的特征与条件；

5）揭示大地幔楔的形成过程及其与陆内壳幔系统演化的关联；

6）构建小地幔楔、大地幔楔形成与演化的理论模型，揭示不同大小地幔楔中岩浆产生机制。

（四）科学挑战 7：俯冲带流体活动与元素迁移

1. 问题的提出

俯冲带流体活动是壳幔之间发生物质和能量交换的重要途径。流体本身也是俯冲过程中发生元素迁移、同位素分馏、矿物反应以及对上覆地幔楔化学交代不可或缺的介质。一般来说，根据流体的状态和性质，俯冲带可能出现的流体可以分为富水溶液、含水熔体和超临界流体三种类型（图 3-6）。超临界流体是一种不同于富水溶液和含水熔体的具有特殊物理化学性质的流体，其形成要求体系的温压条件接近或者超过体系的第二临界端点，这时候流体中的硅酸盐溶质与溶剂之间呈现连续完全互溶的状态（Zheng et al.，2011）。

一般来说，富水溶液含有小于 30% 硅酸盐溶质，含水熔体溶解小于 30% 的水，这样溶质和含水量介于二者之间的中间成分流体最有可能是超临界流体（Hermann et al.，2006；Zheng et al.，2011；Ni et al.，2017）。但是，超临界流体在物理化学性质上最关键的特征就是溶质与溶剂之间达到完全混溶

（Zheng et al.，2011）。因此，如果溶质与溶剂之间未达到完全混溶，即使在成分上介于二者之间也还不是超临界流体（Zheng，2019）。这涉及超临界流体形成热力学和动力学的双重控制：在特定的温度压力下，只要溶质与溶剂之间达到完全混溶即成为超临界流体，与溶质和溶剂之间的相对含量无关；在进入超临界流体稳定域之后，即使溶质和溶剂含量介于两个 30% 之间，只要溶质与溶剂之间未达到完全混溶就还不是超临界流体。

俯冲带最初释放出来的低温低压流体溶解离子的总量一般较少，并且其金属离子的溶解能力主要受控于水溶液中阴离子化学配体。富水溶液主要来源于俯冲岩石的脱水作用，成分上以水为主，并含有 CO_2、N_2、S 以及其他类型的挥发分；其产生以及成分变化与俯冲岩石中含水矿物的稳定性密切相关（郑永飞等，2016），同时受俯冲带温压结构的控制。即使在高温高压条件下富水溶液的溶解物质总量也只是在 5 wt%～15 wt%，并且主要溶解的是 Si、Al 以及碱金属元素。微量元素主要富集流体活动性元素（如大离子亲石元素），相对亏损高场强元素（Hermann and Rubatto，2009；Hermann and Spandler，2008；Hermann et al.，2013）。相比于富水溶液，超临界流体可以迁移高得多的大离子亲石元素和轻稀土元素（图 3-7）。对俯冲带流体地球化学传输的研究，是理解岛弧型和洋岛型镁铁质岩浆岩微量元素组成的关键（图 4-12）。

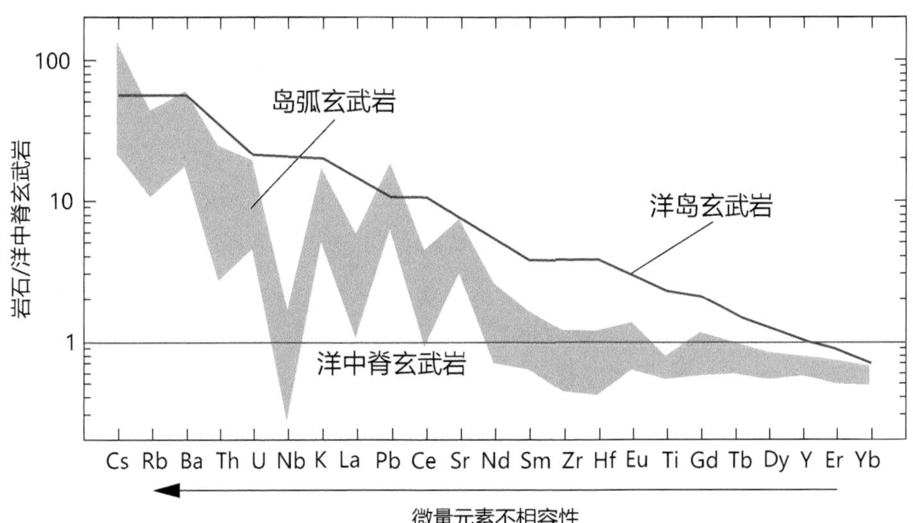

图 4-12　岛弧玄武岩和洋岛玄武岩相对于洋中脊玄武岩微量元素分布图（修改自 Zheng，2019）

横坐标箭头方向指地幔橄榄岩部分熔融过程中微量元素不相容性增大的方向；纵轴数据无单位，是指岩石中微量元素含量与洋中脊玄武岩微量元素含量的比值俯冲带从板片俯冲到折返整个过程，至少涉及两方面的流体活动以及与之相关的元素迁移。一方面，在俯冲过程中，岩石由于脱水作用会在不同深度产生含有不同元素种类和元素含量的流体，这些流体的产生和迁移会导致俯冲板片岩石本身的元素组成发生变化。另一方面，俯冲板片产生的流体会交代上覆地幔楔，从而使上覆地幔楔岩石不仅发生地球化学特征的改变，而且地幔交代岩的固相线温度也相应降低。一旦这些地幔交代岩发生部分熔融，就可以在80～160km的弧下深度产生岛弧型岩浆作用，在大于200km的后弧深度产生洋岛型岩浆作用（Zheng，2019）。在岛弧岩浆演化的晚期，岩浆热液流体发生出溶并携带金属元素富集成矿。正是这些普遍存在的流体迁移、交代和出溶及再沉淀，使得俯冲带成为元素迁移和再分配的大型工厂，也使得金属元素迁移和局部富集成为可能，并最终形成俯冲带之上广泛发育的热液金属矿床。

2. 面临的挑战

关于俯冲带流体地球化学以及相关的元素迁移研究已经取得了很大的进展，并且随着实验分析方法的进步，如热液金刚石压腔等实验方法的建立，其为理解流体成分和矿物溶解度提供重要数据。新型地球化学示踪工具的开发，如轻元素（如Li、B）同位素体系，以及更高精度分析技术，为研究俯冲流体和元素迁移及其地质效应提供了新的手段。但总体来说，在这一领域还存在较多需要解决的问题，面临的挑战和问题主要表现在以下方面。

1）目前对俯冲带超临界流体的研究还十分有限，阻碍了对俯冲带物质循环过程和机制的深入认识。这方面工作面临的挑战主要包括如下几点。①超临界流体在岩石学和特定地球化学意义上与富水溶液和含水熔体有何准确的区分指标？不同岩石体系的湿固相线位置、第二临界端点位置如何？由于超临界流体的产生受体系组成、温压条件和其他诸多地质因素的影响，目前对于俯冲带超临界流体出现的深度、岩石学证据、元素和同位素组成特征，以及超临界流体出现导致的地质效应等尚缺乏清晰和统一的认识。②如何准确界定自然界条件下含水熔体、富水溶液和超临界流体各自稳定的温压范围？虽然在实验室条件下对于这几种流体形成的温压条件有较严格的界定，但是对于自然状况下的俯冲带而言，具体情况要复杂得多。同时，在温压降低时，自然界超临界流体会与围岩反应从而失去其超临界性质，并分离成为含水熔体和富水溶液，这样自然界就很难保存原始的超临界流体信息。③含水熔体和超临界流体元素迁移的关键控制因素是什么？目前已知弧岩浆富集大离子亲石元素，但不同的岩浆弧这些元素的富集程度存在很大差别。尽管弧岩浆总体显示高场强元素负异常，但这些元素在弧岩浆中的含量是高度变化的，表明板片释放的流体并非都是超临界流体，或者超临界流体的元素迁移能力随物理化学条件的不同有所变化，因此揭示流体元素迁移能力的关键控制因素是至关重要的。

2）俯冲带元素的迁移能力主要由元素本身受不同矿物相与流体之间分配系数的直接控制，而分配系数是温度、压力、矿物和溶质组成、氧逸度等物理化学参数的函数，这些因素（特别是温度、压力）势必同时影响流体能够携带溶质迁移的距离。俯冲板片脱水形成残留矿物 + 流体共存体系，因而流体的组成和微量元素含量受到残留矿物的缓冲。大部分微量元素没有自己的独立矿物，它们在流体中的含量实际由矿物与流体之间的分配系数控制；Ti 和 Zr 是例外，在榴辉岩中常形成独立副矿物金红石和锆石，高度富集高场强元素，因此高场强元素的迁移主要受金红石和锆石等在流体中的溶解度控制。此外，有关俯冲带氧逸度变化对变价元素的分配和迁移行为的控制所知有限。在板块俯冲过程中，脱水反应可能会导致残留相向氧逸度降低的方向演化，而在折返过程中可能由于水化的因素导致板片内部的氧逸度升高。氧逸度的变化会导致变价元素及与其地球化学性质相近的其他元素分配系数的变化，这是一个目前为止所知有限的领域，尚有待深入研究。

3）在浅俯冲条件下，板片在弧前深度脱水往往形成的是稀的富水溶液，其溶解元素的能力很弱；而在深俯冲条件下，由于溶质组分随温度和压力增加而增加，板片在弧下深度脱水和熔融可以形成含水熔体乃至超临界流体。对于俯冲带岩石 -H_2O 体系，富水溶液、含水熔体和超临界流体的形成条件（温度和压力稳定域）由固相线及其临界端点界定。因此，限定弧下深度橄榄岩 -H_2O 体系、玄武岩 -H_2O 体系和沉积物 -H_2O 体系的固相线及其临界端点是揭示俯冲带流体性质的前提。俯冲带温压结构研究（Syracuse et al.，2010）表明，俯冲带弧下板片顶部压力（深度）和温度是 2.5～6 GPa（80～180 km）和 700～900℃。以往俯冲板片岩石 -H_2O 体系的固相线及其临界端点限定多采用多顶砧开展实验，但实验结果相互冲突。因此，要准确限定俯冲带岩石 -H_2O 体系超临界流体的形成条件，亟待开发新的实验方法或技术途径。

4）矿物中的包裹体和微观结构，如熔体包裹体、流体包裹体、晶体包裹体以及出溶体等，由于受到主矿物的保护，相比于全岩和整体矿物颗粒来说更容易记录多期次流体活动的地质过程。流体包裹体是保留古俯冲带流体的唯一直接样本。俯冲带岩石中保存有代表不同俯冲阶段流体成分的流体包裹体，但是目前对于这些微小的流体包裹体（10～20 μm）的成分分析多限于传统的冷热台的方法，利用现代微束分析技术来分析单个流体包裹体以直接获得其定量成分信息仍然是一个需要开发、改进和具有挑战性的任务。

5）详细了解俯冲带流体活动引发的特定元素迁移与成矿的关系是未来工作的一个方面，特别是中酸性岩浆的流体出溶与元素迁移的关系。地幔楔部分熔融导致硫化物溶解形成含 H_2O、S 和亲铜元素的镁铁质弧岩浆；这些岩浆常常在下地壳岩浆房演化形成长英质岩浆，最后流体出溶形成含矿流体，形成引人瞩目的斑岩型 Cu、Au 矿床。这个过程涉及硅酸盐矿物 - 硫化物 - 岩浆熔体 - 流体之间的元素分配，控制 S 和亲铜元素在各相中的分配行为。已有的实验研究表明，Cu、Au 在硅酸盐矿物中是高度不相容的，它们的迁移行为由硫化物和流体控制。因此，流体 - 熔体和流体 - 硫化物元素分配系数对理解 Cu、Au 在岩浆热液过程中的富集行为尤为重要。斑岩型矿床的形成中涉及流体和成矿元素迁移的问题是：Cu、Au 如何从深部迁移到浅部成矿？这一问题存在的原因，主要是由于以往大部分实验研究只关注成矿发生场地（＜5 km）的成矿过程和机制研究，对深部岩浆过程，特别是下地壳岩浆 - 热液过程的研究几乎是空白。大陆弧长英质岩浆主要起源于幔源铁镁质岩浆在下地壳的结晶分异（Annen et al.，2006），原始弧岩浆含 2 wt%～6 wt% 的 H_2O（Plank et al.，2013），产生长英质岩浆要求镁铁质岩浆经受大于 70% 的结晶分异。尽管在下地壳压力下岩浆中的 H_2O 溶解度可高达 15 wt%～20 wt%，但是原始弧岩浆 6 wt% H_2O 经历 70% 的结晶分异完全有可能演化出含 H_2O 高达 20%～30% 的长英质岩浆，因此在下地壳条件下长英质岩浆完全有可能发生流体出溶。

3. 重点研究方向

针对以上俯冲带流体活动与元素迁移研究面临的问题和挑战，在未来有必要加强以下几个重点方向的研究。

1）俯冲带超临界流体相关的元素迁移过程和机制的研究，特别是通过革新性实验和分析技术在俯冲带岩石样品研究方面的应用，揭示自然界不同种类流体元素迁移能力的关键控制因素。

2）揭示温度、压力和流体化学成分以及氧逸度变化对流体 - 矿物微量元素分配系数，以及对副矿物在流体中溶解度的影响，是研究富水溶液和超临界流体元素迁移关键控制因素的核心内容。

3）在高温高压实验方面，未来应加强限定弧下深度板片脱水产生的流体性质及超临界流体形成条件的研究，开发新的实验方法或技术途径。

4）针对矿物内部的微区结构，如流体包裹体、晶体包裹体以及出溶体等的各种原位分析技术的开发，将为定量了解俯冲带不同流体的元素迁移能力提供极大帮助。同时，如何通过直接提取与俯冲相关的流体包裹体中的 H、O 等同位素的信息，也是俯冲带流体研究的一个值得努力的重要方向。

5）下地壳岩浆房流体出溶以及流体－熔体和流体－硫化物 Cu、Au 分配，可能是定量理解这些元素如何从深部迁移到浅部并成矿的关键所在。因此，从浅部岩浆热液体系转向深部岩浆热液体系的岩浆热液铜金成矿机制实验研究应当是未来一个重要的研究方向。

4. 应对策略

1）在天然样品中对不同流体进行识别和成分特征的反演，是进一步理解俯冲带流体活动的基础。流体包裹体保存了自然界古地质流体的直接证据。同时，与流体包裹体相补充，近年来发现在深俯冲的变质岩石中存在较多的不同类型的晶体包裹体（以多相为主，包含两种或两种以上结晶矿物），它们很可能代表了超高压变质作用峰期之后的熔体组成。因此，在未来加强俯冲带天然样品中各类型流体和晶体包裹体的定量研究，并利用各种新的微区分析手段定量探讨其中的元素含量，是深刻理解俯冲带相关流体特别是超临界流体活动的重要手段。

2）高压－超高压变质脉体被普遍认为是俯冲板片中流体活动的产物。尽管这些脉体中的矿物组合更多代表的是流体活动残留下来的组分，但是其全岩成分及同位素信息还是在很大程度上反映了当时流体活动的特征。因此，加强对各种俯冲带内生流体相关的不同成因脉体的系统的岩相学、矿物学和地球化学研究，对于了解俯冲带流体与元素迁移具有重要意义。

3）高温高压实验研究一直是了解俯冲带流体与元素迁移关系的重要手段。在目前的实验手段及实验条件下，直接通过平衡实验来准确测量流体－矿物间的元素分配系数非常困难，但是前人已在这一方面做出了大量的工作（Brenan et al.，1995；Kessel et al.，2005）。尽快开发对于高场强元素等流体中溶解度低的元素的分配系数的测量手段，将对定量化了解俯冲带流体与元素迁移的关系提供重要帮助。

4）通过分子动力学模拟了解俯冲带流体与元素迁移的关系，是对高温高压实验和天然样品研究的重要补充手段。

5. 预期目标

通过对以上面临问题和挑战以及重点方向的深入研究，可以在定量理解和准确区分俯冲带含水熔体、富水溶液、超临界流体方面取得重要进展。这些进展将对理解俯冲带流体释放的深度、流体成分随温压条件的变化，各种元素包括挥发分（如 F、Cl 和 S 等）的迁移通量及机制，以及不同种类流体的元素迁移能力和关键控制因素等几个方面提供重要限定。

三、关键科学问题之三：俯冲带产物

（一）科学挑战 8：洋壳俯冲带变质岩

1. 问题的提出

高压－超高压变质岩的出露遍布全球汇聚板块边缘，它们代表了地质历史上地壳（洋壳和陆壳）岩石俯冲到下地壳到岩石圈地幔深度经历了高压－超高压变质作用，而后又折返回地表的过程。对这些岩石的研究可以揭示俯冲带长时间跨度的地球动力学过程和物理化学状态（Chopin，2003；Liou et al.，2009；Zheng，2012）。相对于陆壳俯冲带高压－超高压变质岩，洋壳俯冲带高压－超高压变质岩出露相对稀少，这可能归因于基性成分的洋壳岩石在经历超高压变质后形成了具有较大密度的榴辉岩，这样就难以折返回到浅部地壳层位。不过，洋壳的属性决定了其在理解和发展板块构造理论方面所发挥的作用。洋壳俯冲带形成的典型变质岩是蓝片岩和榴辉岩（图 4-13），它们比陆壳俯冲带同类型变质岩具有更特殊的地质意义。

研究洋壳俯冲带变质岩对深入理解俯冲带深部过程、壳幔相互作用和地球动力学过程具有重要的科学意义。通过对天然样品的研究，结合高温高压实验、相平衡模拟、理论模拟计算、现代地球化学测试技术，对洋壳俯冲带温压结构、俯冲地壳变质脱水和部分熔融、俯冲带流体活动与元素迁移等的研究已经取得了一系列重要进展（Syracuse et al.，2010；Hacker and Gerya，2013；Penniston-Dorland et al.，2015；Zheng，2019）。但是作为俯冲带研究的重要对象，随着俯冲带科学的发展，洋壳俯冲带变质目前也面临着诸多关键的科学问题，需要开展进一步的研究工作。

2. 面临的挑战

（1）洋壳俯冲与板块构造

俯冲带变质岩石是在板块俯冲过程中形成的，特别是俯冲带高压－超高

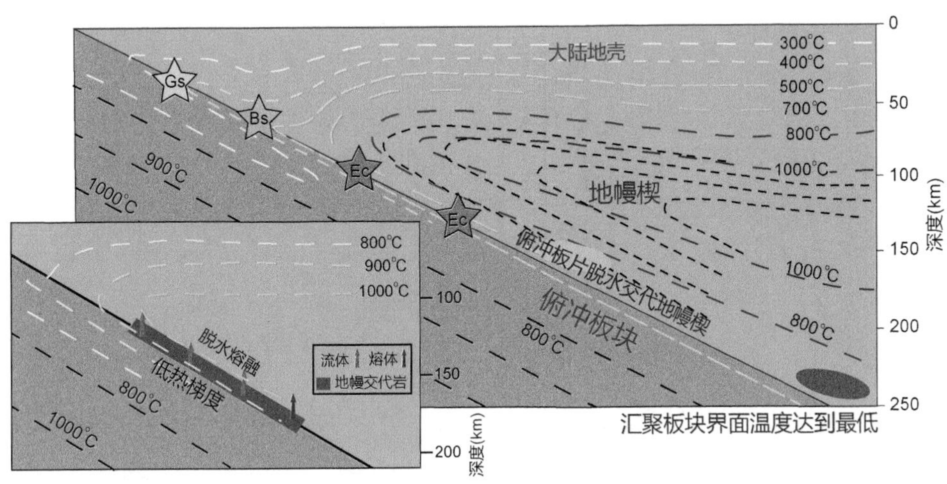

图 4-13　大洋俯冲带蓝片岩相和榴辉岩相变质岩形成与板块俯冲关系示意图（修改自 Zheng and Chen，2016）

阿尔卑斯型变质相系只会形成于板块俯冲早期阶段的低地热梯度条件下，在进入弧岩浆作用的晚期阶段俯冲带温压条件已经显著升高到中等到高的地热梯度，形成的是巴肯型变质相系（Zheng and Chen，2017）。变质相缩写：Gs- 绿片岩相；Bs- 蓝片岩相；Ec- 榴辉岩相

压变质岩，如蓝片岩和榴辉岩，是目前鉴别板块俯冲带的主要标志岩石（李继磊，2020）。在显生宙以来和现今正在进行的冷板块俯冲带都发现了大量的蓝片岩和榴辉岩，是研究现代板块俯冲作用和构造演化的主要岩石类型。此外，对于前寒武纪时期蓝片岩和榴辉岩的识别和深入的岩石学研究，也是判断古代板块起始的重要依据。目前，世界范围内发现的蓝片岩带主要集中在显生宙以来的洋壳俯冲带上，确切的前寒武纪蓝片岩比较稀少（Maruyama et al.，1996）。发育完好的前寒武纪蓝片岩是中国新疆的阿克苏蓝片岩，它出露在塔里木板块的西北缘阿克苏市附近约 300km^2 的区域，其上为震旦纪磨拉石建造不整合覆盖（张立飞等，1998）。阿克苏蓝片岩主要岩石类型为基性绿帘石蓝片岩 - 长英质蓝片岩和多硅白云母片岩，其峰期变质温压条件是 320～410℃和 680～870 MPa，变质时代为 805～769 Ma，这是目前最古老的蓝片岩之一（Xia et al.，2019）。目前发现的最老榴辉岩属于古元古代，规模最大的是俄罗斯白海活动带榴辉岩（Yu H L et al.，2017；Liu F L et al.，2017），因此认为现代板块在古元古代就已经出现（Xu et al.，2018）。板块构造体制在地球早期什么时候启动是固体地球科学研究的重要问题（Zheng，2021a）。洋壳俯冲带高压 - 超高压变质作用形成的蓝片岩和榴辉岩是板块俯冲带最直接的产物，对于建立地球早期板块构造机制具有重要意义，也是目前俯冲带

变质岩研究的前沿领域。

（2）洋壳榴辉岩的折返机制

高压－超高压变质岩石抬升折返机制一直是俯冲带变质作用研究的关键科学问题之一（Li et al.，2016；刘贻灿和张成伟，2020；张立飞和王杨，2020；Zheng，2021d）。从超高压变质岩石发现以来，最大的困惑就是俯冲到弧下深度的地壳岩石是如何抬升折返到地表的？对于大陆型超高压变质地体来说，其折返机制通常被认为是由于陆壳密度较低可以通过浮力折返。俯冲变质洋壳主要由以高密度榴辉岩为代表的铁镁质岩石所组成，由于其密度比周围的橄榄岩还大，因此其折返过程及机制一直存在争议。目前提出来的洋壳榴辉岩折返机制主要有逆掩推覆模式、板底垫托模式、俯冲隧道模式、板块后撤模式等（Hacker and Gerya，2013；Warren，2013）。在众多模式中，密度演化无疑是最重要的影响因素之一。利用数值模拟方法，建立变基性岩、变沉积岩和蛇纹岩等洋壳主要岩石类型在 *P-T* 空间内的三维密度演化，探讨变质脱水作用及俯冲带温压结构对大洋岩石的密度演化和折返过程的控制，一直是大洋型超高压变质岩石折返研究的前沿方向。

张立飞和王杨（2020）依据密度演化将现今折返到地表的洋壳榴辉岩分为两类：一类是具有自折返能力的榴辉岩，岩石在自身浮力的驱动下发生折返（Chen et al.，2013）；另一类是携带折返的榴辉岩，岩石具有负浮力，只能依靠低密度的变沉积岩和蛇纹岩的携带作用才能发生折返（Li et al.，2016）。数值模拟计算结果表明，几乎所有的洋壳硬柱石榴辉岩都属于具有自折返能力的洋壳榴辉岩，而绝大部分的绿帘石榴辉岩则属于携带折返的洋壳榴辉岩（图 4-14）。此外，张立飞和王杨（2020）还得出了洋壳榴辉岩抬升折返过程中将会出现“莫霍面停滞”的结论，即莫霍面是洋壳榴辉岩抬升折返过程的阻碍面，洋壳榴辉岩折返到这个界面处时，其密度会远高于周围岩石的密度，结果出现明显的折返速度下降，直至发生折返停滞。停滞在莫霍面附近的榴辉岩则需要外力作用才能折返到地壳表面。因此，高压－超高压变质地体的折返是一个受多因素影响的地球动力学过程，目前仍存在诸多尚未解决的问题。

3. 重点研究方向与应对策略

（1）洋壳俯冲带变质岩温压条件的准确限定

对俯冲带温压结构的有效恢复和计算模型的地质验证，都需要对洋壳俯冲带变质岩温压条件的准确限定。为解决这一问题，应聚焦：①代表性高压－

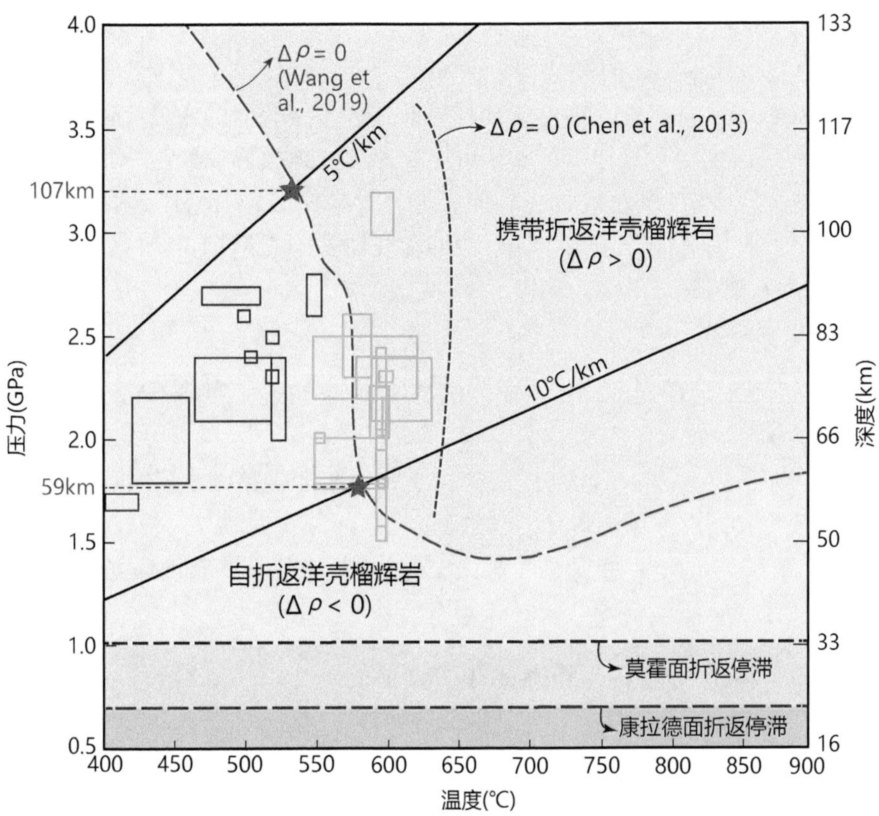

图 4-14 洋壳榴辉岩的 P-T-$\Delta\rho$ 演化图解（张立飞和王杨，2020）

$\Delta\rho$ 为特定温度压力条件下榴辉岩与围岩橄榄岩的密度差

超高压变质地体的不同岩石体系的相互验证；②有效的相平衡模拟及温压计算手段的发展；③岩石折返过程中热叠加效应的有效剔除；④计算动力学模拟的多参数集成与优化。

（2）洋壳俯冲带变质岩的流变性模拟

更深入地理解大洋和大陆俯冲带岩石的俯冲－折返过程，需要从地质学和数值模拟结合的角度开展更深入的研究。虽然目前数值模拟方法还存在很多不确定因素，但随着地质学观察的不断深入，岩石学和地球化学数据的不断增加，应用数值模拟方法定量地再现和评估高压－超高压变质地体抬升折返的地球动力学过程势在必行。

（3）全球尺度的洋壳蓝片岩／榴辉岩对比研究

对地球演化的理解离不开对全球尺度的整体把握和考量。对洋壳蓝片岩－榴辉岩的系统分类和解析首先需要建立完善的蓝片岩数据库，包含全球洋壳蓝片岩/榴辉岩的产出位置、围岩类型、原岩属性、年龄、温度、压力、T/P 比值、板片年龄、俯冲速率、俯冲角度等一系列重要参数的统计。基于这些统计数据，可以建立地质时期洋壳蓝片岩－榴辉岩的成分变化规律、俯冲带的地热梯度变化规律等信息；可以反映板片地热梯度、俯冲角度和俯冲速率的改变对岩石矿物组合、矿物成分、含水量、岩石密度和黏滞度的影响；还可以根据对应于俯冲早期、中期和晚期不同类型洋壳蓝片岩－榴辉岩的特征来反演俯冲带的生命历程。

4. 预期目标

1）准确限定洋壳俯冲带变质岩的精细 *P-T-t* 轨迹，厘清不同洋壳俯冲带变质岩的构造含义；

2）岩石学特征与高温高压实验相互验证俯冲带精细变质过程，地质学和数值模拟共同反演俯冲带动力学过程；

3）系统建立洋壳蓝片岩－榴辉岩数据库，提炼具有地质意义的变化规律。

（二）科学挑战9：陆壳俯冲带变质岩

1. 问题的提出

20世纪80年代以来，在大陆地壳中陆续发现超高压变质矿物柯石英（Chopin，1984；Smith，1984）和金刚石（Sobolev and Shatsky，1990；Xu et al.，1992），表明低密度的大陆地壳岩石可以俯冲至大于80 km和大于120 km的地幔深度（图3-3），引起了板块构造的一场革命（Chopin，2003；Liou et al.，2009；郑永飞等，2015）。随着研究的深入，其他特殊矿物或特殊的矿物出溶结构在这些超高压岩石中被不断发现，如 α-PbO_2 结构的金红石（Hwang et al.，2000），橄榄石出溶钛铁矿和铬铁矿（Dobrzhinetskaya et al.，1996），石榴子石出溶单斜辉石和斜方辉石（van Roermund and Drury，1998），石榴子石出溶单斜辉石（Ye et al.，2000；Zhang et al.，2003），榍石出溶柯石英（Ogasawara et al.，2002），石英出溶Al和Fe氧化物（Liu L et al.，2007）等。对这些出溶结构的深入研究证明，部分陆壳曾俯冲到至少200 km深的地幔并折返到地表。

科学家已经在全球发现了30多条含柯石英或金刚石的超高压变质带（Chopin，2003；Liou et al.，2009，2014），大陆俯冲带超高压变质岩石的研究成为地球动力学研究的重要领域之一。通过对大陆深俯冲和超高压变质的深入研究，已经在陆壳俯冲的深度、超高压岩石的全球分布情况、深俯冲陆壳的变质时代、产出规模和 *P-T-t* 轨迹、大陆俯冲带流体活动、大陆深俯冲的机制和动力学过程等方面取得了一系列明确的认识（Coleman and Wang，1995；Liou et al.，2009；Zheng，2012；Hermann and Rubatto，2014）。

但是，俯冲带大陆深俯冲变质岩研究仍然面临以下方面的关键科学问题：①深俯冲陆壳的俯冲深度和命运；②大陆深俯冲变质岩石的形成条件；③变质过程中矿物组合及矿物地球化学和岩石学变化特征；④大陆地壳深俯冲和折返阶段的精确时间以及全岩主微量元素和同位素组成变化及其控制因素等。

2. 面临的挑战

（1）深俯冲陆壳的俯冲深度和命运

进入 21 世纪以来，人们把在大洋俯冲带研究中提出的俯冲隧道模型拓展到大陆俯冲带，提出大陆俯冲隧道模型（Gerya et al.，2002；Guillot et al.，2009；郑永飞等，2013；张建新，2020），强调大陆俯冲过程的实质是下伏俯冲大陆板片与上覆大陆板片在一定的空间发生的物质运动，及伴随的这些板块界面之间的相互作用。随着两个刚性板块之间的汇聚，俯冲隧道内的物质形成韧性剪切带，表现为“动态三明治结构”，因此出现同一造山带不同岩片之间的构造拆离以及同一个超高压岩片中不同部位记录的差异性 *P-T-t* 演化轨迹（Zheng et al.，2019a），从而解释了大陆碰撞造山带中出现的各种微观和宏观岩石构造单元以及深俯冲陆壳的折返机制。大陆隧道模型的提出是大陆动力学的重要发展。

高温高压实验岩石学研究表明，大陆地壳深俯冲至 250 km 以上的深度后，其密度会大于周围地幔岩石的密度，很难由于浮力作用折返至地表而滞留在地幔过渡带（Wu et al.，2009）。关键科学问题包括：如何确定深俯冲大陆陆壳的极限俯冲深度，大陆超深俯冲的动力来源如何限定，如何识别未折返的大陆地壳在地幔中的命运并确定它们对地幔成分不均一性的贡献等。

（2）大陆深俯冲变质岩石形成条件的精细限定

大陆深俯冲变质岩石的温压条件及以此为基础构建的 *P-T* 轨迹，对探讨大陆深俯冲的动力学过程具有非常重要的意义。目前，用于变质岩石 *P-T* 条件计算的方法主要是传统的矿物温度计和热力学视剖面模拟计算，其中前者需要找到峰期及不同阶段的变质矿物组合并准确判断哪些矿物或矿物域处于化学平衡，后者则需要明确有效的全岩组分。目前学者应用不同的方法计算了全球大陆俯冲带的温压条件，并构建了 *P-T* 演化轨迹。实际上，陆壳深俯冲形成的超高压岩石记录的峰期压力大于 2.5 GPa，并可高达 7 GPa，峰期温度条件为 500～1200℃（Carswell et al.，1997；Krogh Ravna and Terry，2004；Wei and Clarke，2011；Hermann and Rubatto，2014）。

但是，这些超高压变质岩大多经历了近等温降压甚至降压升温的退变质过程，部分岩石在折返过程中还经历了部分熔融。因此，前进和峰期矿物的保存以及在有效全岩成分的精确确定上都存在误差，这些给限定岩石的精细 *P-T* 条件带来了不确定性。另外，大陆俯冲隧道的温度与现代大洋俯冲隧道相似（图 3-5 和 4-6），大陆俯冲隧道中地壳岩石在俯冲过程中发生的脱水作用应与大洋俯冲隧道类似（郑永飞等，2016）。但是，大陆俯冲带之上没有出现同俯冲大陆弧岩浆作用（郑永飞等，2015）。

大陆俯冲带之上虽然缺乏弧火山作用，但是不等于说俯冲带内部缺乏流体活动（Zheng，2009，2012；Zheng and Hermann，2014）。大陆俯冲带流体活动依然存在，加快了变质反应进程、导致了不同性质元素分异、引起了壳幔相互作用等。特别是在深俯冲陆壳折返的过程中，由于含水矿物的脱水分解、名义上无水矿物中结构水和分子水出溶以及流体包裹体爆裂等，流体活动显著，可形成高压脉体或引起部分熔融（Zheng，2009；Zheng and Hermann，2014）。在大陆地壳俯冲过程中，虽然在浅部释放的流体极为有限，但是在进入弧下深度之后，含水矿物也会发生分解，产生流体交代地幔楔，却没有引起同俯冲岩浆作用，其原因是大陆俯冲带温度较低，未能引起地幔楔部分熔融（郑永飞等，2016）。

目前，已确定在大陆俯冲带存在富水溶液、含水熔体和超临界流体，并认识到它们在俯冲带对元素具有明显不同的分异作用，且不同岩性之间的流体活动和元素分异行为也存在很大差异（Zheng et al.，2011；Zheng and Hermann，2014；Zheng，2019）。如何精细限定前进 - 峰期 - 退变质不同阶段的精确 *P-T* 条件，准确限定陆壳深俯冲岩石俯冲 - 折返的动力学过程，建立流体活动与岩石 *P-T* 轨迹之间的准确对应关系，并确定流体的性质及其对壳幔相互作用的影响，是下一步研究需要关注的关键科学问题。

（3）特征矿物组合及有效地球化学记录的准确提取

大陆深俯冲形成的超高压岩石从俯冲到折返至地表的过程中，原岩性质和 *P-T* 条件的差异，导致不同矿物在不同阶段发生结晶和 / 或重结晶，俯冲和折返的不同阶段形成特定的矿物组合和地球化学成分特点。但是，这些超高压岩石在折返过程中经历了明显的退变质改造，使得很多前进和峰期变质矿物的记录仅仅保存于局部岩石或少数矿物包裹体中。特征变质矿物对元素的稳定性具有决定性影响，如石榴子石是重稀土元素的主要载体、金红石主要赋存 Ti-Nb-Ta 等。有效恢复大陆深俯冲超高压变质岩石不同变质矿物组合和成分特点，不但可以揭示这些岩石演化的精确历史和演化过程，还有助于理解俯冲过程中元素和同位素的迁移转化（Rubatto and Hermann，2003；Gao et al.，2011；Guo S et al.，2014）。关键科学问题包括：如何准确构建前进和峰期变质矿物组合，如何有效识别特征矿物的出现和消失及其影响，如何确定不同阶段出现矿物的成分和影响因素，如何限定关键矿物对特征元素迁移转化的影响和不同矿物组合对全岩成分的影响等。

（4）不同造山阶段岩石组合特征及其动力学意义

大陆深俯冲的超高压岩石在峰期变质后一般会经历两阶段折返过程（Zheng，2021d）。碰撞造山带中的高压－超高压变质岩在第一阶段会通过岩片逆冲方式折返，岩石在峰期变质之后出现等温降压作用，结果在下地壳深度发生高压麻粒岩相叠加。接下来这些岩石在第二阶段以穹窿式隆起方式折返，首先是造山带根部拆沉减薄，软流圈地幔上涌加热减薄的岩石圈地幔，结果上覆地壳发生脱水熔融引起混合岩化作用，所形成的富水溶液和含水熔体向上迁移，形成角闪岩相变质岩和长英质脉体，留下的属于低压麻粒岩（Wallis et al.，2005；Labrousse et al.，2011；Gordon et al.，2012；Zheng and Chen，2017，2021；Zheng and Zhao，2017；Zheng，2021b）。这个过程引起了造山带形成之后根部岩石的改造，混合岩化本身也会改变岩石的流变学性质，对造山带的演化具有非常重要的意义。

许多碰撞造山带出现的麻粒岩相变质作用和混合岩化作用都发生在超高压岩石折返到造山带下地壳之后，与碰撞造山过程本身没有直接关系，而与造山后山根垮塌过程有关。例如，大别造山带超高压变质岩经历了两阶段折返：第一阶段是深俯冲大陆地壳在弧下深度拆离后沿俯冲隧道以岩片逆冲的方式折返到地壳的不同深度，而后第二阶段是那些位于下地壳底部的岩石在造山带岩石圈减薄后发生部分熔融使其以穹隆侵出方式折返（Zheng et al.，2019a；Zheng，2021d）。显然，第一阶段的岩片逆冲式折返发生在晚三叠世，属于大陆碰撞的晚期阶段；第二阶段的穹隆侵出式折返发生在早白垩世，已经是超高压岩片折返到下地壳层位的约 100 Ma 之后（Wu et al.，2007）。关键科学问题包括：造山带折返时麻粒岩相叠加的时间和机制，超高压岩石多期深熔作用的时空分布特征及其与造山带演化的关系，混合岩与花岗岩形成和陆壳演化的关系，造山后山根垮塌、麻粒岩相变质和混合岩化的动力学过程与机制。

（5）精确限定大陆地壳深俯冲和折返阶段的时间

精确测定大陆深俯冲超高压岩石不同阶段的变质年龄，对于确定这些超高压岩石的俯冲折返历史及其动力学过程和演化机制具有非常重要的意义。目前，确定超高压变质岩石的峰期变质时代主要依靠矿物封闭温度较高的同位素体系，包括石榴子石和其他矿物的 Sm-Nd 和 Lu-Hf 同位素等时线法（Li et al.，1993，2000；Cheng et al.，2008），副矿物 U-Th-Pb 定年法等（Liu and Liou，2011）；确定后期退变质演化历史主要依靠多硅白云母和角闪石等矿物 Ar/Ar 定年法以及封闭温度较低的副矿物（如金红石和磷灰石）U-Th-Pb 定年方法等。结合多同位素体系的定年结果，可以很好地查明造山带各个阶段的构造热演化历史和动力学过程（Li S G et al.，1993，2000）。关键科学问题包括：变质过程中不同定年矿物特别是副矿物的差异性响应行为，不同同位素体系封闭温度的准确确定及其所得年龄与 *P-T* 关系的准确制约，如何确定前进和峰期变质作用不同阶段的年龄。

（6）全岩主微量元素及同位素组成变化及其控制因素

俯冲大陆地壳以长英质片麻岩为主，夹有少量榴辉岩和橄榄岩，它们具有与俯冲洋壳不一样的岩石组合和全岩组分（Zheng，2012；Hermann and Rubatto，2014）。深俯冲成因超高压岩石的全岩组成特别是榴辉岩成分可以很好地用来限定原岩性质，从而确定超高压变质地体的性质。但是，由于俯冲过程中受到变质作用特别是流体组分迁移等的影响，超高压岩石的元素和同位素组成会受到不同程度的改造（Zhao et al.，2007；Zheng and Hermann，2014），影响这些组分对原岩性质的限定。通过对在变质过程中稳定的原岩矿物（如锆石）和变质作用不活动元素（如重稀土元素和高场强元素等）的研究，可以进行原岩构造环境的有效甄别。关键科学问题包括：超高压变质作用如何影响变质岩原岩的元素和同位素组成，流体活动对超高压岩石的元素和同位素如何改造，如何获得超高压岩石准确的原岩元素和同位素组成，如何利用有效的岩石学和地球化学判别指标来准确识别榴辉岩的原岩特征并判断其构造环境，不同矿物和副矿物的生长和分解对变质岩成分的影响。

3. 重点研究方向

拟开展的重点研究方向包括：俯冲大陆地壳的俯冲深度极限和命运；深俯冲大陆地壳的 *P-T-t* 演化轨迹；大陆深俯冲和折返过程中的岩石组合变化特征及其对造山带演化的意义；深俯冲岩石不同阶段的矿物组合特点及其对俯

冲带物质循环的影响；深俯冲陆壳的岩石地球化学特征及其控制因素等方面。

4. 应对策略

通过岩石矿物学研究识别深俯冲大陆地壳的最大俯冲深度，通过实验岩石学研究揭示深俯冲大陆地壳的折返极限深度和俯冲极限深度，通过深源岩浆岩的研究有效识别深俯冲陆壳对地幔岩石的影响；通过数值模拟，确定大陆超深俯冲的动力来源。拓展传统岩石学 *P-T* 轨迹研究方法，使其能够适用于开放体系和动态变化过程的 *P-T* 轨迹研究；加强同位素年代学和岩石学的结合，进行岩石年代学研究，特别是加强不同阶段形成的岩石和副矿物定年与示踪，精细限定大陆深俯冲岩石各阶段的时间和流体活动；对大陆碰撞造山带中常常伴生的榴辉岩－麻粒岩－混合岩－花岗岩进行系统研究，确定不同岩石类型的成因、联系及其转化规律，为大陆碰撞带的生长－演化和消亡过程提供准确限定，并建立造山带变质作用和岩浆作用的联系。对不同变质条件下的岩石进行系统的岩石地球化学和同位素研究，限定陆壳深俯冲不同阶段岩石的元素和同位素特征，结合微区分析手段确定岩石元素和同位素特征的动态演化过程，更好地限定大陆深俯冲过程中的元素和同位素行为，并为俯冲带元素和同位素循环提供重要制约。

5. 预期目标

认识深俯冲陆壳的俯冲和折返极限，以及陆壳深俯冲过程中的壳幔相互作用过程；精确确定大陆深俯冲的 *P-T-t* 演化轨迹；深刻认识大陆深俯冲过程中的元素和同位素循环及其对壳幔相互作用的影响；发展板块构造理论，拓展其从大洋俯冲到大陆碰撞和碰撞后演化阶段。

（三）科学挑战 10：俯冲带镁铁质岩浆岩

1. 问题的提出

与俯冲带系统演化相关的最重要的岩石学问题之一是俯冲带岩浆的起源。对于俯冲带岩浆岩的研究最早可追溯到 19 世纪末期。20 世纪早期已有科学家对日本大陆边缘弧开展了火山岩岩石学的研究。20 世纪岩石学研究中一个最重要的进展，就是揭示玄武岩是来源于地幔的部分熔融产物，并揭示了水在岛弧玄武岩成因中的重要作用：不仅可以降低地幔橄榄岩的固相线温度，还可以改变熔体的成分（Tatsumi et al.，1983；Kushiro，1968）。

Taylor 和 White（1965）最早提出安山质岩浆形成于俯冲洋壳熔融，而并非来自高铝玄武质岩浆的分异或玄武质与流纹质岩浆的混合。该观点得到了实验岩石学结果的支持，即与地幔平衡的石英榴辉岩部分熔融可以产生原始安山质岩浆（Green and Ringwood，1968；Green，1972）。后来的研究认识到，虽然石英榴辉岩发生部分熔融可以产生安山质岩浆，但单独的低钾玄武质洋壳部分熔融不可能产生在大陆弧主体广泛出现的中–高钾安山质岩浆岩（Gill，1974；Ratajeski et al.，2005）。因此，一些研究提出，这种中–高钾安山质岩浆可能来自富集蚀变洋壳的部分熔融（Morris et al.，1990；Hernández-Uribe et al.，2019）。

另外，一些学者通过对大洋沉积物的 B 和 Be 同位素研究（Morris et al.，1990；Ishikawa and Nakamura，1994）以及对玄武岩橄榄石斑晶中熔体包裹体的 H_2O/K_2O 研究（Sobolev and Chaussidon，1996），提出大洋沉积物中的水参与了所有弧岩浆的形成，并且水在地幔楔中以独立的流体形式存在。随后对全球橄榄石斑晶中的熔体包裹体数据分析表明，俯冲带玄武岩原始熔体的平均含水量约为 1.7 wt%（Sobolev and Chaussidon，1996），最高可达 14 wt%（Krawczynski et al.，2012）。

实验岩石学研究结果进一步表明，蚀变洋壳和沉积物中的含水矿物（如角闪石、蛇纹石、绿泥石、绿帘石和硬柱石等）将随着俯冲深度的增加而依次逐渐脱水分解（Schmidt and Poli，1998），向地幔楔释放出水。由此一些研究者提出，来自俯冲板片的水流向较热的地幔楔后引发了橄榄岩部分熔融，产生的幔源熔体因充足的水可以不被地幔橄榄岩再平衡（Grove et al.，2002，2012）。另一种观点认为，北美喀斯喀特（Cascades）弧原始岩浆具有高的固相线温度 1350～1450 ℃（70 km 深），可能指示其来源于地幔楔的降压熔融（Lee et al.，2009）。

经过多年研究，一个基本的共识就是（图 4-15）：俯冲带镁铁质岩浆岩的形成与俯冲板片流体对弧下地幔的交代作用、壳幔相互作用以及弧下热演化密不可分（Zheng，2019）。作为俯冲岩浆体系中最重要的岩石类型，俯冲带镁铁质岩浆岩已经成为揭示俯冲带地幔富集机制、壳幔相互作用、物质循环以及深部动力学过程最重要的“岩石探针”之一（Zheng et al.，2020）。但是，有关流体性质与组成特征、弧下混杂岩形成与熔融过程、俯冲带镁铁质岩浆岩的形成机制等仍然存在激烈的争论。

另外，原始玄武质或安山质岩浆在上升过程中经历一系列壳内分异过程形成长英质岩石。这涉及两种类型的岩石学过程。第一种过程是部分熔融和岩浆混合：原始岩浆底侵到下地壳底部，形成热带（Annen et al.，2006），诱发下地壳物质发生部分熔融，并经历熔融–同化–储存–均一过程（即

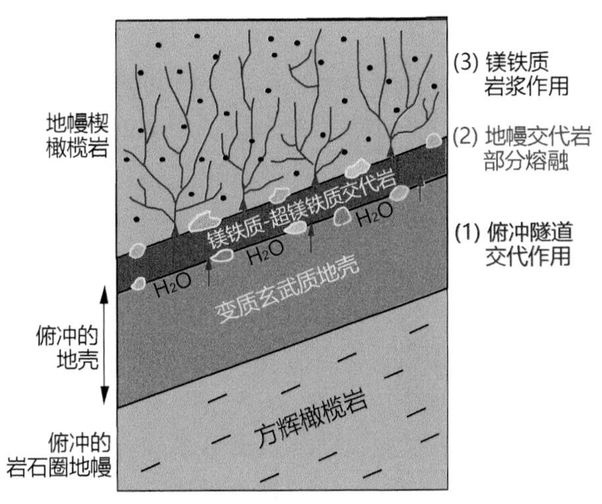

图 4-15 大陆俯冲带镁铁质岩浆岩形成过程中的地幔源区交代作用
（修改自郑永飞等，2015）

MASH 过程）（Hildreth and Moorbath，1988）。幔源岩浆和混合岩浆在不同地壳层次的岩浆房或在岩浆上升过程中发生分离结晶和同化混染，形成长英质岩石（Sisson et al.，1996）。实验岩石学研究表明，当下地壳含水镁铁质岩石在超过 1.0 GPa 发生部分熔融时，将产生长英质岩石，残留体的矿物组合以石榴子石 + 单斜辉石为主（榴辉岩相）（Saleeby et al.，2003）。第二种过程是岩浆分异和岩浆混合：在相对高的压力和含水量条件下（相对于低压下拉斑玄武岩结晶出橄榄石的条件），原始岩浆发生辉石 ± 石榴子石的结晶分离，形成堆晶高镁辉石岩（单斜辉石为主），产生低镁富铝派生岩浆（Lee et al.，2006）。此时由于高镁辉石岩的 SiO_2 与初始岩浆近似，该阶段的分异不会改变派生岩浆的 SiO_2 含量，仅使得其 MgO 显著降低。如果这些低镁富铝派生岩浆继续在较高压力下发生石榴子石 + 单斜辉石的结晶分离，则形成堆晶低镁辉石岩；如果这些派生岩浆侵位到中地壳（压力更低），则形成矿物组合为橄榄石 + 辉石 + 斜长石的堆晶辉长岩（Lee et al.，2006）。此时由于低 SiO_2 含量的石榴子石和橄榄石的结晶，派生岩浆将向高硅方向演化，形成长英质岩石。实验岩石学研究也表明，原始玄武质岩浆在 0.7GPa 条件下，通过结晶分离可以产生酸性钙碱性岩浆，矿物结晶分离顺序为橄榄石→单斜辉石→斜长石 + 尖晶石→斜方辉石 + 角闪石 +Fe-Ti 氧化物→磷灰石→石英 + 黑云母（Nandedkar et al.，2014）。

2. 面临的挑战

（1）不同类型俯冲带镁铁质岩浆岩形成的控制因素

大洋弧与大陆弧和陆缘岛弧在岩浆岩组合和地球化学特征上存在一系列异同点（徐义刚等，2020）。大洋弧岩浆岩以拉斑玄武岩和钙碱性玄武岩为主，伴有极少量的橄榄玄粗质玄武岩和少量安山岩-英安岩，有时还有特殊的高镁安山岩类和埃达克岩（Kelemen et al.，2003）。与此相比，大陆弧岩浆作用以安山质岩石为主体，陆缘岛弧以玄武质岩石为主体。不过，这三类弧岩浆岩都具有典型的岛弧型微量元素分布特征，其中大洋弧玄武岩不太富集不相容元素和放射成因同位素。另外，一些富 Nb 玄武岩和洋岛型玄武岩也在俯冲带有所出现，指示其地幔源区可能受到了深俯冲洋壳衍生熔体的交代（Wang et al.，2007；Zheng，2019；徐义刚等，2020）。

大陆俯冲带岩浆作用通常缺乏同俯冲岩浆岩，发育大量碰撞后花岗质岩石，含有少量钾质和高钾钙碱性镁铁质岩。碰撞后镁铁质岩大致分为两类（图 4-12）：一类具有洋岛型微量元素分布和亏损的放射性成因同位素，记录了先前俯冲古洋壳物质再循环；另一类具有岛弧型微量元素和富集的放射性成因同位素，来自大陆地壳熔体交代的地幔源区（Zhao et al.，2013；Zheng，2019；许文良等，2020）。

大洋和大陆俯冲带出露的镁铁质岩浆岩，在岩石类型和成分上均存在一定差异，其受控于诸多因素，如俯冲带温压结构、俯冲板片和上覆地幔楔的成分、上覆岩石圈的厚度、板片在不同深度析出流体的成分和比例、地幔部分熔融深度和程度、岩浆上升过程中的演化过程等。

（2）俯冲带地幔熔融与镁铁质岩浆的产生、演化过程

地幔橄榄岩精确的湿固相线温度的确立对揭示弧玄武质岩浆的产生机制非常关键。但是，不同实验得到的橄榄岩湿固相线温度差别很大，从低可达约 800℃到高可达 1000～1100℃（Grove et al.，2009；Green et al.，2012；Till et al.，2012）。这种固相线温度的巨大差异会影响对岛弧玄武质岩浆产生机制的认识，包括地幔熔融深度、地幔熔融机制、地幔熔融的合适时间等（Till et al.，2012；郑永飞等，2016）。因此，精确确定不同温压下地幔橄榄岩湿固相线的温度是一个亟待解决的重要科学问题。

计算地球动力学研究提出了混杂岩熔融形成弧岩浆的新模型（Marschall and Schumacher，2012）。该混杂岩熔融模型只是从地球化学成分上论证了二元混合的可行性，但是没有从地球动力学上说明地幔楔如何在超高压混杂岩

未发生部分熔融的情况下就产生裂隙允许固态物质进入地幔楔。此外，大洋俯冲带弧后拉张与俯冲板片回卷有关，这时一方面板片－地幔楔界面受到侧向流入的软流圈地幔加热发生部分熔融，另一方面回卷板片上方产生拉张环境引起上覆板块拉张。但是，混杂岩熔融模型没有说明在板片回卷之时那些超高压混杂岩是否还能保持固体状态。因此，混杂岩熔融模型能否成立仍然值得推敲。

弧系统玄武质岩浆的产生与弧下深度动力学过程、热演化密切相关。除了板片回卷和软流圈地幔角流之外，板片撕裂、板片断离或岩石圈拆沉、扩张洋脊俯冲、大洋高原俯冲等过程在弧岩浆岩的形成中也发挥了一定的作用（王强等，2020；徐义刚等，2020）。但是，在实际的研究中，要识别这些深部动力学、热演化过程，仍旧十分困难。

镁铁质岩浆产生后，其在上升穿过地幔、进入地壳并在最终喷出地表的过程中，要经历一系列的演化过程，如幔源玄武质岩浆作用上升途中发生结晶分异－同化地壳或熔融－同化－存储－均一过程。地壳内的岩浆储库大部分时间以晶粥体形式存在，而非传统认为的富熔体相岩浆房（Bachmann and Bergantz，2008；Cashman et al.，2017）。因此，针对喷出到地表玄武岩的成分演化，需要综合考虑地幔源区和喷发前岩浆的演化过程。

（3）俯冲带流体对地幔的交代作用

板片释放的流体交代上覆地幔楔曾被认为是形成弧岩浆作用的重要机制。板片流体的加入可以降低地幔橄榄岩的固相线温度，但立即引发地幔楔熔融的传统认识与大洋俯冲带温压结构研究结果相矛盾（Manning，2004；Grove et al.，2009；Syracuse et al.，2010；郑永飞等，2016）。这意味着，地幔楔不会因为富水溶液的加入立刻发生部分熔融，而需要后期软流圈地幔的进一步加热才会熔融形成镁铁质熔体（Zheng，2019）。根据现代大洋俯冲带温压结构随时间变化的性质，更为可行的机制是俯冲洋壳在弧下深度脱水交代地幔楔之后，俯冲板片回卷使地幔楔底部加热才发生部分熔融（Zheng and Chen，2016）。软流圈地幔的降压熔融可以出现在上覆岩石圈减薄的弧后伸展背景下，结果产生弧后盆地洋脊型玄武岩；如果是火山弧岩石圈减薄引起弧后盆地打开，结果就是岛弧型玄武岩与洋脊型玄武岩的先后产出。

洋壳深俯冲还可以产生硅酸盐熔体，熔体－橄榄岩反应形成的地幔交代岩发生部分熔融形成俯冲带镁铁质岩浆岩。含水熔体相比富水溶液，携带熔体活动性元素甚至高场强元素的能力更强。越来越多的研究也发现，俯冲带中富 Nb 玄武岩和洋岛型玄武岩的源区具有洋壳熔体的贡献（Wang et al.，

2007; Zheng，2019; Sobolev et al.，2007; Zheng et al.，2020)。

除了富水溶液、硅酸盐熔体外，含 CO_2 流体或碳酸盐熔体对地幔也有交代作用，其对弧下地幔交代形成弧岩浆岩的源区（Zheng，2012，2019)。然而，学界对俯冲板片的脱碳机制以及对弧玄武岩的贡献还存在较大争议（Kepezhinskas and Defant，1996; Blundy et al.，2010; Sano and Williams，1996; Dasgupta，2013; Thomson et al.，2016)。因此，弧下含水或 CO_2 流体、硅酸盐或碳酸盐熔体对地幔的交代作用及其与物质循环的关系仍然是一个重要的前沿科学问题。

俯冲板片还可能在高压环境下形成超临界流体（Kessel et al.，2005; Kawamoto et al.，2012)。超临界流体具有比富水溶液和含水熔体高得多的元素迁移能力，甚至能够迁移高场强元素，对俯冲带镁铁质岩浆岩的形成具有重要意义（Zheng et al.，2011; Zheng，2019)。然而，Kessel 等（2005）确定的超临界流体产生压力大于 6.0 GPa，因此俯冲带超临界流体在向上运移过程中很可能超出其稳定温压范围，只能瞬时存在。厘清俯冲带超临界流体与镁铁质岩浆岩的成因联系，将是今后重要研究方向之一。

大陆俯冲带普遍缺乏同俯冲弧岩浆作用，过去认为是由于陆壳相对较干、缺乏俯冲脱水作用（Rumble et al.，2003; Zheng et al.，2003)。然而，俯冲陆壳也含有相当含量的含水矿物，如角闪石、多硅白云母和硬柱石（Zheng，2009; Zheng and Hermann，2014)，因此另一种解释是陆壳流体交代地幔后，大陆俯冲带的低地热梯度不能使地幔楔熔融形成弧岩浆作用（Zheng and Chen，2016)。大陆俯冲带的另一特点是发育大量碰撞后岩浆岩，其中绝大多数镁铁质岩具有岛弧型微量元素分布型式和富集的放射性成因同位素组成，记录了俯冲 / 折返时陆壳衍生熔体的交代作用（Zhao et al.，2013；许文良等，2020)。在大陆俯冲带也发现少量镁铁质岩具有洋岛型微量元素分布型式和亏损的放射性成因同位素组成，记录了先前俯冲古洋壳衍生熔体的交代作用（Zheng，2019)。对于大陆俯冲带之上的地幔楔，曾经历了先前古洋壳到随后大陆地壳来源的不同性质流体的叠加交代，由此部分熔融所产生的镁铁质岩浆岩会表现出什么特点？今后研究中需要进一步识别和区分。

（4）俯冲带镁铁质岩浆作用与大陆地壳形成和演化

大陆地壳的形成和演化是通过地壳增生和再造来完成的，这也一直是国际地学界长期关注的重大基础科学问题。大陆地壳的许多微量元素地球化学特征与俯冲带岩浆岩十分相似（Rudnick and Gao，2003)，这使得众多的研究者认为大陆地壳产生于大洋弧和大陆弧环境（Rudnick，1995；Tang et al.，2017），并逐渐演变为成熟的大陆地壳。尽管俯冲带通常被认为是陆壳形成的主要场所，但是以玄武质组分为主的弧岩浆无法直接解释以花岗岩为主的上地壳（Cawood et al.，2013）和全球陆壳安山质成分特征（Rudnick and Gao，2003），这也被称为陆壳成分悖论（Rudnick，1995）。由于构成地壳的长英质岩石无法与地幔达成化学平衡，因此解决陆壳成分悖论的主要方式有两类。一是通过幔源弧岩浆或弧地壳的演化和改造来形成安山质成分，如分离结晶和基性 - 超基性弧根的拆离、沉积物的大量底辟、混杂岩的熔融和化学风化等。二是寻找替代或补充弧岩浆外其他能够形成大规模安山质岩浆的源区和动力学环境（Niu et al.，2013；Zhu et al.，2019），如大陆碰撞或后碰撞造山环境。但是，上述两类模式仍存在激烈争论。

尽管大洋俯冲、大陆碰撞带的大陆地壳生长与演化被广泛关注，但是一些重要的问题仍然没有很好地解决，主要有三个方面。①地壳垂向生长机制。弧环境下岩浆岩成分往往呈周期性变化，这可能代表了大陆地壳垂向生长的非均一性（Tang et al.，2017）。但是，导致这种周期性垂向地壳生长的机制究竟是俯冲大洋板片后撤或者洋中脊俯冲作用，还是造山带岩石圈拆沉作用，存在不同的认识。这种弧周期性垂向地壳生长的关键控制因素是与全球板块动力学过程有关，还是受区域的构造动力学过程所主导，并没有定论。②大陆地壳净增生长量的估计。大陆地壳净增生长量的估计是俯冲带研究的一个难点。一方面，由于岛弧往往被沉积物覆盖或者经历了强烈的剥蚀，很难确定岛弧岩浆岩的时空分布，尤其是火山岩产生的量很难确定。另一方面，计算新生地壳的比例，需要精确确定岩浆岩的成因，尤其是长英质岩浆岩中新生地壳的比例。由于俯冲带兼具较大的地壳生长速率和破坏速率（Hacker et al.，2011），俯冲带地壳净增生长量也被认为可能极为有限（Niu et al.，2013）。但是，具有低侵蚀速率的大陆碰撞带是地壳再造的场所，不是陆壳净生长的地方（郑永飞等，2015）。因此，俯冲带和碰撞带大陆净生长量的正确估计仍然是个亟待解决的难点问题。③太古宙大陆地壳生长与板块构造启动。大量的研究将太古宙的岩浆岩同现代弧岩浆岩（玄武岩、高镁安山岩、玻安岩、埃达克岩等）进行对比，提出板块构造活动（即洋壳俯冲）在太古宙已经出现，其启动时间争议巨大，大陆地壳的形成与大洋岩石圈的俯冲有关（王强等，2020；徐义刚等，2020）。但是，上述研究推论还需要大地构造、沉积岩和变质岩等其他证据的支持。因此，大洋岩石圈俯冲与太古宙板块构造的启动、大陆地壳的生长，仍旧是当前国际地球科学界尚未解决的一个热点问题。

3. 重点研究方向

1）不同类型俯冲带镁铁质岩浆岩形成的控制因素：不同类型俯冲带镁铁质岩浆岩时空分布及其控制因素；俯冲带镁铁质岩浆岩特征及其源区物质的识别。

2）俯冲带地幔熔融与镁铁质岩浆的产生、演化过程：不同压力下地幔橄榄岩湿固相线温度的精确确定；混杂岩的熔融与弧岩浆岩的产生；俯冲带镁铁质岩浆产生的动力学机制及岩浆储库的形成与演化。

3）俯冲带流体对地幔的交代作用:多同位素体系联合示踪岩浆源区组成；利用矿物及流体包裹体的原位分析识别原始岩浆成分及源区组成；利用天然样品结合高温高压实验来约束地幔交代作用；弧下地幔交代作用及弧岩浆源区形成。

4）俯冲带岩浆岩的形成与大陆形成、演化：俯冲带和碰撞带大陆地壳生长机制；俯冲带和碰撞带大陆地壳再造机制；俯冲带和碰撞带大陆地壳净增生长量的估计；太古宙大陆地壳生长与板块构造的启动。

4. 应对策略

要深入研究俯冲带镁铁质岩浆作用，应对的策略包括：①加强国际合作，选取现代大洋俯冲带（如环太平洋和澳大利亚－东南亚之间的印度洋新生代大洋俯冲带）和特提斯－青藏高原碰撞带岩浆岩特别是镁铁质岩浆岩作为重要研究对象，选取典型区域进行解剖，建立新生代不同类型俯冲带的镁铁质岩浆岩时空分布规律和岩石成因模型；②加强古俯冲带与现代俯冲带之间的镁铁质岩石的对比研究，揭示古俯冲带的镁铁质岩石的成因及形成机制；③开展多学科交叉，除了对天然样品进行矿物学、岩石学、地球化学和年代学研究外，还需要结合地质学、地球物理资料、高温高压实验和数值模拟，特别需要高温高压和数值模拟来研究板片流体、熔体组分的性质与特点及其对地幔的交代作用；④采用大数据结合岩浆岩的时空分布和体量开展综合研究，揭示镁铁质岩浆形成的深部动力学过程。

5. 预期目标

预期目标主要包括：①查明不同类型俯冲带镁铁质岩浆岩时空分布及其控制因素，阐明俯冲带镁铁质岩浆岩特征及其源区物质组成；②精确确定不同压力下地幔橄榄岩湿固相线温度，建立混杂岩的熔融产生与弧岩浆岩模型，揭示俯冲带镁铁质岩浆产生的动力学机制以及弧镁铁质岩浆储库的形成与演化；③揭示俯冲板片的流体、熔体组分特征及其起源，应用高温高压实验和数值模拟来约束俯冲板片流体、熔体特征，探究弧下地幔交代作用、弧岩浆源区形成及其动力学过程；④揭示俯冲带和碰撞带大陆地壳生长、再造机制，估算俯冲带和碰撞带大陆地壳净增生长量，探究太古宙大陆地壳生长与板块构造的启动可能性。

（四）科学挑战 11：俯冲带长英质岩浆岩

1. 问题的提出

长英质岩石构成了大陆上地壳的主体，是造成大陆地壳平均成分为安山质到英安质（Rudnick and Gao，2003；Hacker et al.，2011）的重要原因，这也是地球有别于太阳系其他行星的独特特征。因此，大陆地壳如何形成一直是国际学术界长期关注的重大基础科学问题（Lee et al.，2007；Kelemen and Behn，2016）。早期研究认为，大陆地壳主要形成于大洋弧和随后的弧陆碰撞以及大陆弧岩浆作用，弧岩浆过程对大陆地壳的形成发挥了关键作用（Taylor and McLennan，1995）。但是，大洋弧环境下的大洋地壳成分是玄武质，明显不同于安山质到英安质的平均大陆地壳成分。

根据对全球代表性俯冲带的深入研究，尤其是基于对出露非常完好的大陆弧地壳剖面详细的野外地质调查、岩石学和地球化学工作，已经分别提出了拆沉模型和刮垫模型来解释上述不一致性。拆沉模型（图 4-16）认为，在大陆弧下地壳底部，幔源玄武质岩浆发生结晶分异形成堆晶岩，因其密度大，拆沉再循环进入地幔，导致大陆弧从镁铁质地壳转换为长英质大陆地壳（Kay and Kay，1991；Ducea et al.，2015a）。刮垫模型（图 4-17）强调，在沉积物俯冲、俯冲侵蚀、大洋弧俯冲和大陆俯冲过程中，镁铁质岩石转变成榴辉岩沉入地幔，长英质岩石转化为片麻岩进入地幔楔顶部的莫霍面附近形成长英质成分的大陆地壳，导致镁铁质地壳转换为长英质地壳（Hacker et al.，2011；Kelemen and Behn，2016；Maierova et al.，2018）。

很早就有学者将长英质岩石（尤其是花岗岩类）与造山带过程联系起来，并根据已知不同构造背景花岗岩的微量元素特征区分为火山弧、同碰撞、板内、碰撞后等不同类型（Pearce et al.，1984；Harris et al.，1986）。俯冲板片部分熔融产生含水熔体仅仅发生在暖俯冲条件下。这种情况在显生宙相对稀少，这也是环太平洋俯冲带只有极少数地方出现埃达克岩的基本原因。但是，在太古宙时期，板片俯冲基本上是在中等地热梯度下（Zheng and Zhao，2020），这样汇聚板块边缘加厚洋壳的部分熔融就能够产生成分上与埃

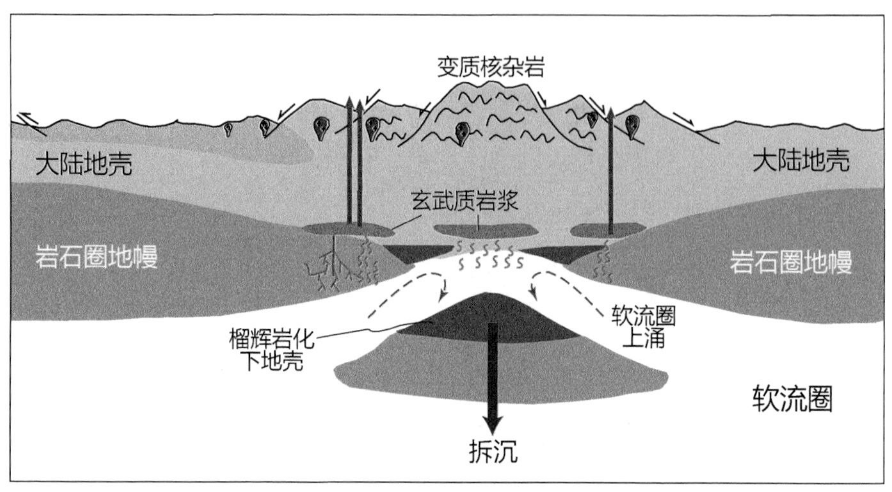

图 4-16　大陆弧岩石圈镁铁质－超镁铁质堆晶下地壳拆沉模型

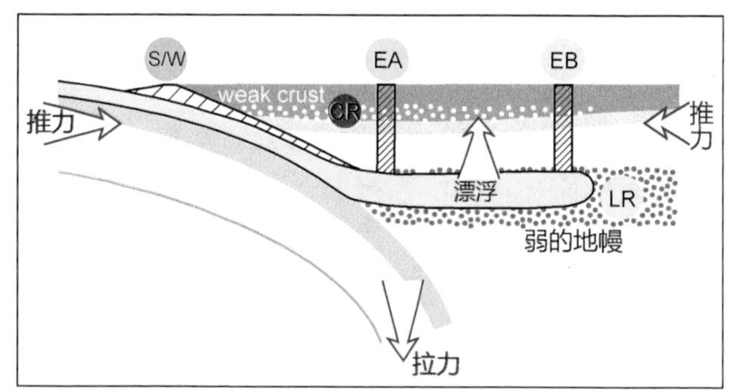

图 4-17　大陆碰撞带长英质地壳底辟进入莫霍面位置的刮垫模型
（修改自 Maierova et al.，2018）
weak crust- 流变强度较弱的地壳; LR- 岩石圈之下刮垫; CR- 壳内刮垫; S/W- 俯冲隧道和增生楔; EA- 大陆弧位置折返; EB- 弧后位置折返

达克岩相似的英云闪长岩－奥长花岗岩－花岗闪长岩（Tonalite-Trondhjemite-Granoditite，TTG）（Zheng，2021b）。由于板片断离，上覆地幔和岩石圈因减压和升温发生大规模熔融，形成大规模岩浆活动。在进入碰撞后张裂造山作用阶段，岩石圈减薄和山根垮塌过程中软流圈地幔上涌引发造山带地壳发生部分熔融，从而形成广泛的长英质岩浆岩（Zheng and Zhao，2017; Zhao et al.，2017a，2017b）。该机制可能是汇聚板块边缘长英质岩浆岩形成的主导机制（Zheng and Gao，2021）。

根据板块运动的主要驱动力（即俯冲大洋板片下沉引起的板片拖曳力）

和岩浆产生的三种主要机制（即加水、升温和降压），有研究将印度－亚洲大陆碰撞带上盘冈底斯造山带岩浆作用表达为四个阶段（Zhu D C et al., 2016, 2019；朱弟成等，2017）。①特提斯洋壳俯冲期，岩浆作用以俯冲大洋板片衍生熔体交代地幔楔，地幔楔中玄武质交代岩部分熔融形成的正常钙碱性安山质岩浆为主。②初始大陆碰撞期，这时因仍存在洋壳俯冲，岩浆活动类似于大洋板片俯冲期，仍然以正常钙碱性安山质岩浆为主。③同碰撞期，包括两种情况：一是在俯冲大洋板片回卷引起的伸展背景下，地幔楔之下的软流圈地幔对流增强，板片脱水带向海沟方向扩展，地幔楔继续受到交代作用并向海沟迁移；二是在峰期碰撞挤压后的伸展条件下，岩石圈发生降压熔融形成碱性岩浆作用，其中包括小规模壳源过铝质岩浆作用。④碰撞后，同样包括两种情况：一是俯冲板片在洋陆板片过渡带发生断离，结果俯冲大洋板片的拖曳力完全消失（Duretz et al., 2012），大陆俯冲/碰撞作用急剧减弱甚至停止，软流圈上涌引起断离板片边缘部分熔融；二是大陆碰撞带加厚岩石圈发生拆沉减薄作用，引起软流圈上涌加热减薄岩石圈之上的地壳部分熔融，俯冲带转入陆内演化阶段。

2. 面临的挑战

（1）俯冲带长英质岩石的源区

俯冲带长英质岩石有俯冲板片、俯冲沉积物和上覆地壳等多种可能的来源，在大陆弧则与地幔楔来源原始玄武质或安山质岩浆的结晶分异有关。前者对应于碰撞造山带，后者对应于增生造山带（Zheng, 2021b）。面临的挑战是，如何准确识别这些造山带的属性及其可能的源区，以及不同源区的相对贡献大小，尤其是地壳沉积物的贡献。

（2）俯冲带长英质岩石的成因

全球岩浆岩的全岩地球化学数据显示，俯冲带岩浆岩的分异趋势（从原始的玄武质到长英质）无法区分，而大陆张裂带岩浆岩的成分是分散的。对于大陆弧来说，玄武质岩浆结晶分离作用是长英质岩石产生的主要机制，作为地壳分异残余物的镁铁质堆晶岩，在成分上与被抽取熔体是互补的，通过结晶分离和随后镁铁质堆晶岩的丢失（拆沉）产生安山质地壳是可能的（Keller et al., 2015）。在这种背景下，面对的挑战是，俯冲带长英质岩石究竟是由源区地壳物质经历部分熔融而来，还是由原始幔源岩浆经历结晶分离作用产生？这些原始岩浆究竟是安山质的还是玄武质的？

（3）俯冲带长英质岩石与造山过程之间的关系

如何界定地质历史上的同碰撞、碰撞后等造山过程，学术界并未取得一致认识（Liegeois，1998；Zheng et al.，2019a；许文良等，2020）。同时，长英质岩石的地球化学成分主要取决于源岩的矿物组成、化学成分和熔融时的物理化学条件（包括温度、压力和挥发分）以及随后的岩浆过程（如结晶分离、岩浆混合和同化混染），在多数情况下与碰撞造山过程的关系不大。因此，要建立长英质岩石与造山带过程之间的联系，仍然是一个极具挑战性的课题。

（4）俯冲带大陆地壳的形成机制

岩浆弧根部高密度下地壳的拆沉模型，一直是解释大陆地壳从镁铁质演化到长英质平均成分的主流模型。但近年来一些学者发现，即使发生了下地壳拆沉作用，相同深度大陆弧地壳的微量元素含量仍然明显不同于大陆内部地壳。这种差异性促使了刮垫模型的提出。但有关刮垫作用观点的两个最重要例子，即印度大陆上地壳物质底垫到拉萨地体之下（Chemenda et al.，2000）和松潘-甘孜三叠系复理石在侏罗纪沿金沙江缝合带逆冲到羌塘地体之下的解释（Kapp et al.，2003），并未得到现今地球物理结果（Gao et al.，2016）和后续地质研究的支持（Zhu et al.，2013）。因此，究竟是哪种或哪些机制控制了俯冲带大陆地壳的形成，仍然有待于进一步研究。

3. 重点研究方向和应对策略

（1）俯冲带长英质岩石的源区

确定俯冲带长英质岩石的岩浆源区，是精确限定其形成过程和形成机制的前提。这需要在传统岩石学和地球化学研究的基础上，结合新的测试分析技术获取诊断性的地球化学指标（如锆石O同位素、全岩或单矿物K同位素、B和Be同位素等）进行联合示踪。

（2）俯冲带长英质岩石的成因

准确约束俯冲带长英质岩石成因的关键在于，识别不同时期的岩浆源区和岩浆作用过程。面临的挑战在于，能否根据岩相学特征和地球化学成分演变趋势准确鉴别矿物的结晶顺序，能否利用矿物-熔体平衡法恢复原始岩浆成分并模拟矿物结晶顺序，能否有效鉴别部分熔融、结晶分离、岩浆混合、同化混染等不同过程在长英质岩石形成过程中的作用，能否通过野外观察、地球化学和实验岩石学查明岩浆弧下地壳不同构造演化阶段的部分熔融（或结晶分离）条件、部分熔融程度（或分离结晶程度）和熔体成分。建立变质-深熔-岩浆作用之间的联系，将为揭示长英质弧岩浆岩的成因提供关键约束。

（3）俯冲带长英质岩石与造山带过程之间的关系

岩浆的产生和演化是地幔动力学和壳幔相互作用的自然结果，造山带过程与深部地幔动力学密切相关。因此，在理论上将岩浆成分与造山带过程联系起来是可能的。综合考虑俯冲带板块运动驱动力、岩浆产生机制和岩浆作用过程（部分熔融、结晶分离和同化混染等）等多种因素，以岩浆成分的时间演化为主线，对构造环境已知的长英质岩石进行大数据分析，可能是未来将俯冲带长英质岩石与造山带过程关联起来的努力方向和有效手段。

（4）俯冲带大陆地壳的形成机制

镁铁质岩石的拆沉作用和长英质岩石的刮垫作用都可以解释部分观察现象，如拆沉作用可以很好地解释北美科迪勒拉岩浆弧的长英质地壳成分，而刮垫作用形成的下地壳与地球物理数据吻合。无论哪种机制，都会改变大陆下地壳的成分和结构（如地壳厚度）。因此，能否查明岩浆弧的岩性剖面结构和成分特征（尤其是下地壳岩性和物质组成），能否找到两种机制的确切证据、论证两种机制的效率和发生条件，或者提出其他可能的机制等，可能是这一主题未来需要关注的重要研究方向。对北美科迪勒拉活动大陆边缘岩浆弧的研究表明，其岩浆岩的平均成分是安山质到英安质的（Ducea et al.，2015a），非常类似于平均大陆地壳成分（Rudnick and Gao，2003；Hacker et al.，2011）。因此，要解决俯冲带大陆地壳的形成机制问题，最有效的应对策略是聚焦俯冲带大规模长英质岩浆活动的成因。

4. 预期目标

准确限定俯冲带长英质岩石的源区，查明俯冲带长英质岩石的成因，特别是不一致熔融对岩浆成分变化的贡献，建立俯冲带长英质岩石与造山带过程之间的成因联系，提出能够经得起地质学、地球化学和地球物理学检验的俯冲带大陆地壳形成理论。

（五）科学挑战 12：俯冲带热液矿床

1. 问题的提出

矿床学研究的核心内容是揭示成矿机制并建立成矿模式，其对找矿勘查的指导作用也是立足于建立准确成矿模式的基础之上。经过国内外矿床学家百余年的努力，已经建立了覆盖各个矿种的数十种重要成矿模式（陈华勇和吴超，2020；王瑞等，2020），包括俯冲带成矿中最为重要的斑岩铜矿模式（图 1-4 和图 3-10）。这些模式在相应找矿勘查工作中也起到了巨大的推动作用。

对特提斯俯冲带热液成矿机制，Zheng 等（2019b）根据张洪瑞和侯增谦（2018）对欧亚大陆南缘新生代热液矿床时空分布特点及其形成条件的研究，将金属成矿元素的迁移富集概括为三个阶段：①大洋俯冲阶段，洋壳玄武岩和沉积物在弧下深度脱水交代上覆地幔楔，在高氧逸度情况下使金属从硫化物中释放出来进入地幔楔得到初步富集；②大陆碰撞阶段，大陆俯冲/碰撞引起上盘构造拉张，地幔楔在弧下深度发生部分熔融形成镁铁质熔体，其中水和成矿元素得到进一步富集，然后上升侵位到地壳不同层位，其中上升到浅部地壳的岩浆结晶分异就会形成同碰撞热液矿床；③碰撞后再造阶段，位于下地壳层位的镁铁质岩石在拉张背景下发生部分熔融产生长英质熔体，在高氧逸度下经过充分的结晶分异使水和成矿金属元素进入高分异的长英质岩浆，结果在花岗斑岩中发生成矿作用。

目前，国内外研究相对成熟的斑岩型铜矿床，已经基本确立了俯冲和碰撞两种不同构造背景的成矿体系，揭示了不同环境下斑岩铜矿的成因机制（Sillitoe，2010；侯增谦等，2012；Zheng et al.，2019b）。然而，对于斑岩铜矿的成矿实验模拟、斑岩型矿床后期保存改造以及成矿宏观控制要素确定（如地壳性质、物质来源、深部过程等）等重要科学问题的研究还相对欠缺，而这些方面的创新研究将会显著拓展和完善已有的斑岩铜矿成矿模式。因此，对于如何将成矿实验模拟、成矿精细过程刻画、矿床后期保存、矿床类型、尺度、成矿规律（控制因素）等多维度结合，综合研究和丰富完善已有的成矿模式（包括可能发现新的成矿模式）将是未来俯冲带热液成矿作用研究最为关键的科学问题。

矿床尺度成矿模式的确立一般是建立在典型矿床解剖的基础上，主要是通过矿床地质特征、成矿流体演化、构造岩浆控矿等几个方面的研究工作来限定成矿动力－物质来源和成矿过程，从而建立相应矿床或成矿类型的成矿模式。成矿带（或成矿域）尺度成矿模式的建立多集中于不同矿床地质和地球化学特征对比与总结，从而得出具有鲜明地域特色的区域成矿模式。这些成矿模式的建立工作在21世纪初达到高峰，近年来主要是围绕已有成矿模式进行进一步精细化工作，从而加强对成矿机制的认识。

2. 面临的挑战

（1）缺乏成矿作用实验模拟系统研究

作为地质过程恢复和探索的重要手段，实验模拟已经在岩石学和地球动力学等众多地质分支领域得到广泛应用，然而对于重要成矿作用的反演和机理分析却鲜见具有针对性的实验模拟工作。即使对于俯冲带中最为重要的斑岩型矿床，尽管通过天然样品的获得已经对成矿热液出溶和热液蚀变矿化过程有了较为全面的认识，但这些过程的具体物理化学参数却一直未能准确限定，一些重要的地质现象得不到合理解释（如钠化与钾化的关系；钾化主成矿与绿泥石-绢云母主成矿差异原因等），这主要与成矿作用的实验模拟要求较高相关，如对于实验母体的选择要求高（特殊热液配置等）、实验条件的相对苛刻（反应时间普遍较长，尤其是中低温实验）、实验产物较难分析（较难形成新的矿物相，不能原位实时测试等）。

（2）较难进行成矿过程的精细厘定

成矿过程研究是矿床解剖研究的核心内容，也是揭示成矿机制的关键。虽然已有的矿物微区微量元素分析、流体包裹体温度与成分测定、同位素定年和同位素示踪等多种手段，已经对成矿过程的精细研究起到极大的推动作用，但对于俯冲带矿床中（如斑岩、夕卡岩和 VMS 矿床）经常出现的“瞬时成矿爆发”与“多期叠加成矿”等重要成矿作用的成矿机制依然缺乏清晰的认识。这要求在未来研究中多利用更新的方法手段（如硫化物高精度 Re-Os 定年、单颗粒锆石高精度定年、NanoSIMS 成分测试等）以及更多的学科交叉融合（如结合材料学方法进行成矿元素在矿物中的赋存形式和富集机制研究）来进一步实现成矿过程的高精度厘定，从而为成矿机制的准确认识和成矿模式的完善提供支撑。

（3）长期忽略成矿后保存机制研究

作为矿床特征中的重要一部分，矿体在形成后的保存和变化等过程的揭示对认识已有矿床的“真实面目”，修订已建立的“成矿模式”，以及对该地区相似矿床的勘查都具有非常重要的意义。然而，这种“非传统”的“成矿后”研究工作却长期受到忽视。作为俯冲带最为典型的斑岩型矿床，形成深度一般较浅（3～5 km）且极易剥蚀，但不同地质时代的俯冲带中却大量存在（如古生代中亚造山带和中-新生代特提斯造山带等）。显然，这些斑岩型矿床形成后的保存和变化机制一定有其“独特之处”，这使得斑岩型矿床得以保存，并常以暴露于地表的方式被人类发现利用。

（4）缺少大尺度控矿因素深入分析

对于斑岩型矿床，虽然已有成矿机制相对清晰，成矿模式相对成熟，然而，对于成矿物质来源（如 Cu-Au-Mo-S 的来源与分异）、成矿构造控制（如挤压与拉张环境的控矿机制）、成矿母岩岩石成因（如埃达克岩对成矿的控制）、成矿动力学背景（如洋脊俯冲对成矿的制约机制）等诸多成矿规律方面的认识还相对模糊，并长期存在争论。这不仅体现在成矿类型与尺度上，对于区域不同矿床类型综合成矿规律也缺乏深入的认识。

3. 重点研究方向

1）成矿作用实验模拟：如何克服实验技术困难，选择重要成矿类型中关键成矿过程进行巧妙的实验设计，并尝试在已有的成矿模式和成矿机理推导中增加相对准确的物理化学反应参数，将是未来俯冲带成矿研究中最为基础和非常关键的一环。

2）成矿过程精细厘定：未来，应利用多种方法综合的方式（如利用低温年代学中的裂变径迹和 U-Th-He 定年法，并结合区域地质分析），大力加强俯冲带热液矿床保存和变化的研究工作，揭示俯冲带中矿床能被保存并分布于合理开采深度的机制，并同时为矿产勘查提供重要指示。

3）成矿后保存机制：如何利用多种技术手段，并与地质因素结合，对矿床保存的特殊条件进行深入分析，从而进一步揭示矿床保存机制和区域成矿规律，并为勘查应用提供科学指导。

4）大尺度控矿机制：在未来俯冲带热液成矿研究中，需要与俯冲带其他领域研究成果紧密结合，并要重视利用大数据和人工智能手段进行大尺度控矿规律的探索。

4. 应对策略

成矿模式的理论创新研究，首先需要加强细致的矿床解剖工作，这也是目前国内矿床学者较为忽视的一点，已有成矿模式的建立无不是来源于准确精细的矿床解剖研究。只有准确精细的矿床解剖研究才能进一步校正已有的成矿模式，并为发现和建立新的成矿模式提供可能。在这个核心工作的基础上，研究人员应该突破传统矿床研究的界限，从不同维度围绕成矿机制进行思考，由此拓展和完善已有成矿模式。例如，对于很多俯冲带中重要的成矿系统，如斑岩型铜矿、VMS 型矿床、浅成低温型矿床等，虽然已有很多精细的矿床解剖工作且建立了相对成熟的成矿模式，但成矿相关的实验模拟工作却非常缺乏，导致很多成矿理论的建立只能通过其他学科的简单实验体系来大致推导，降低了矿床学成矿理论研究的科学基础性。再如，当前矿床学研究往往注重于成矿机制研究，成矿模式中也很少涉及矿床形成后的保存和改造作用，而“源、运、储、变、保”是矿床“来世今生”的全过程，缺一不可。

在汇聚板块边缘产有大量热液矿床，不仅种类多而且储量大、品位高。一般认为，这些矿床形成于增生造山带，往往忽视了碰撞造山带的成矿潜力。出现这种认识偏差的原因主要在于，前人对碰撞造山带与增生造山带之间在结构和成分的继承和叠加关系缺乏分辨，对大陆碰撞带下盘与上盘在板块汇聚过程中的地位缺乏认识。例如，在印度－亚洲大陆碰撞带，下盘是喜马拉雅碰撞造山带，上盘是冈底斯增生造山带，两者在新特提斯洋关闭过程中处于不同的构造背景，成矿元素在其中的迁移富集具有不同路径和机制。

为了澄清汇聚板块边缘造山作用与成矿作用之间的时空关系，Zheng 等（2019b）通过解析成矿元素在板块俯冲带发生富集直至成矿的三阶段机制，强调了以下四点：①大陆碰撞带有矿，甚至有大矿、富矿；②大陆碰撞造山之前的增生造山作用可以成矿，碰撞之后的张裂造山作用也可以成矿，但是碰撞造山过程对成矿的贡献更值得研究；③在碰撞造山作用期间有矿形成，成矿出现在大陆俯冲带上盘岩石圈中成矿元素预富集区受到构造拉张改造的部位（不是在受到构造挤压改造的部位）；④如果板块俯冲带上盘岩石圈中成矿元素预富集区在拉张构造作用下发生再活化，就能在碰撞造山带形成热液矿床。因此，确定大陆碰撞与成矿作用在时间上是否处于相同点，是认识造山带成矿作用的关键；认识不同类型造山作用与成矿作用之间在时间和空间上的关系，是解决矿床成因问题的关键。

随着低温年代学和微区分析的技术突破，对于矿床的保存和改造研究正进入一个新的发展阶段，但目前这些方面的优秀研究成果还相对较少。除此之外，更大尺度的成矿要素分析也是未来需要加强的研究方向。例如，利用大数据手段对全球斑岩三大成矿带（特提斯、中亚、环太平洋）进行地质－地球化学分析对比，确定地壳性质、厚度以及岩浆氧逸度、含水量等对斑岩铜矿及其他不同类型矿床的控制作用，目前这些工作有了较为成功的实例（Richards et al.，2012；张洪瑞和侯增谦，2018），但显然还需要更多的研究。

5. 预期目标

未来对于俯冲带热液矿床成矿模式和成矿机制的理论创新研究，不仅要立足于精细成矿过程的核心工作，还要在更为基础的成矿实验模拟、更接近现实的矿床“变、保”过程和更为宏观的成矿控制要素方面加强研究力量，以期建立更为全面和牢固的成矿模式和理论体系。

四、关键科学问题之四：俯冲带动力学

（一）科学挑战 13：俯冲地壳再循环形式

1. 问题的提出

自将岛弧玄武岩地球化学成分与俯冲大洋地壳变质脱水和部分熔融过程

联系起来以后，一般认为俯冲地壳再循环的形式是流体，包括富水溶液、含水熔体和超临界流体［图 4-18（a）］。虽然洋岛玄武岩地球化学成分中也含有地壳信息，但是对进入洋岛玄武岩地幔源区的地壳组分是以流体还是固体的形式未加区分。

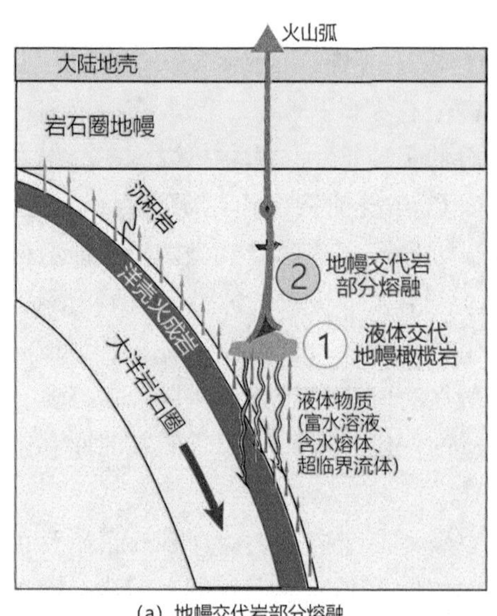

（a）地幔交代岩部分熔融

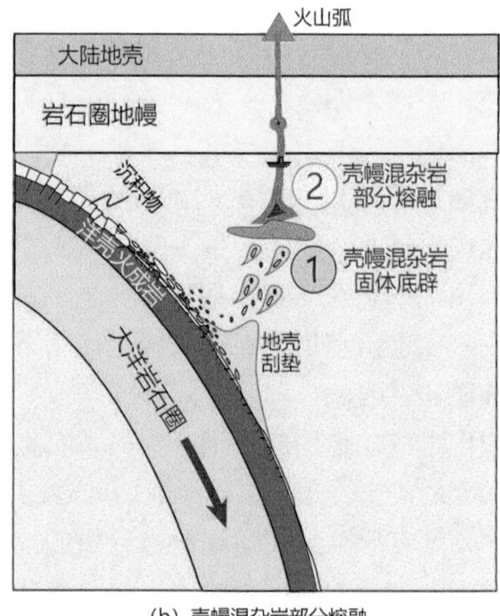

（b）壳幔混杂岩部分熔融

图 4-18　俯冲带板片物质迁移的两种端元模型（修改自 Nielsen and Marschall，2017）

此外，俯冲进入弧下深度的岩石圈地壳会受到上覆地幔楔的刮削作用而与下伏岩石圈基底发生拆离，所拆离的地壳物质会作为“冷柱”底辟进入地幔楔内部（而不是沿俯冲隧道折返到弧下深度），在受到地幔楔加热后发生部分熔融并与橄榄岩反应，由此成为镁铁质弧火山岩的岩浆源区［图 4-18（b）］。该模拟结果意味着，俯冲地壳再循环形式是固体而不是流体。由于流体只溶解有少量硅酸盐物质和大量不相容元素，它与地幔楔橄榄岩反应后形成的蛇纹石化和绿泥石化橄榄岩以及辉石岩和角闪石岩依然具有超镁铁质成分，因此其部分熔融产物可以具有玄武岩的成分。对于固体而言，其熔融产物包含大量硅酸盐物质（其中也富集不相容元素），它与地幔楔橄榄岩反应的产物可能不再具有超镁铁质成分。因此，究竟俯冲地壳是化学分异产生流体上升还是物理分离产生固体上升还有待进一步研究。

俯冲带是地壳物质再循环和地球各圈层发生物质交换的核心场所（Stern，2002；Tatsumi，2005；Zheng and Chen，2016）。俯冲地壳物质的再循环是引起地幔组成不均一性、地壳生长、碳 - 氢 - 氮 - 硫等挥发性元素循环、火山作用以及许多矿产资源形成的重要因素（Stern，2002；Bebout et al.，2018；Zheng，2019）。俯冲板片物质，包括洋壳、陆壳、大陆和大洋岩石圈以及俯冲侵蚀的地壳等，都可能发生再循环（Bebout，2014；Zheng and Chen，2016）。板片俯冲伴随的物理化学条件改变，导致俯冲地壳在不同深度释放出富水溶液、含水熔体以及超临界流体等多种类型俯冲带流体（Kessel et al.，2005；van Keken et al，2011；Spandler and Pirard，2013；Schmidt and Poli，2014；Ni et al.，2017；Zheng，2019）。

数值模拟结果揭示，俯冲隧道中以沉积物为主的混杂岩还会在浮力作用下发生固相底辟，在进入地幔楔内部后与橄榄岩一起部分熔融形成安山质岩浆（Behn et al.，2011；Marschall and Schumacher，2012），从而实现地壳物质的再循环［图 4-18（b）］。在弧前和弧下深度经历脱水的残余板片会继续俯冲进入深部地幔（Spandler and Pirard，2013；Pirard and Hermann，2015）。需要回答的问题是：俯冲地壳物质是以液体还是固体的形式发生再循环？如何区分这两种物质形式？是否两者之一具有占据主导作用的再循环方式？俯冲地壳物质在迁移过程中与俯冲板片本身和地幔楔之间发生过什么样的相互作用？

2. 面临的挑战和重要研究方向

（1）俯冲带液相物质再循环

大洋弧和大陆弧玄武岩中熔体活动性元素和放射成因同位素的富集，表明其地幔源区存在俯冲地壳来源流体和熔体的交代（Tatsumi，2005；Spandler and Pirard，2013；Bebout，2014，Zheng，2019）。俯冲带中富 Nb 玄武岩和洋岛型玄武岩的出现，被认为是洋壳硅酸盐熔体交代橄榄岩的结果（Ringwood，1990；Zheng，2019）。俯冲带部分洋岛型玄武岩表现出低的 Mg 和 Ca 同位素组成，低 Ti/Eu、高 Ca/Al 和 Zr/Hf 值，指示其来自含碳熔体交代的碳酸盐化地幔源区（Dai et al.，2017；Zheng，2019）。在火山弧之下的板片深度，俯冲地壳来源的流体已经进入超临界流体的稳定域（Kessel et al.，2005；Kawamoto et al.，2012；Ni et al.，2017；Zheng，2019），但是这些流体是否已经成为超临界流体还有待确定。

地幔楔橄榄岩的研究识别出俯冲地壳来源的不同性质流体的交代作用（郑建平等，2019）。高温高压实验和理论模拟也显示，不同性质流体－橄榄岩反应会形成不同类型的地幔交代岩，它们进一步发生熔融能解释观察到的俯冲带镁铁质岩浆岩组成（Zheng et al.，2020）。俯冲地壳释放的流体在来源、成分及物理化学性质上存在巨大差异，它们会导致流体交代形成的地幔源区具有岩石学和地球化学不均一性，并最终传递到俯冲带镁铁质岩浆岩。如何有效识别俯冲带镁铁质岩浆岩地幔源区中的不同地壳组分，特别是估算俯冲带地壳物质通量是当前研究的一个难点。进一步，这些再循环地壳物质能否直接诱发地幔楔部分熔融，还是一个具有极大争议的问题（Manning，2004；Grove et al.，2009；郑永飞等，2016）。如果交代地幔楔不会发生立即熔融，其再次熔融从而完成地壳物质再循环的机制是什么（底辟加热 vs 拖曳加热 vs 后撤加热）？这一机制在不同构造背景下有何区别（俯冲早期 vs 俯冲晚期 vs 俯冲后）？尽管地幔楔橄榄岩和超高压变质岩的研究为俯冲地壳流体-橄榄岩相互作用提供了直接制约，但是仍然不清楚现今观察到的地幔交代岩和折返变质岩在多大程度上能够指示弧下板片-地幔楔界面所发生的地壳物质再循环。

未来需要重点开展的研究领域包括：①俯冲地壳释放的超临界流体和含碳流体相的识别；②地幔楔橄榄岩与俯冲地壳来源流体的交代反应；③地幔楔镁铁质/超镁铁质交代岩在俯冲期间和之后的行为；④俯冲地壳和地幔楔熔融过程中元素的地球化学分异；⑤俯冲带不同阶段流体活动性元素和流体不活动性元素在流体相、交代岩和镁铁质熔体中的分异行为；⑥不同温压条件下俯冲地壳、流体相、地幔橄榄岩和交代岩之间的元素质量收支。

（2）俯冲带固相物质再循环

数值模拟研究提出，在俯冲板片与地幔楔之间的接触界面会发生蚀变洋壳、沉积物、蛇纹石化橄榄岩与地幔楔橄榄岩之间的物理混合，从而形成超高压壳幔混杂岩［图 4-18（b）］。这种混杂岩在浮力作用下会底辟上升进入地幔楔内部，在那里受到加热与周围橄榄岩一起发生部分熔融，从而形成大陆弧安山质岩浆（Behn et al.，2011；Hacker et al.，2011；Marschall and Schumacher，2012）。计算地球动力学和地球化学模拟结果表明，这种混杂岩加橄榄岩部分熔融形成的熔体具有与实际观测的大陆弧安山岩相似的地球化学特征（Marschall and Schumacher，2012；Nielsen and Marschall，2017；Cruz-Uribe et al.，2018）。然而上述模型仍然存在如下问题需要考虑：在俯冲带热体制上，冷地幔楔上部如何获得高温？在俯冲带构造体制上，受到挤压的地幔楔如何裂开？在板片-地幔楔耦合关系上，板片-地幔楔接触界面如何从挤压转换到拉张？由于混杂岩熔融过程中主量与微量元素配分存在差异，如何在与橄榄岩混合的过程中同时满足质量平衡？

未来需要重点开展的研究领域包括：①壳幔混杂岩如何在固态条件下物

理传输进入地幔楔？②壳幔混杂岩在地幔楔内部是否能够以及如何获得足够热量发生熔融，进而发生熔体－橄榄岩反应形成俯冲带岩浆？③如何区分固相物质底辟体与液相流体对弧岩浆成分的贡献？

（3）大陆俯冲带和大洋俯冲带俯冲地壳再循环异同点

大陆俯冲带普遍缺乏同俯冲弧岩浆作用，一般认为是由于陆壳相对较干、缺乏同俯冲脱水作用（Rumble et al.，2003；Zheng et al.，2003）。然而，超高压条件下稳定含水矿物的发现、名义上无水矿物中微量水的存在、超高压变质岩中高压－超高压脉体以及流体包裹体的发现，指示大陆俯冲带同样存在显著的流体活动（Zheng，2009；Hermann and Rubatto，2014）。根据陆壳和洋壳中含水矿物的比较，推测在弧下深度，大陆俯冲带存在与大洋俯冲带类似的流体活动（郑永飞等，2016）。对大陆俯冲带出露的地幔楔橄榄岩的研究也证实，大陆俯冲带与大洋俯冲带一样存在板片来源流体对地幔楔的交代作用（Malaspina et al.，2006；Chen R X et al.，2017；郑建平等，2019）。因此，大陆俯冲带弧岩浆作用缺乏的一种解释是陆壳流体交代地幔后，大陆俯冲带冷的地热梯度阻碍了地幔楔熔融导致弧岩浆不发育（Zheng and Chen，2016）。

此外，大陆俯冲带广泛发育碰撞后岩浆岩，其中岛弧型镁铁质岩浆岩记录了俯冲/折返陆壳来源流体在弧下深度对地幔楔的交代作用（Zhao et al.，2013；许文良等，2020），而洋岛型镁铁质岩浆岩记录了先前俯冲古洋壳来源流体在后弧深度对地幔楔的交代作用（Zheng，2019）。由此可见，大陆俯冲带上覆地幔楔经历了从俯冲古洋壳到大陆地壳来源的不同性质流体的叠加交代，它们部分熔融所产生的镁铁质岩浆岩会表现出不同的特点。此外，虽然大陆和大洋俯冲带在流体活动上存在许多相似之处，但是二者在俯冲板片和上覆地幔楔组成上存在一系列差异，由此必然导致两种类型俯冲带在俯冲地壳再循环过程上的差异。

未来需要重点研究的领域包括：①大洋和大陆俯冲带温压结构差异及其对地幔交代岩的影响；②大陆俯冲带碰撞后基性岩中俯冲大陆和大洋地壳来源组分的识别和区分；③造山带橄榄岩中俯冲大陆和大洋地壳来源介质交代作用的识别；④俯冲地壳再循环作用在洋－洋俯冲带、洋－陆俯冲带以及陆－陆俯冲带的异同点。

（4）俯冲地壳深部再循环与板内玄武岩成因

洋岛玄武岩和大陆溢流玄武岩通常被认为是远离现代大洋俯冲带的板内岩浆活动产物（Zindler and Hart，1986；Hofmann，1997；Kelemen et al.，2014）。在这一成因信条下，洋岛玄武岩中大离子亲石元素和轻稀土元素的富集被解释为源自核幔边界的地幔柱所取样的下地幔源区，而放射成因 Sr-Nd 同位素的富集则被归因于地壳再循环（Hofmann and White，1982；Allègre and Turcotte，1986）。再循环的地壳组分被认为来自蚀变洋壳及上覆沉积物、大陆上地壳、大陆下地壳、交代大洋岩石圈地幔或交代大陆岩石圈地幔等（Zindler and Hart，1986；Hofmann，1997；Workman et al.，2004）。一些研究认为，洋岛玄武岩的地幔源区存在俯冲的榴辉岩（Sobolev et al.，2005）。然而主微量元素的质量平衡计算指示，俯冲地壳物质并不直接进入镁铁质岩浆岩的地幔源区（Zhao et al.，2013；Xu et al.，2017）。相反，地壳岩石熔融过程中元素分异作用的研究发现，深俯冲洋壳在大于 200 km 的后弧深度、金红石不稳定条件下部分熔融衍生熔体交代地幔楔形成的交代岩部分熔融就可以产生洋岛型玄武岩（Ringwood，1990；Zheng，2019）。造山带橄榄岩的研究也表明，俯冲带交代作用在后弧深度可以发生（Scambelluri et al.，2008）。由此可见，板内玄武岩的地幔源区也可能形成于俯冲带深部壳幔相互作用（Zheng，2019）。

但是，目前对洋岛玄武岩地幔源区深度的确定仍存在很大困难。虽然许多研究猜想壳幔相互作用的深度在地幔过渡带之上的 200～400 km，但很少有研究尝试定量确定这一相互作用发生的具体深度（郑永飞等，2018）。如果洋岛玄武岩的地幔源区确实是通过俯冲板片在后弧深度熔融交代地幔形成，那么就要求地幔楔随着板片俯冲从弧前经由弧下到后弧深度，最终在地幔过渡带形成大地幔楔（Zheng，2019）。但是，目前并不清楚俯冲板片什么时候发生回卷，需要多长时间才能产生大地幔楔。俯冲洋壳进入深部地幔产生的石榴辉石岩被认为是具有同位素富集特征的玄武岩地幔源区中富化组分的主要来源（Allègre and Turcotte，1986），但是已有研究对它们在大洋俯冲隧道中板片－地幔界面的溶解过程并没有说明。由此引出的问题是：这些超高压辉石岩是否可能由俯冲洋壳来源长英质熔体与地幔楔橄榄岩在弧下乃至后弧深度的交代反应形成？地幔中的富集组分是否可能是由正在俯冲的大洋岩石圈而不是古老的俯冲大洋岩石圈溶解形成？

未来需要重点研究的领域包括：①板内玄武岩中不同地壳组分的识别；②俯冲带壳幔相互作用的深度；③弧下和后弧深度地幔中交代介质的来源；④板内玄武岩地幔源区深度；⑤大地幔楔的形成机制。

3. 应对策略和预期目标

俯冲地壳再循环形式的发展目标是找到区分不同形式地壳再循环的判定

指标；实现对不同构造背景下俯冲地壳物质再循环的形式及其相对权重的总体把握；尝试对具有特殊意义的元素（如碳、氢、氮、硫等）的再循环进行局部和区域性的定量化约束。推进俯冲地壳再循环形式研究，需要从以下几个方面着手。

1）造山带橄榄岩和镁铁质岩浆岩为俯冲地壳再循环提供了直接和间接记录。以它们为主要研究对象，综合利用多种示踪地壳物质再循环的同位素体系，考察地壳物质再循环的岩石学和地球化学效应；通过对橄榄岩和镁铁质岩浆岩成分的协同分析，定量化约束再循环地壳物质的成分特征，并通过定量模拟探讨其与不同性质流体的亲缘性。将地球化学示踪（镁铁质岩石成分及其空间分布规律）与区域构造演化，特别是造山带温压结构的时空变化紧密结合起来，通过物理和化学的双重约束，探讨特定再循环地壳物质形成的可能性。综合考察地幔包体、造山带橄榄岩、折返变质岩和镁铁质岩浆岩所给出的地壳再循环证据，找出共同点，查明这些记录之间出现差异的基本原因。

2）开展高温高压实验岩石学研究，查明俯冲带超临界流体的形成条件和成分，获取元素在不同岩石体系下的分配系数，确定各种地幔交代岩发生熔融的条件，揭示不同性质流体 - 橄榄岩相互作用机制、过程及产物，为天然样品中再循环地壳组分识别提供依据，为定量模拟提供基础参数。

3）定量模拟从变质作用到交代作用再到最终岩浆形成整个过程中的元素和同位素行为，加深对地壳物质在俯冲板片和地幔楔之间以及幔源岩浆在地幔和地壳中迁移等过程所伴随的元素分异和同位素分馏的认识。

4）深化俯冲带过程的数值模拟工作，尽可能再现最接近自然观察的深部过程。完善俯冲带温压结构模型，结合高温高压实验得到的热力学参数和对天然样品的观测结果，建立地壳物质再循环和俯冲带温压结构演化之间的内在联系。

5）鉴于俯冲地壳再循环形式的多样性和深部地幔样品获取困难的特点，需要有机结合高温高压实验、理论计算、天然样品观测、地球物理探测等多种研究手段来研究俯冲地壳再循环形式。

（二）科学挑战 14：俯冲带板片再循环机制

1. 问题的提出

俯冲板片（包括大洋岩石圈和大陆岩石圈板片）再循环进入软流圈地幔的不同深度，是地球物质循环的重要表现形式。主要受到来自俯冲带大洋岩石圈板片下沉拖曳力（Forsyth and Uyeda，1975）的牵引，俯冲带先后经历大洋板片俯冲、大洋板片断离（Davies and von Blanckenburg，1995）或拆离（Soesoo et al.，1997）、大陆板片俯冲（郑永飞等，2015）和岩石圈拆沉（DeCelles et al.，2009）等不同方式再循环进入地幔（图 4-19）。

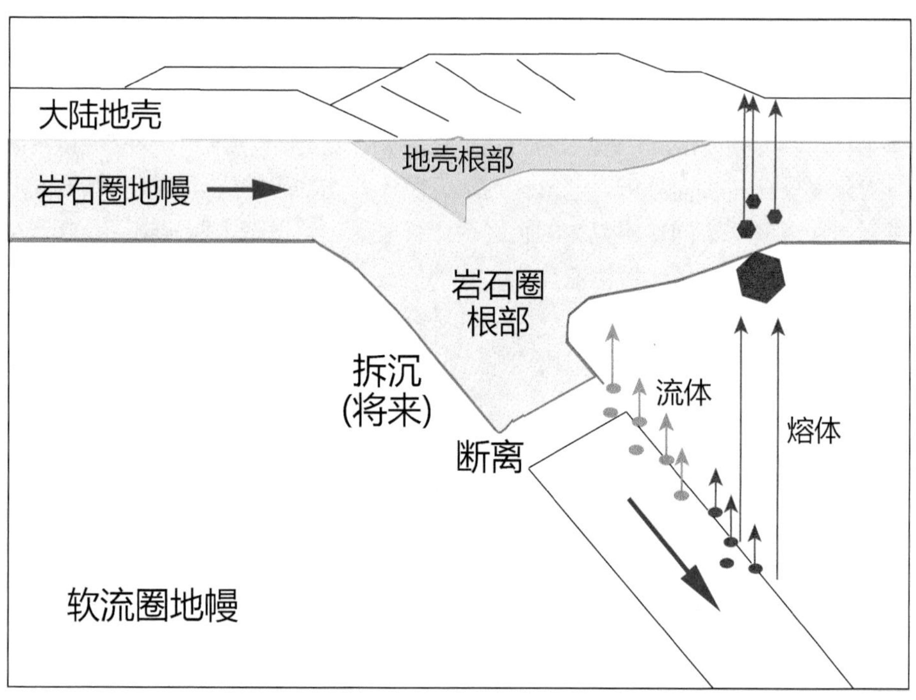

图 4-19　汇聚大陆边缘大洋板片断离与残留大陆板片将来拆沉关系示意图（修改自 Schott and Schmeling，1998）

俯冲大洋板片从上到下主要由大洋玄武岩和辉长岩、方辉橄榄岩和二辉橄榄岩等主要岩性单元组成，在从弧下深度到达地幔过渡带底部之前，俯冲板片的整体密度大于周围地幔（Irifune，1993；Duesterhoeft et al.，2014）。当俯冲大洋板片到达上地幔底部时，一些板片的俯冲角度将变小乃至变平并滞留在过渡带，形成滞留板片，如中国东部下方地幔过渡带的太平洋滞留板片；而另外一些板片可穿过地幔过渡带（410～660 km）进入下地幔，如北美法拉隆板片（Goes et al.，2017）。在有被动大陆边缘卷入的洋－陆俯冲带，随着俯冲作用的进行，如果被动大陆边缘的下地壳黏度较大，大陆板片被先期俯冲的大洋板片牵引俯冲到不同深度（Chopin，1984；郑永飞等，2015）。由于大陆板片相对浮力较大，当其俯冲到岩石圈 / 软流圈边界的弧下深度发生超高压变质时俯冲阻力达到最大，大洋板片与大陆板片之间的过渡带将会发生板片断离（Davies and von Blanckenburg，1995；Magni et al.，2013）。在大洋两侧均发育活动大陆弧的洋－陆俯冲带，随着俯冲作用的进行，两侧正在俯冲的大洋板片的下沉拖曳力，将导致大洋板片与上覆陆壳发生拆离（Soesoo et al.，1997）。

板片断离后，大陆板片在浮力作用下发生折返，而大洋板片在重力牵引下继续俯冲进入软流圈地幔。对正在进行的板片俯冲，板片断离后软流圈物质会沿着板片窗上涌，降压熔融形成玄武质熔体，同时也加热正在折返的大陆地壳形成超高温变质岩，出现超高温变质作用叠加在超高压变质岩之上的现象，这两种极端变质的时间间隔较小（Zheng et al.，2020）。随着板块汇聚过程的持续进行，由于岩浆底侵和构造缩短，汇聚板块边缘的岩石圈会增厚。加厚的造山带根部岩石圈由于重力不稳定而发生拆沉作用，其结果是将部分岩石圈地幔（有时可能包括上覆的镁铁质下地壳）带入软流圈地幔。该过程一般发生在古俯冲带，那里一旦造山带发生去根作用，汇聚板块边缘的岩石圈就会发生减薄，接下来就会出现大陆张裂作用（Zheng and Chen，2017，2021；Zheng and Gao，2021）。即使软流圈上涌不会降压熔融形成玄武质熔体，也会加热减薄的岩石圈地幔使上覆大陆地壳发生巴罗型高温－超高温变质作用，并且引起高温－超高温变质作用叠加在阿尔卑斯型高压－超高压变质岩之上，不过这两种极端变质的时间间隔较大（Zheng，2021c）。

准确鉴别地质历史时期的古俯冲带究竟以何种机制（断离、拆离、拆沉）实现板片再循环，是汇聚板块边缘研究持续关注的重大科学挑战之一。地震层析成像和俯冲带应力状态研究表明，俯冲板片下插到地幔过渡带时，既可以滞留在过渡带，也可以穿透过渡带进入下地幔，这为俯冲带板片再循环方式提供了最直观和最重要证据（Goes et al.，2017）。在上地幔和岩石圈层次，俯冲带震源分布的不连续性促使了板片断离概念的提出（Isacks and Molnar，1969），这种断离被认为是陆－陆碰撞后低密度大陆岩石圈难于俯冲造成的（McKenzie，1969）。板片断离这一概念真正引起学术界的广泛关注，是在利用其解释阿尔卑斯造山带碰撞阶段（始新世）大规模岩浆作用和高压变质岩的折返现象（Davies and von Blanckenburg，1995）之后。但近年来通过高精度地球物理观测发现，阿尔卑斯造山带深部俯冲板片保留了从阿尔卑斯特提斯洋俯冲到陆内汇聚阶段俯冲消减的岩石圈，具有很好的连续性，表明阿尔卑斯造山带可能至今仍未发生板片断离（Zhao et al.，2016）。这使得一些研究者对板片断离模型提出了质疑（Niu，2017；Garzanti et al.，2018）。在青藏高原南部的印度－亚洲碰撞带，也存在俯冲板片是连续的（Zheng H et al.,2007）或不连续的（Replumaz et al.,2010；Liang et al.,2016）等不同看法，直接导致新特提斯大洋板片没发生断离（Niu，2017）和发生了断离（Zhu et al.，2019）两种截然不同的解释。因此，要回答古老的俯冲板片以何种方式再循环进入地幔面临的挑战是，能否利用高精度地球物理观测数据准确识别俯冲的大陆板片和大洋板片，确定俯冲板片是否具有连续性。

2. 面临的挑战

（1）俯冲板片再循环机制的识别

板片断离能够形成成分多样的同期岩浆，包括软流圈地幔来源的玄武质岩浆（Davies and von Blanckenburg，1995；Ferrari，2004），岩石圈地幔来源的钙碱性、碱性和超钾质熔体（Davies and von Blanckenburg，1995；van de Zedde and Wortel，2001），以及持续时间更长的壳源长英质熔体（van de Zedde and Wortel，2001；Zhu D C et al.，2015）。在无被动大陆边缘卷入的洋-陆俯冲带，大洋板片的拆离将导致弧陆碰撞带因没有经历明显的大陆深俯冲而缺乏超高压变质岩，发生持续时间较长且具幔源同位素指标的大规模岩浆活动（Soesoo et al.，1997；Zhu et al.，2016）。

汇聚板块边缘增厚的岩石圈发生拆沉作用之后，软流圈地幔绝热上涌到浅部，降压熔融的比例增大，岩浆成分将发生明显改变，如位于北美板块南侧的墨西哥火山岩带规模相对较小的拉斑质玄武岩被解释为岩石圈拆沉作用的产物（Mori et al.，2009）。同时，板块汇聚边缘的岩石圈拆沉作用，将导致地表发生大约 2 km 的抬升。在方法学上，这些时代可知的沉积、构造、岩浆、变质、地表抬升和剥露记录，既提供了鉴别俯冲板片再循环进入地幔的地质证据，又为限定俯冲板片再循环的时间提供了可能。

（2）俯冲板片再循环过程的数值模拟

从时间演化的角度来看，俯冲板片的再循环最初可能发生在大洋内部的薄弱带（洋中脊、转换断层、破碎带等），然后可能发生在洋-陆过渡带（被动陆缘），最后是发生在成熟的汇聚板块边缘。近年来的计算地球动力学模拟结果显示，正常板块构造推动力可以促使年轻的大洋板片（≤30Ma）在被动陆缘发生俯冲起始，但却无法驱动大洋板片在相对古老的大西洋型被动陆缘发生俯冲起始（Zhong and Li，2019）。在较大的边界应力和弱化的被动陆缘条件下，大洋板片的再循环可以由地体碰撞拼贴所触发，从而提供了通过地质证据约束地体拼贴时代来限定大洋板片再循环开始的时间（晚于地体拼贴，＜10Ma）（Zhu et al.，2013；Zhong and Li，2020）。

数值模拟结果表明，碰撞背景下的板片断离不会在碰撞之后马上发生，而是发生在初始碰撞之后的 5～30 Ma（多数集中在 10～15Ma）（Macera et al.，2008；van Hunen and Allen，2011），这不同于从综合多学科地质资料角度提出的印度－亚洲碰撞带板片断离可能发生在初始碰撞之后约 2 Ma 的认识（Zhu D C et al.，2015）。在这种背景下，如何基于地质观察，针对特定俯冲带开展俯冲板片再循环过程的数值模拟研究，对合理解释特定俯冲带演化历史具有极为重要的意义。

（3）俯冲板片再循环的时间

在特定地区很难保存有完整的各类地质证据，而单个地质证据在多数情况下又是多解的，这就为如何限定俯冲板片再循环（如板片初始俯冲、断离和拆离）的时间带来了巨大挑战。大洋板片的初始俯冲标志了俯冲板片再循环的开始。一些弧前蛇绿岩的年代学信息，如阿曼塞迈尔蛇绿岩变质底板的年龄（Guilmette et al.，2018）、伊豆－小笠原－马里亚纳岛弧弧前蛇绿岩的基底熔岩（Arculus et al.，2015）和辉长岩年龄（Reagan et al.，2019），可用来限定大洋板片俯冲再循环开始的时间。在大洋板片的俯冲晚期，在有被动大陆边缘卷入的洋－陆俯冲带，陆－陆碰撞后将发生板片断离，导致高压－超高压岩石的剥露，并引发上覆岩石圈板块发生区域变质、伸展和快速抬升（地表抬升幅度可达 4 km）以及邻近盆地中剥蚀产物的沉积作用等（von Blanckenburg and Davies，1995）。

3. 重点研究方向和应对策略

（1）俯冲带板片再循环方式的高精度地球物理探测

环太平洋和阿尔卑斯－喜马拉雅造山带经典的地球物理研究结果，凸显了高精度地球物理探测在揭示俯冲带板片再循环方式方面的重要性。因此，未来研究应重点以典型俯冲带为对象，开展高精度地球物理探测，结合古地理重建，识别出地幔中俯冲的大洋板片、断离或拆离的大洋板片以及拆沉的岩石圈，准确甄别大洋和大陆板片，既可为俯冲带板片的再循环方式提供确凿证据，又可为重建俯冲带的俯冲消减历史提供关键约束。

（2）俯冲带板片再循环时间的多学科综合限定

地球物理探测是确定俯冲带板片再循环的关键证据，但地球物理方法只能给出现今的地壳和地幔结构，很难限定古俯冲板片的再循环时间。大洋板片初始俯冲的地质记录主要保存在弧前蛇绿岩的不同岩石单元中，对保存完整的弧前蛇绿岩剖面，尤其是弧前玄武岩以及从弧前玄武岩过渡为受俯冲影响明显的玄武－安山岩组合，开展系统深入的地质和地球化学研究，是限定大洋板片何时发生再循环的关键。聚焦典型碰撞造山带，鼓励综合使用地质和地球化学证据，查明岩浆活动的时空迁移规律和岩浆源区性质，精确限定高压–超高压岩石剥露的时间以及俯冲带上覆岩石圈板片发生区域变质、伸展和快速抬升剥露的时间，可靠约束邻近盆地剥蚀产物的快速沉积时间，从多学科角度分析俯冲带是否发生了板片断离、板片拆离和岩石圈拆沉，并利用各种现代定年方法精确限定俯冲带板片再循环的时间。

（3）俯冲带板片再循环过程的地质学和数值模拟交叉研究

数值模拟是重现俯冲带板片再循环过程的有效手段，能够直观地显示从俯冲起始到板片断离、拆离等再循环过程。尽管已有大量俯冲带板片再循环的数值模拟研究，然而多数数值模拟仅仅是证明了再循环能够发生和在什么条件下发生，模拟结果与特定俯冲带的实际地质观察的差距比较大。板片断离是一个持续过程，涉及从边部撕裂还是通过颈化逐渐断开、从开始软化到彻底断离所需时间、形成板片窗的大小和速度、俯冲板片之下的软流圈地幔物质从板片窗上涌进入上覆地幔的速率等问题。这些过程直接影响了上覆岩石圈的岩浆、变质、构造和剥露事件。因此，综合考虑各种地质边界条件，对特定俯冲带的板片再循环过程进行精细的数值模拟，是未来值得关注的重要研究方向。

4. 预期目标

以典型俯冲带为重点研究对象，通过高精度地球物理观测获取不同岩石圈和地幔层次的俯冲板片形态，建立起鉴别俯冲带板片再循环方式的岩石学和地球化学方法，查明俯冲带板片再循环的开始时间和方式，利用基于地质观察的数值模拟，再现俯冲带板片再循环过程。

（三）科学挑战 15：俯冲带地热梯度演化

1. 问题的提出

对碰撞造山带变质岩的研究发现，虽然也存在双变质作用，但是两者在时间和空间上呈现叠加，指示俯冲带在地热梯度上发生了变化，早期形成于低地热梯度的变质岩受到晚期高地热梯度变质作用的叠加（Zheng and Chen，2017，2021；Zheng，2021c）。一方面，是否俯冲板片回卷也是导致地幔楔底部受到软流圈加热发生部分熔融引起弧岩浆作用的机制，已经成为俯冲带弧岩浆作用动力学研究的焦点和前沿。另一方面，是否碰撞加厚的造山带根部岩石圈在拆沉之后就发生大陆张裂作用，导致软流圈上涌加热减薄的岩石圈地幔使上覆大陆地壳发生超高温变质作用，结果引起超高温变质作用叠加在超高压变质岩之上，已经成为造山带变质作用动力学研究的焦点和前沿。

一般来说，俯冲板块边缘在低的地热梯度下表现出刚性行为，表壳岩石能够俯冲到大陆岩石圈地幔深度；在中等地热梯度下即可表现出韧性行为，表壳岩石只能俯冲到大陆下地壳深度。由此可以区分出两种范式的板块构造（Zheng and Zhao，2020），其中现代板块构造以冷俯冲为特征，典型产物是阿尔卑斯型蓝片岩相－榴辉岩相变质系列；古代板块构造以暖俯冲为特征，典型产物是巴罗型角闪岩相－麻粒岩相变质系列。在太古宙广泛出现的是暖俯冲，在古元古代开始出现局域性冷俯冲，到新元古代起才出现全球性冷俯冲。另外，软流圈上涌将热能从地幔传输到地壳，不仅引起巴肯型高温低压变质作用，而且使下部地壳发生部分熔融产生长英质熔体。这些高温低压过程是在先前俯冲带的基础上发展起来的，由此所产生的大陆张裂带构成古板块边缘的活动带，其性质取决于软流圈地幔上涌的效率。太古宙地幔温度较高，俯冲地壳在弧下深度可以发生显著部分熔融；显生宙地幔温度较低，俯冲地壳在弧下深度只能少许脱水熔融。

板块俯冲下沉进入上地幔时，俯冲板块的地热梯度在不同的俯冲带具有很大的不同，从接近垂直的陡俯冲到俯冲角度较缓的平板俯冲都有观测结果（Goes et al.，2017）。俯冲板块的地热梯度与俯冲角度对地表地质和构造造山过程有重要影响，上地幔中俯冲板块形态的一个典型特征是板块的平板俯冲，另一个特征是板块形态的张裂、断离和形态扭曲（Thorkelson，1996；Schellart et al.，2007；Andrews and Billen，2009；Groome and Thorkelson，2009）。俯冲板块的地热梯度演化过程主要受板块的负浮力、地幔中大范围的流场以及板块自身的流变强度和板块边界处的耦合等因素的控制，一般主要使用数值模型结合地质观测结果进行研究。

板块俯冲一般分为两个阶段（图 4-20），早期阶段是低角度俯冲，汇聚板块边缘地热梯度较低；晚期阶段变成高角度俯冲，汇聚板块边缘地热梯度升高（Zheng，2019，2021b）。前人在大洋俯冲带发现双变质作用，靠近海沟一侧出现蓝片岩－榴辉岩组合带，形成于低的地热梯度（小于 10℃ /km）；而在靠近岛弧一侧出现角闪岩－麻粒岩组合带，形成于高的地热梯度（大于 30℃ /km）。对于正在进行的板块俯冲，地球动力学数值模拟结果显示，俯冲板片回卷引起弧后位置岩石圈拉张减薄，导致弧后盆地打开，结果软流圈地幔发生上涌，不仅发生降压熔融形成玄武质熔体，而且加热地壳岩石形成高温－超高温变质岩（Zheng and Chen，2017，2021）。

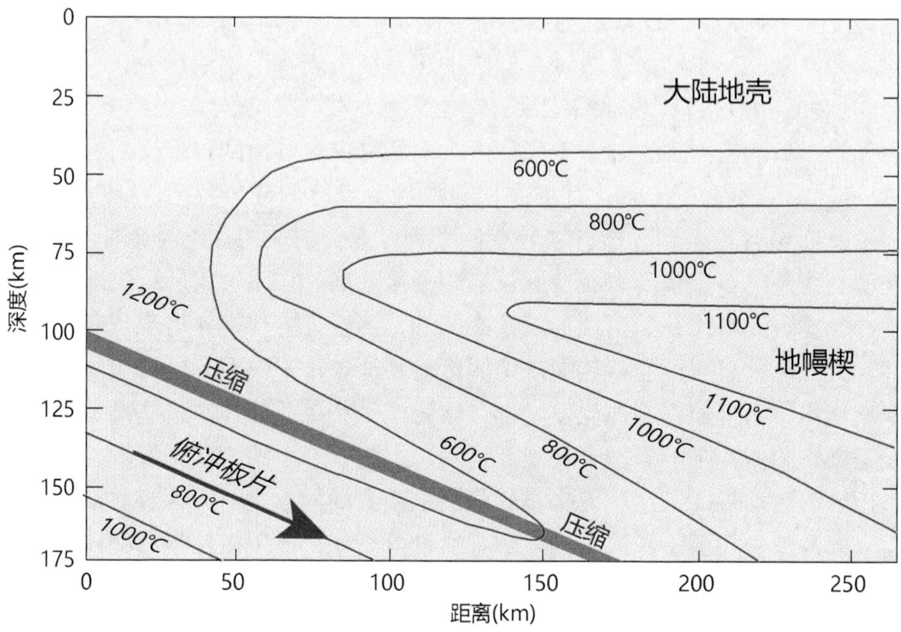

（a）早期阶段低角度俯冲产生低热梯度压缩环境

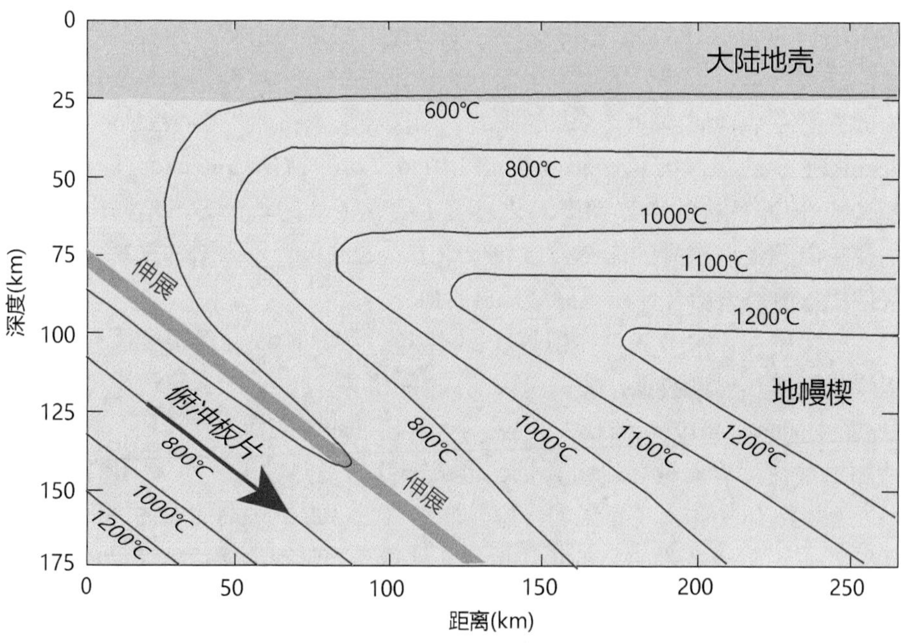

（b）晚期阶段板片回卷导致高角度俯冲产生高热梯度伸展环境

图 4-20　一个板片向另一个板块之下俯冲的温度结构演化示意图（修改自 Zheng and Chen，2016）

2. 面临的挑战

采用数值模拟方法，可以对俯冲板块在上地幔的地热梯度演化进行系统研究，结果能够较好地解释通过地质和地球物理方法观测到的平板俯冲、板块扭曲和断离等板块地热梯度的动力学成因及其地表响应。然而，随着更多观测结果的获得和计算能力的提高，对俯冲板块地热梯度演化提出了新的问题和挑战。

1）相对于大洋俯冲带，大陆俯冲带俯冲板块的性质具有很大的不同，如大陆俯冲板块的地壳厚度较大、浮力较大等。这种板块性质的不同会造成大陆俯冲带与大洋俯冲带结构的不同。但是由于全球俯冲带的主要形式为大洋俯冲，目前对大陆俯冲带地热梯度演化的相关研究开展得较少。随着对大陆俯冲带的研究持续深入，通过各种岩石和地球化学方法得到的大陆俯冲带和大陆俯冲隧道过程中形成的高压和超高压变质岩的数据不断增多（Zheng，2012；Zheng and Chen，2016），为通过变质岩的温压特征约束大陆俯冲带的地热梯度演化提供了可能。

2）Penniston-Dorland 等（2015）收集了全球范围内已知的高压和超高压变质岩的分布，并将变质岩的温压数据与 Syracuse 等（2010）对现代大洋俯冲带温压结构进行数值模拟所得到的板片表面温度进行了比较。结果显示，对于土耳其－希腊－伊朗造山带、亚洲大陆造山带以及阿尔卑斯造山带的高压变质岩，所得到的变质温度比现代俯冲板片表面温度高 100～200℃。解决这种现代大洋俯冲带温压结构与自然界变质温压数据之间的差异是理解大陆俯冲带地热梯度演化的关键，需要在未来的研究中通过建立更符合实际情况的大陆俯冲带模型进行重点探讨。

3）目前对于全球俯冲动力学模型，尚未解决的一个关键问题是如何在板块自身重力驱动下，使得地表板块的运动速度与通过板块重构模型得到的板块运动历史相一致。当前的动力学模型中，如果仅仅使用板块自身重力驱动模型，可以大致吻合现今各板块的运动速度（Conrad and Lithgow-Bertelloni，2002；Stadler et al.，2010），但是很难得到与板块重构模型结果相一致的地表板块运动历史。进一步的研究应深入探讨如何提高板块边界处的分辨率以及改进岩石圈的黏弹塑性设置和断层模拟方法，逐步解决这一问题。

3. 应对策略和预期目标

在未来研究中，首先需要将原有以大洋俯冲板块为研究对象的地热梯度演化拓展到既包括大洋俯冲带，又包括大陆俯冲带。同时，应深入研究如何将通过高压和超高压变质岩得到的温压特征与通过数值模拟研究得到的温压特征进行对比并互相验证，使得两种方式得到的结果一致。此外，应大力发展新一代数值模拟方法，构建在动力学上自洽的、在自身板块重力驱动下的俯冲板块模型，使得实际得到的地表速度场能够吻合通过板块重构得到的速度场演化历史。通过这些策略，可以打开俯冲带地热梯度演化领域的数个全新研究方向，大大拓展本领域的覆盖面，使得对于俯冲带地热梯度的演化提升到国际前沿水平。

（四）科学挑战 16：俯冲带构造体制

1. 问题的提出

板块汇聚边界由外缘隆起、海沟（或缝合带）、岛弧、弧前盆地、弧后盆地、地幔楔、俯冲板片等构造单元组成。板块汇聚边界的构造增生和侵蚀与俯冲板块和上盘板块的物质组成和流变学性质密切相关。根据汇聚边界两侧板块的类型，可区分出不同的俯冲带类型：洋－洋俯冲带（如沿马里亚纳海沟太平洋板块向菲律宾海板块俯冲），洋－陆俯冲带（如沿秘鲁海沟纳斯卡板块向南美板块俯冲），陆－洋俯冲带（如大陆属性的南沙地块俯冲到婆罗洲的巽他板块东侧洋壳之下），洋－弧俯冲带（太平洋板块向具有陆壳性质的日本岛弧俯冲，弧后扩张形成沟－弧－盆体系），陆－陆俯冲带（如印度板块向欧亚板块俯冲，形成喜马拉雅造山带），弧－陆俯冲带（如东亚陆缘向吕宋岛弧俯冲，澳大利亚板块的被动陆缘向 Banda 大洋弧俯冲），弧－弧碰撞带（如印度尼西亚 Sangihe 岛弧和 Halmahera 岛弧碰撞）。

Uyeda 和 Kanamori（1979）发现，不是所有的大洋俯冲带都伴生活动的弧后盆地。这在环太平洋俯冲带最为典型。根据上盘板块是否发育正在打开的弧后盆地，可以分成智利型和马里亚纳型两类俯冲边界。对于东太平洋的智利型俯冲边界，纳斯卡板块以缓倾角沿智利海沟俯冲到南美板块之下，俯冲板块与上盘板块强耦合，可发育 8 级以上的逆冲断层型大地震，弧后区处于挤压应力场，不发育弧后盆地，弧岩浆作用的产物为安山岩－英安岩－流纹岩系列，安山岩十分发育。对于西太平洋的马里亚纳型俯冲边界，太平洋板块以陡倾角沿马里亚纳海沟俯冲到菲律宾海板块之下，俯冲板块与上覆岩石圈板块解耦或者弱耦合，无地震，海沟向大洋方向后撤，弧后区处于拉张应力场，形成弧后盆地，弧岩浆作用的产物以玄武岩为主，安山岩较少。他们提出了两种模型来解释俯冲边界差异性的形成机制：①俯冲带的阶段演化模型（这两类俯冲边界代表了俯冲过程的不同阶段）；②固定板片模型（俯冲板片和海沟的位置相对于软流圈固定，上盘板块向大陆方向的运动速率决定了弧后盆地是否打开）。这一研究把俯冲带的几何形态与上盘岩石圈的构造演化和岩浆活动联系起来，引起了广泛关注。

2. 研究现状

Davies 和 von Blanckenburg（1995）为解释阿尔卑斯新生代岩浆活动、始新世超高压变质岩折返以及阿尔卑斯地形隆起等一系列地质过程，通过数值模拟提出大陆碰撞过程中大洋岩石圈最终会断离。自此，板片断离成为包括阿尔卑斯－喜马拉雅造山带在内的诸多造山带演化的关键一步（Xu et al.，2008；Zhu D C et al.，2009，2015）。然而在大多数情况下，板片断离常缺失地质学证据（Zhao et al.，2016；Niu，2017）。地球物理观测数据只是部分解释了超高压变质带下方的板片结构，有很多地质学和地球动力学问题依然有待讨论（Zhao et al.，2016）。

俯冲板片在上地幔中的平板俯冲和板块断离对地表的动力地形和火山活动有显著影响。例如，Liu 等（2008）通过使用在时间域上反演的地幔对流模型模拟重建了自 100 Ma 以来法拉隆板片的俯冲历史，模型给出的动力地形与实际观测到的北美大陆地表沉积层指示的晚白垩世沉积环境的演变基本一致。Flament 等（2015）使用全球动力学演化模型研究了南美洲地表长波长地形随板块俯冲历史的演化。他们发现板块的平板俯冲会造成海沟到俯冲前缘之间的地形隆升以及随俯冲前缘向内陆迁移的地形下降过程，模型结果能够有效解释南美洲典型区域观测到的地表沉积层变迁和亚马孙河流域系统的转变。Liu 和 Stegman（2012）建立了法拉隆板片断离过程的模型，结果显示哥伦比亚河区域的大规模玄武岩喷发的时空分布与模型预测一致，为该区域火山岩的成因提供了一种可能的新机制。基于以上这些模型结果，使用地表的地形变化和火山岩分布并结合板块运动历史，是发现和识别板块平板俯冲和断离的重要手段。

数值模拟显示，当俯冲板块携带着较轻的块体俯冲时，由于该块体的浮力作用，俯冲板块的倾角会显著减小并形成平板俯冲（van Hunen et al.，2002，2004）。对于美国西南部，Coney 和 Reynolds（1977）发现地表的岛弧火山岩的位置在 110Ma 内先从板块边界向内陆迁移了 1000 km 以上，之后又回到板块的边界处。他们认为，该过程的形成原因是该区域下方的俯冲板块的俯冲角度经历了从高角度俯冲到平板俯冲再到高角度俯冲的转化。基于这个认识，Bird（1988）使用数值模型模拟了从中生代晚期到新生代早期北美洲岩石圈下方的平板低角度俯冲对地表地质构造演化的影响。模型结果能够很好地解释拉勒米（Laramide）造山活动所产生的地壳在水平方向缩短的方向和大小，以及与之伴随的沿西南向东北方向的大量下地壳物质的迁移过程。English 等（2003）通过进一步的热动力学模拟研究发现，北美洲下方的平板俯冲可能仅存在于美国的下方，而在加拿大和墨西哥地区并不存在。Liu 和 Currie（2016）的模拟结果则显示，北美洲板块向西的运动加速可能是造成该区域平板俯冲的主要原因。

除了北美洲，其他区域也发现了可能的平板俯冲。例如，对于南美洲的安第斯山脉，地震活动性和层析成像结果显示上千千米宽的区域都受到平板俯冲的影响（Gutscher et al.，2000）。Capitanio 等（2011）的三维数值俯冲模拟结果显示，俯冲的纳斯卡板块的年龄沿海沟存在侧向变化，造成了板块厚度的变化，俯冲板块的厚度变化导致局部区域形成平板俯冲，是导致当前观测到的安第斯山脉地表形态分布的主要因素。这是目前观测到的平板俯冲对地表形态影响的最典型例子之一。

尽管东太平洋板块目前是低角度俯冲在美洲大陆之下，但是该低角度俯冲与中生代岩浆作用之前是否存在成因联系还有待确定，否则有可能属于“关公战秦琼”式的错位解释（Zheng，2021b）。在中国的华南区域，Li Z X 和 Li X H（2007）应用地球化学和沉积学数据建立了一个平板俯冲模型，提出在 250～190 Ma 华南陆块之下受到过古太平洋平俯冲的影响，结果引起宽度达 1300km 的造山带从近海岸线向内陆的迁移，并伴随着前海盆地和岩浆活动的产生。但是，这个时期的岩浆岩同位素年龄和地球化学深入研究并不支持该模型（Wang et al.，2013）。此外，古特提斯大洋板片在三叠纪时期不仅俯冲到华南陆块南缘之下而且俯冲到华北南缘之下，由此阻挡了古太平洋板块的西向俯冲。因此，对于东亚大陆岩石圈之下是否出现过古太平洋板块平板俯冲这个假设，目前还缺乏确凿的地质学证据。

此外，当洋中脊随板块俯冲抵达海沟后，进一步的俯冲会使得洋中脊两侧的板块进入上地幔后逐渐裂开，形成板片窗（Thorkelson，1996）。Groome 和 Thorkelson（2009）使用一系列三维数值模型探讨了洋脊俯冲和板片窗形成的演化过程，结果展示了板片窗对地表的热流和板块内部应力具有重要影响。Andrews 和 Billen（2009）的二维模型结果显示，当上地幔为非牛顿黏性流体时，俯冲板块能够发生断离。对屈服应力大的强板块，断离缓慢发生在深部地幔；而对屈服应力小的弱板块，断离快速发生在浅部地幔。Schellart

等（2007）采用板块自由俯冲的三维数值模型进行计算，发现俯冲板块自身的宽度对其后撤速度和板块俯冲形态有重要影响，模型给出的不同俯冲板块对应的海沟曲度与板块的宽度相互关联并且与观测结果基本一致。板块俯冲过程中如果遭遇了上升的地幔热柱，其与地幔热柱发生相互作用，会影响海沟的迁移速度，并可能造成板块撕裂和板片窗的形成（Betts et al.，2012）。

3. 面临的挑战

近年来，不断积累的观测数据和动力学模拟为俯冲带构造体制的变化提供了新的约束。新生代大多数俯冲板块的运动速率是 5～10 cm/a，运动速率随俯冲板块在海沟处的年龄增加而变快，但是法拉隆板块的运动速率超过 15 cm/a，与 50 Ma 之前的印度板块向北的运动速率类似（Müller et al.，2008；Goes et al.，2011；Zahirovic et al.，2015）。约 70% 的俯冲带在新生代都存在海沟向大洋方向后撤现象，大多数海沟的运动速率在 –1 cm/a（前进）和 +2 cm/a（后撤）之间，而汤加海沟、伊豆 – 小笠原海沟、日本海沟在新生代的后撤速率显著高于其他海沟，可达 +10 cm/a（Goes et al.，2011；Williams et al.，2015），因此固定板片模型不符合观测结果，智利型和马里亚纳型俯冲边界应该代表了俯冲带演化的不同阶段。

如图 4-21 所示，海沟移动的速率由俯冲板片的密度和强度控制，与板片的几何形态或倾角无关（Goes et al.，2017）。由于板块年龄的增加会提高板

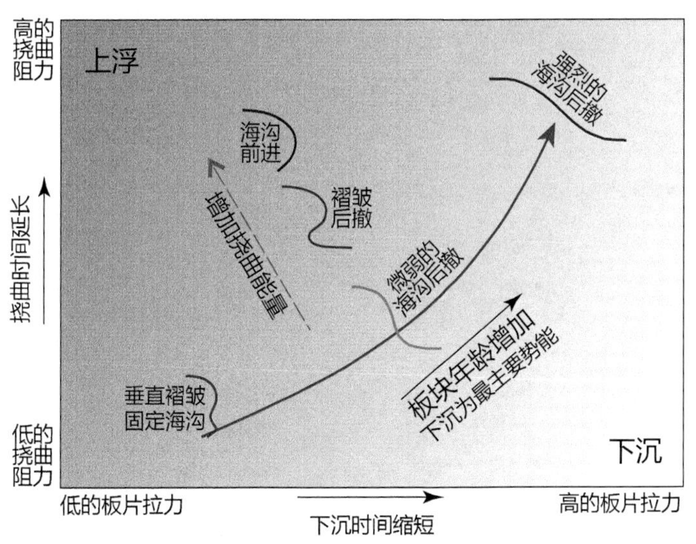

图 4-21 板片几何形态与板片拉力和挠曲阻力之间的关系（Goes et al.，2017）

片密度和强度，古老大洋岩石圈的板片拉力和挠曲阻力很高，俯冲板块与上盘板块强耦合，更易于发生海沟后撤。大多数在地幔过渡带的滞留板片在新生代都经历了大于 2 cm/a 的海沟后撤，固定海沟只发育在板块挠曲阻力和板片拉力都非常小的情况下，即俯冲板块与上盘板块的力学耦合很弱。但是，在智利海沟缓倾角俯冲的纳斯卡板块的年龄是始新世，而在马里亚纳海沟陡倾角俯冲的太平洋板块的年龄是侏罗纪，和传统模型预测的板块年龄与俯冲倾角的关系不一致。这可能涉及东太平洋板片俯冲角度在地质历史时期发生过从低到高再到低的演化过程（Zheng，2021b），弧岩浆作用和弧后盆地打开只与高角度俯冲相对应。

在自然界，板块俯冲一般可分为两个阶段（图 4-22）。早期阶段是上下板块耦合，汇聚板块边缘处于挤压状态；晚期阶段上下板块解耦，汇聚板块边缘变成拉张状态（Zheng，2021b）。在挤压体制下，板块界面具有低的地热梯度，引起蓝片岩相－榴辉岩相变质作用，不会发生弧岩浆作用（Zheng，2021c）。在拉张体制下，板块界面具有高的地热梯度，地幔楔底部受到软流圈加热引起弧岩浆作用，弧后地壳出现角闪岩相－麻粒岩相变质作用和镁铁质岩浆作用（Zheng and Chen，2017，2021）。

A. 早期阶段低角度俯冲产生压缩构造

冷

挤压

缓倾斜板片

无弧岩浆作用
阿尔卑斯型变质

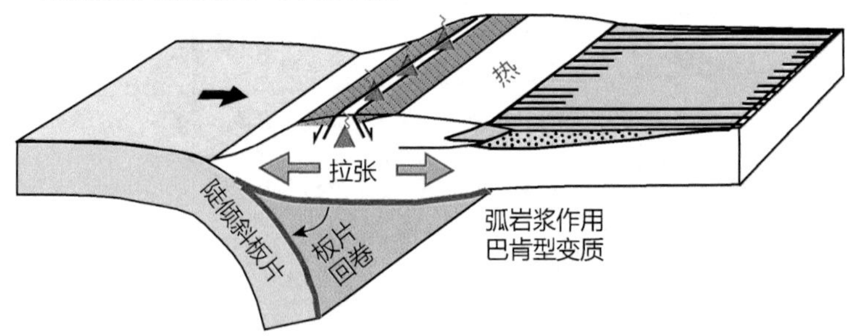

图 4-22　不同地热梯度下的大洋板块俯冲示意图（引自 Zheng and Zhao，2020）

如果碰撞加厚的造山带根部岩石圈发生拆沉作用，减薄的岩石圈就会发生大陆张裂作用，结果软流圈上涌加热减薄的岩石圈地幔使上覆大陆地壳发生超高温变质作用，引起角闪岩相 - 麻粒岩相高温 - 超高温变质作用叠加在榴辉岩相高压 - 超高压变质带之上，伴有花岗质岩浆作用（Zheng，2021b）。在挤压体制向拉张体制转换的过程中，俯冲隧道中的岩石在浮力驱动下发生折返，不仅蓝片岩相 - 榴辉岩相变质岩受到角闪岩相 - 麻粒岩相变质叠加，而且折返的地壳岩石会发生降压熔融引起同折返岩浆作用。因此，俯冲带构造体制的变化导致了俯冲带地热梯度的变化，结果引起了不同类型的变质作用和岩浆作用。如何认识俯冲带构造体制随时间的变化，已经成为俯冲带动力学研究的焦点和前沿。

在俯冲带的四维演化过程中，俯冲带的俯冲倾角、俯冲速率、俯冲板片的形态与汇聚边界两侧板块的类型和俯冲极性有密切关系。由于地球物理观测结果只是俯冲带漫长演化历史中的一瞬间，如何通过对现代俯冲带地质实例分析和动力学模拟，重建俯冲带构造体制随时空的演化是俯冲带研究面临的重大挑战。只有在此基础上，才能更好地理解曾经的俯冲带——蛇绿岩和高压 - 超高压变质带的构造演化历史。

4. 重点研究方向

（1）俯冲带几何形态与海沟后撤的关系

数值模拟结果表明，如果俯冲板片具有高黏度和相对较低密度，将发生海沟前进（Ribe，2010）。此外，上盘板块的强度和俯冲板片与地幔过渡带的相互作用也可能影响海沟后撤。由于大多数动力学模型是二维模拟，而俯冲过程中可以在地幔楔产生平行海沟的角流，影响俯冲板块界面的剪应力。因此，需要与地震学和地质观测结果相结合，开展三维动力学模拟，探索俯冲带的阶段演化与海沟后撤和弧后盆地发育的关系。

（2）俯冲边界的时空演化与上盘岩石圈的响应

板块构造理论根据欧拉旋转定理，用球面上的刚性块体旋转来解释板块间的相对运动。离旋转极越远，板块扩张和俯冲的线速度越快，为了调整板块内不同部位扩张速度的差异，转换断层必然与洋中脊垂直。随着俯冲过程的不断进行，海沟可以变成转换断层，如随着法拉隆板块与太平洋板块之间的洋脊 - 转换断层系统俯冲消失，原来的海沟变成北美洲板块与太平洋板块的边界——圣安德烈斯转换断层（Atwater and Stock，1998）。圣安德烈斯转换断层的走滑作用在加利福尼亚湾形成了一系列拉分盆地，应变比较集中

（Kuiper and Wakabayashi，2018），这与受法拉隆板块平板俯冲导致的地壳伸展减薄形成的盆岭省完全不同（Humphreys，2009）。此外，随着洋盆的不断俯冲，海沟处也可能发生洋脊俯冲（ridge subduction），如约 50 Ma 依泽奈崎－太平洋板块的洋中脊俯冲到东亚陆缘之下（Müller et al.，2016）。洋中脊在俯冲时可能已经停止扩张，但是俯冲到一定深度时，洋中脊作为俯冲带中的软弱带，会再次打开形成板片窗，从而成为软流圈物质上升的通道，导致上盘板块出现高温岩浆活动和变质作用。当具有较厚地壳的洋底高原抵达俯冲带时，俯冲板块的汇聚速率会显著降低，洋底高原可能直接增生成为上覆板块的一部分，导致陆壳生长，海沟跃迁到洋底高原的另一侧，形成新的俯冲带（Mann and Taira，2004）。将俯冲引起的地幔流动与地震波各向异性相联系，为检验不同的构造模型提供了重要约束（Li et al.，2014）。

随着超级计算机和大数据的应用，全球板块重建工作揭示了更多俯冲边界的演化细节，但是这些构造演化模型常存在矛盾，尤其是对中生代以前俯冲边界性质、俯冲极性和俯冲时间的厘定存在很多不确定性。关键科学问题包括：俯冲板块的几何结构、温压结构、年龄和汇聚速率等因素如何影响俯冲带的构造体制转变，如何通过构造地质学、岩石学和地球化学的工作追踪俯冲边界的时空演化，上盘岩石圈的流变学性质对俯冲边界构造演化的影响。对这些问题的解答需要使用现代板块构造系统建立模型，通过正反演确定板块俯冲边界演化的主要控制因素，然后系统梳理地质和地球物理观察结果；通过构造体制变化与变质作用、岩浆作用之间的联系，确定板块汇聚边界的不同构造单元分布，通过动力学模拟追溯早期的俯冲带演化历史。

（3）板块构造的起始时间和机制

对板块构造的起始时间还存在很多争议，其中一个关键点就是俯冲的识别。古代板块构造运动可能早在 40 亿年前就已经开始（Korenaga，2013；Cawood et al.，2018），到约 540Ma 时转变成现代板块构造（Zheng and Zhao，2020）。虽然在冥古宙时期出现的是停滞层盖构造，但是岩石圈层盖经常受到陨石撞击、地幔柱冲击和层盖滴落的影响。早期地球的地热梯度与显生宙时期相比，上地幔的温度和地热梯度都更高，意味着岩石圈的强度显著降低，现在的大洋板片冷俯冲模式可能不适用于地球早期构造体系。以地幔上涌占主导的垂直运动和以板块构造占主导的水平运动可能从太古宙起就开始运行，引起地质历史上的跨圈层物质循环，但是两者之间的转换是否意味着地球演化历史上的重大转折还有待进一步确定（Zheng，2021a）。由于早期地质记录的保留非常有限，因此关键科学问题包括：早期板块汇聚边界的构造单元组成，早期俯冲的规模、时间和形成机制，太古宙构造体制转换与俯冲带类型。

5. 应对策略

对上述问题的研究需要把构造地质学、岩石学、地球化学和同位素年代学相结合，对前寒武纪高压 - 超高压变质岩、高温 - 超高温变质岩、蛇绿岩带进行系统研究，确定这些岩石形成的构造背景和变质历史，厘定板块汇聚边界。此外，由于俯冲带演化过程中岩浆、变质和构造记录会通过剥蚀 - 沉积作用保留在地层中，厘定与俯冲阶段对应的沉积盆地类型并建立从源到汇的物质迁移链对重建俯冲带构造体制具有重要意义。以地质观察为基础建立初始模型，通过三维动力学模拟，可以通过不同边界条件下的俯冲带演化过程，发现不同构造体制下控制俯冲带几何学和运动学的关键因素。

6. 预期目标

对俯冲带构造体制的深入研究将厘定古板块汇聚边界单元，确定俯冲带几何形态与海沟后撤的关系，揭示俯冲边界的时空演化与上盘岩石圈的响应，制约板块构造的起始时间和机制，从而促进对板块构造理论的认识，为全球板块重建提供可靠依据，提出具有全球视野的国际大科学计划。

（五）科学挑战 17：俯冲带动力来源

1. 问题的提出

板块运动的驱动力问题一直都是地学领域内的研究热点和争论焦点。在板块构造理论建立之初，一般将地幔对流作为板块运动的驱动力。随着研究的深入，人们逐渐认识到，俯冲的大洋岩石圈在进入弧下深度后玄武质地壳发生榴辉岩化，由于榴辉岩的密度比橄榄岩大，重力成为板块俯冲的主要驱动力。另外，在海底扩张处软流圈地幔在浮力驱动下发生降压熔融，形成洋中脊玄武岩，洋中脊推力也是引起板块俯冲的动力之一。在一段时间内，许多科学家支持洋中脊推力为板块运动的主要动力。但是自 20 世纪末开始，越来越多的学者支持 20 世纪 70 年代后期就已经提出的俯冲驱动力模型。近 50 年来，相关学者针对俯冲带动力来源提出了很多模式（Forsyth and Uyeda，1975；Bott，1993；Lithgow-Bertelloni and Richards，1998），如板片拉力、海沟吸力、洋脊推力、相变驱动、密度差产生的浮力等（图 4-23）。

俯冲洋壳向下俯冲后不久，含水玄武岩和辉长岩将随着温度压力的增加通过变质脱水反应释放出大量的水，在 60～100km 深度将洋壳岩石转变为高

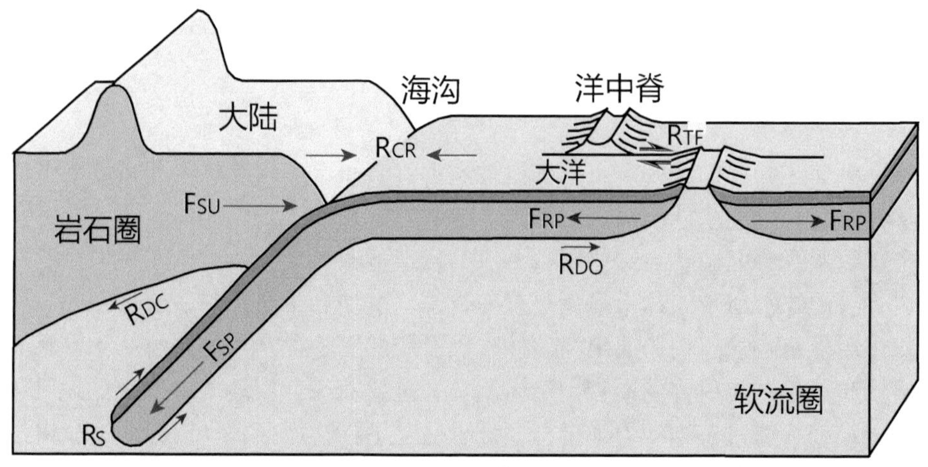

图 4-23　板块运动作用力示意图（修改自 Forsyth and Uyeda，1975）

F_{SP}- 俯冲板片拖拽力；F_{RP}- 洋中脊推力；F_{SU}- 使上覆板块向海沟运动的吸力；R_{DO}、R_{DC}- 大洋板块和大陆板块与软流圈地幔界面处的剪切牵引力；R_S- 板片俯冲阻力；R_{CR}- 碰撞阻力；R_{TF}- 转换断层阻力

压 - 超高压榴辉岩，结果变质洋壳的密度可增加到超过 3500 kg/m^3（Ahrens and Schubert，1975；Kirby et al.，1996）。伴随俯冲深度的增加，变质大洋板片与周围地幔之间的密度差（负浮力）进一步增加，由此拖曳整个板块继续向下俯冲，直到进入 440～660 km 的地幔过渡带。Tassara 等（2006）的研究推测，纳斯卡板块之下软流圈的平均密度差为 90 kg/m^3；Ganguly 等（2009）提出，在古老板块的情况下，深部软流圈上方的过渡区这个密度差更大。以目前对俯冲板片进入地幔之后发生的物理化学变化的认识，俯冲一旦开始，在板片的负浮力作用下，俯冲将持续进行，直到俯冲板片受到堵塞而停止。因此，现今大多数地球科学家都接受这样一种观点，即俯冲板片的自身重力是驱动板块运动的主要驱动力（Anderson，2001，2007；Conrad and Lithgow-Bertelloni，2002）。

俯冲板片重力驱动理念的建立，强化了俯冲带动力来源探索的核心科学内涵。但是迄今对俯冲带动力的本质并未得到很好的认识。为此，瞄准俯冲带动力这个根本科学问题，针对争议的焦点，有必要开拓新思路以攻克解决该问题。在三维空间中，由于俯冲板块宽度有限，在板块迁移过程中（俯冲板片的回卷），能够导致地幔围绕俯冲板块边缘或凸出部位进行水平环形运动，即环向流。其发育规模及强度与地幔的黏度和俯冲板片拖曳力的大小密切相关，而与俯冲板片年龄和倾角关系不大。这两种地幔流动方式能够解释很多与俯冲相关的特殊现象。例如，极向流与俯冲带弧后地区的伸展构造有关，能够驱动上覆板块张裂直至弧后盆地打开（Schellart and Moresi，2013）。环向流能够促使俯冲板片弯曲向下或者向上，结果影响了俯冲带的温压结构，由此引起板片和地幔楔的部分熔融。需要强调的是，无论是环向流还是极向流，都是随俯冲板块的演化而发生变化的。要确定在不同地质历史时期内地幔楔内部和外部的两种流动方向和速度大小及其对上覆板块构造变形的影响，不仅需要精细的地球物理数据和高分辨率数值模型的定量计算，而且需要区分俯冲早期阶段和晚期阶段板片和地幔楔在温压结构上的变化趋势。

在板块俯冲过程中，除了由负浮力导致的板块拉力以外，还有一种能够影响俯冲带动力学过程的板块吸力（Chase，1978）。板块吸力是在地幔对流拖曳作用下驱动上覆板块向海沟方向运动的力，其大小及绝对运动方向非常难以测定（陈凌等，2020）。这是因为能够影响俯冲带地幔楔对流过程的因素很多，包括：①俯冲板片的几何形态、年龄、俯冲速度、板片撕裂及海沟迁移；②仰冲板块的性质，特别是地壳和地幔顶部的流变学和浮力特征；③地幔楔流变学特征，其受温度、压力、应变速率及组成成分（含水矿物、水及熔融物质的配比）等影响。三维地幔循环的计算地球动力学模拟显示，俯冲带地幔对流过程可分解为垂向的极向流和环绕俯冲板块的环向流。其中，极向流位于俯冲板块上部的地幔楔中，它的形成与板块下沉过程有关，影响因素包括俯冲速度和角度、俯冲板片的形态以及地幔楔流变学性质。地震层析成像结果显示，俯冲的太平洋板块在地幔过渡带的滞留，导致大地幔楔中可能存在一个大的极向流（Liu et al.，2016）。

考虑到矿物相变的影响，俯冲板片几何形态的差异意味着俯冲板片的物质组成、应力状态以及与地幔的相互作用会随着时间演化而发生变化。俯冲洋壳的榴辉岩化可显著提升大洋板片的密度，由于俯冲带的温度较低，俯冲板片岩石圈地幔的密度会大于周围地幔。在这两种密度增加机制的叠加效应下，俯冲板块的拖曳力导致穿越圈层结构的深俯冲，而地幔的黏性拖曳力以及俯冲板片和上覆板块之间的摩擦 / 黏性耦合力阻止板片俯冲（Hacker et al.，2003；van Summeren et al.，2012）。橄榄石相变的高压相产物的密度和黏度都增加，在 410 km 橄榄石—瓦兹利石的相变是放热反应（克拉伯龙斜率＞0），俯冲板片的相对低温，导致该相变面变浅并产生向下的负浮力，促使板片穿过 410 km 间断面。然而在 660 km 林伍德石→布里奇曼石 + 铁方镁石的相变是吸热反应（克拉伯龙斜率＜0），冷俯冲导致这一相变面变深从而产生浮力，板片俯冲进入下地幔受阻，在过渡带底部或者下地幔顶部发生褶皱（Yu et al.，2008；Goes et al.，2017）。因此，对于俯冲深度小于 300 km 的板片，以平行板片的拉张应力场为主，而对于俯冲深度大于 300 km 的板片，以平行板片的挤压应力场为主（图 4-24）。

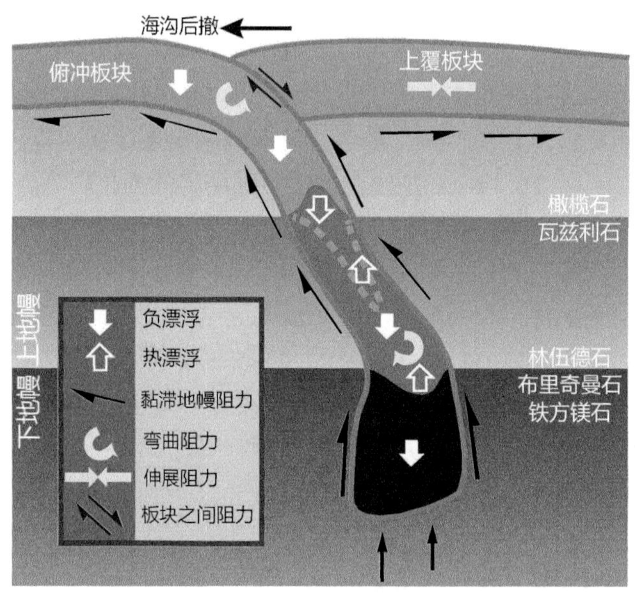

图 4-24　俯冲板片的受力情况示意图（Goes et al.，2017）

目前，对俯冲动力来源主要有两种针锋相对的模型（图 4-25）：一是自下而上（bottom-up）模型；二是自上而下（top-down）模型。自下而上模型认为，地壳运动是由深部地幔的对流过程所驱动的。该模型最早是由 Holmes（1931）和 Hess（1962）提出的，早于板块构造理论。随后，由 Wilson（1963，1973）和 Morgan（1971）进一步修正，并认为深部地幔的对流过程很可能是由来自核 - 幔边界的地幔柱驱动的。在这种自下而上模型中，地幔柱是控制全球地幔对流的根本原因，它在地球早期演化过程中可能起到了控制性作用。但在现今板块构造体制下，地幔柱与板块构造之间是否存在紧密联系已经成为热点和前沿问题，特别是超级地幔柱与超大陆裂解之间。

关于超级地幔柱（或地幔柱群）的形成机制目前有四种观点。①超级地幔柱形成于超大陆的绝热过程（Anderson，1982；Gurnis，1989；Zhong and Gurnis，1993）。这一模型的主要缺陷是对一个超级地幔柱来说热积累不足，无法解释现今太平洋超级地幔柱的形成（Zhong et al.，2007）。②超级地幔柱来源于从地核传导出来的热所引起的一个新形成的超大陆下面的堆积板片的熔融（Maruyama et al.，2007）。③环超大陆板片俯冲的上推效应（Maruyama，

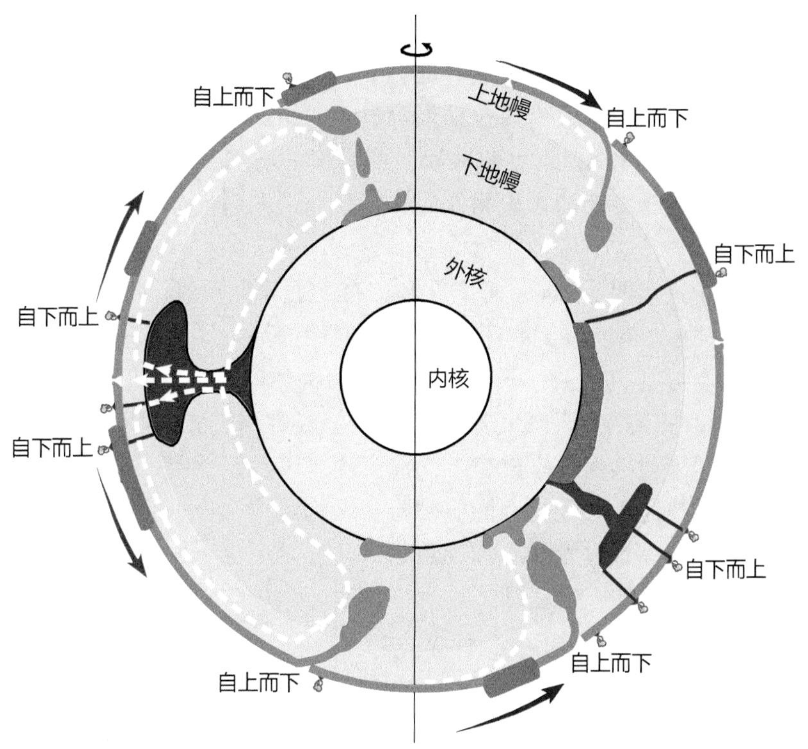

图 4-25　板块运动的“自下而上”和“自上而下”驱动机制（修改自陈凌等，2020）

1994；Li et al.，2004，2008）。④超大陆旋回过程中动态自洽形成的一级地幔对流（在一个半球形成一个主要的上升流系统，而在另外一个半球形成一个主要的下降流系统）或者是二级地幔对流（两个位置相对的上升流系统）（Zhong et al.，2007）。从以上超级地幔柱成因机制的四种观点来看，超大陆的存在是超级地幔柱形成的必要条件，而超大陆的裂解则是地幔柱作用的结果。这一结论，可以从地质历史上一系列超大陆（哥伦比亚、罗迪尼亚和潘基亚超大陆等）裂解后残留下来的大火成岩省的形成时间大体推断出来。显然，这些观点都与板块俯冲过程相关，不同于早期认为地幔柱不受板块动力学所约束的观点。

自上而下模型认为，无论板块运动还是地幔对流，都是岩石圈演化的结果。该种模型最早由 Anderson（2001）、Foulger 和 Natland（2003）提出，并认为板块构造是板块俯冲过程中与上地幔相互作用的结果。之后进一步研究发现，岩石圈的俯冲行为不仅控制了地壳的运动，而且可能决定了地幔对流的样式，甚至地幔柱的形成（Zhong et al.，2007）。在自上而下模型中，板块俯冲是驱动地球深部地幔和浅部地壳动力学过程的根本原因，而板块俯冲的起始则可能来自重力的不稳定性。虽然该观点依然存在争议，但地球物理数据、板块重建数据以及数值模拟结果等显示，俯冲下去的板片比周围地幔更冷，密度差可达 200 kg /m^3。俯冲岩石圈密度的增加主要归因于洋壳在下地壳和岩石圈地幔深度的榴辉岩相变质作用，以及在地幔过渡带上下矿物相变的影响。

由以上论述可见，“自下而上”和“自上而下”两种动力学模型之间存在一种辩证关系。现今地球的板块运动是以俯冲为主导的驱动机制，由垂向俯冲牵引水平漂移，在俯冲板片上方出现两种类型的自下而上的物质流动：一是地幔楔熔融产生弧岩浆上升侵位；二是软流圈物质上涌导致弧后盆地打开。另外，尽管目前对地幔柱来源的深度还难以确定，但是可以明确的是深部地幔物质和热量上升到岩石圈底部对板块构造产生显著的影响。板块汇聚引起超大陆聚合，地幔物质上涌导致超大陆裂解。至于如何区分地幔柱和地幔上涌对板块运动的影响，目前还是一个尚未解决的问题。

2. 面临的挑战和重点研究方向

（1）俯冲带三维应力 – 应变状态分析

对于现今俯冲系统，应发展新的手段，充分利用中国已有的深潜器（蛟龙号、彩虹鱼号等）、海底地震仪、海底电磁仪、深海 Argo 浮标、大量先进海洋科学考察船（东方红 3 号、科学号、嘉庚号等）、海底观测网等，添置便携式电缆立体地形 – 地层 – 地震勘探系统，加速建设中国大洋钻探船，以揭示复杂几何形态的俯冲带应力 – 应变差异，特别是分段性、闭锁性、地震活动性的时空差异调查分析，揭示俯冲板块的弯曲、挠曲、分段、撕裂、拆沉、滞留等多尺度复杂三维几何形态，分析不同大小地幔楔的边界环向流、弧后地幔流等复杂地幔结构和流系分布，建立俯冲板片的综合受力模式。对于古俯冲系统，大陆岩石圈板块无论是在横向还是在纵向上都存在显著的各向异性，具有明显的超越板块构造特征，岩石圈流变学三维不均一性取决于岩石的流变性。中下部地壳岩石所记录的变质、变形及热力学状态是探讨深部地壳流变学结构、揭示壳幔转换关系的关键，深部地壳的流变学性质因为缺乏直接的观察记录而存在不同认识。因此，应加强大陆俯冲带和古大洋俯冲带高压岩石深熔作用与流变学耦合、大陆碰撞、板块俯冲起始与深俯冲和折返的驱动力机制研究。

（2）四维俯冲过程的板块重建

针对陆－陆俯冲、洋－洋俯冲、洋－陆俯冲的不同形式，在进行板块重建的同时，结合多尺度地震层析成像成果，系统反演俯冲带的精细结构及其随时间的演化过程，定量分析俯冲产生的各种地质效应，包括岩浆作用及其资源效应、动力地形及其环境效应、地震分布及其灾害效应等。

（3）多参量多物理场耦合的俯冲动力学模拟

不同区域岩石圈物质成分差异引起的岩石圈密度、浮力的反差可能是俯冲起始的有利和必要条件。当俯冲起始后，下插的大洋板块会沿着俯冲带下沉变重，这种负浮力是板块构造的主要驱动力。无论是对俯冲板片的作用力还是对其他作用力的研究，过去大多是基于简单模型，特别是只考虑密度结构和流变学性质沿径向变化的地球圈层模型来进行的。近年来的计算地球动力学研究显示，地球圈层横向流变学性质的变化会显著影响各种作用力的相对强度及其对板块运动的贡献。考虑不同的板块和地幔黏滞度模型，研究结果会产生对俯冲板片作用力的争议，尤其是对板片吸力的认识差异及对其驱动板块运动作用的争议。当考虑全球岩石圈黏滞度的横向差异时，相对于板片作用力，不能忽视地表地形伴随的重力势能对现今板块运动的作用。显然，对各种作用力的研究和对板块运动驱动力的深入认识，还有待于对地球圈层结构和性质观测的进一步积累和更定量化的细致分析对比。因此，有必要基于现代地球动力学多尺度多场非线性耦合数值模拟，考虑不同黏度、密度、几何结构、成分、温度或地热梯度、受力状态下的各种响应，揭示冷俯冲与热俯冲、大洋与大陆板块俯冲等不同构造背景下的动力过程和机制。

（4）俯冲动力来源和微板块构造体制探讨

主要讨论重力和浮力这两种动力在汇聚板块边缘所发挥的作用。洋壳俯冲、陆壳俯冲的动力来源是什么？深俯冲地壳折返的动力来源是什么？这些动力来源和构造体制、岩石组合（密度差异）有什么联系？微板块构造理论是基于超大陆裂解之后小陆块的漂移而建立的，可以突破传统的大板块构造模型，在许多方面与地体模型相似，因此可以拓展到太古宙时期。虽然微板块俯冲的动力来源可以千差万别，但是其漂移、碰撞和俯冲过程与古大洋板片的牵引密不可分。因此，将不同大小板块综合起来研究，有可能区分不同来源的板块运动动力。

（5）深部过程与地表系统耦合机制

多学科综合模拟分析研究俯冲系统在空间和时间的演化，从多角度验证俯冲系统的动力来源和变化特征。特别是，大洋岩石圈之下的地幔还保存了大量不同地质时期形成的残留板片。这些深部地幔中的残留板片记录了极其丰富的俯冲演化过程，可将俯冲带研究不断向深部深时拓展。在微板块构造理论下，综合多学科的俯冲带四维重建，构建深部过程与地表系统耦合机制，是深化板块构造理论的一个切入点，将会是未来俯冲带研究的主要方向，并以此带动相关深海深时深地观测技术的不断发展。

3. 应对策略与预期目标

在板块构造建立之初，人们将水平运动作为区分板块构造与槽台构造的标志。进入21世纪以来，人们把自上而下的俯冲构造作为识别板块构造起始时间的标志。在板块构造运动体系中，实际上既有自上而下的岩石圈俯冲，也有自下而上的软流圈上涌和洋中脊岩浆上升，后者导致海底扩张。海底扩张和大陆漂移是个联动体系，两者在地球不同位置的相向运动是岩石圈板块保持质量和动量守恒的基础。在海底扩张处，软流圈地幔降压熔融形成洋中脊玄武岩。在大陆张裂带，汇聚板块边缘加厚的造山带岩石圈在减薄后，软流圈也在浮力驱动下上涌（Zheng and Chen，2017，2021；Zheng and Gao，2021）。如果岩石圈减薄速率大于软流圈上涌速率，不仅上涌的软流圈能够发生降压熔融，而且减薄的岩石圈可以发生垮塌作用，导致双峰式岩浆作用。如果岩石圈减薄速率小于或者等于软流圈上涌速率，则上涌的软流圈难以发生降压熔融（相当于超慢速扩张的洋中脊），并且减薄的岩石圈也不会发生垮塌作用；但是，减薄后的岩石圈会受到上涌软流圈的加热，结果在大陆张裂带出现角闪岩相－麻粒岩相高温－超高温变质作用、花岗质岩浆作用、变质核杂岩侵位等组合。因此，在汇聚板块边缘既有重力主导的岩石圈俯冲，也有浮力主导的软流圈上涌（Zheng and Zhao，2020）。两者在地球历史上各自发挥了多大的作用，已经成为板块构造启动机制研究的焦点和前沿。

明确俯冲带动力来源，精细揭示不同深度、不同分段、不同性质、不同演化阶段的四维俯冲过程，多学科融合开展原始创新和全链条集成创新，发展地球系统科学的核心内容，建立深层俯冲系统理论、浅层海沟系统理论，实现地球科学、海洋科学和生命科学交叉，培养国际专业领军人才，造就一批新一代复合型创新人才，参与国际地学竞争。

（六）科学挑战 18：俯冲带时空演化

1. 问题的提出

确定俯冲带的时空演化规律是揭示造山带形成和演化、认识汇聚板块边缘物质和能量传输规律的关键所在（Stern，2002；Zheng and Chen，2016；Zheng，2021b）。俯冲带与造山带之间在结构和成分上具有继承和发展的关系，板块汇聚导致陆缘造山带的形成，板块发散导致陆内造山带的形成（Zheng and Chen，2017，2021；Zheng and Zhao，2017）。在板块汇聚过程中，大洋板块俯冲到大陆板块之下形成增生造山带，典型实例就是东太平洋俯冲带；而大陆板块俯冲到大陆板块之下形成碰撞造山带（图 4-26），典型实例就是阿尔卑斯－喜马拉雅造山带。如果从增生造山带发展到碰撞造山带，则成为复合造山带（郑永飞等，2015），典型实例是喜马拉雅－青藏高原造山带。无论是增生造山还是碰撞造山，陆缘造山带的形成与板块俯冲过程同步，造山旋回属于俯冲进行时，在大陆碰撞情况下有地壳俯冲与折返之分（Zheng et al.，2019a）。确定俯冲进行时形成的陆缘造山带结构、俯冲过程以及变质岩和岩浆岩产物是理解造山带形成和演化的基础。

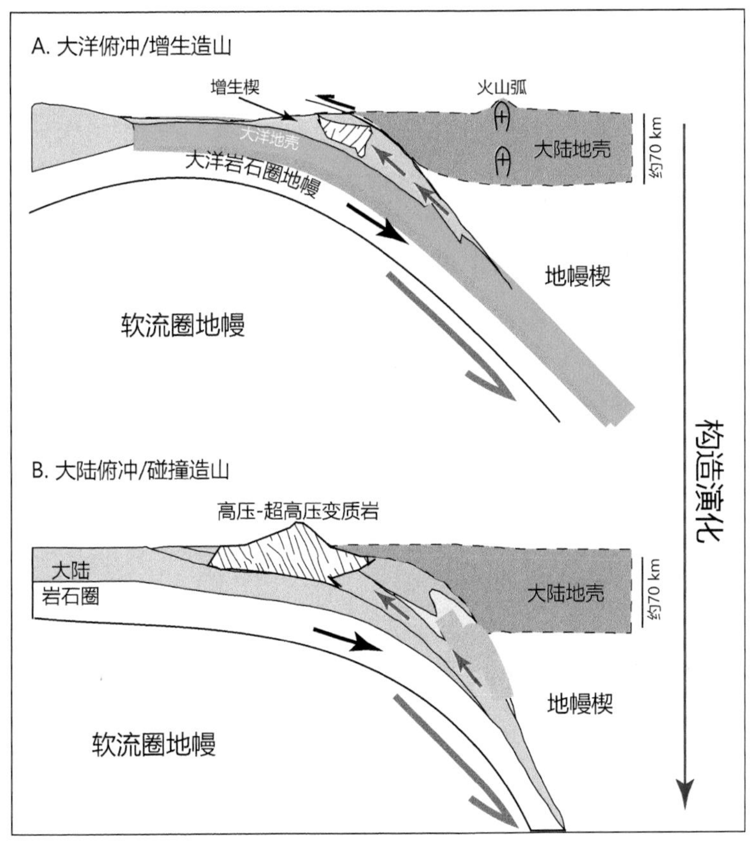

图 4-26　俯冲带构造演化示意图（修改自郑永飞等，2015）

当两个板块不再汇聚之后，板块边缘会处于不活动状态，直到碰撞加厚的造山带根部在重力作用下发生拆沉，或者受到软流圈地幔的对流侵蚀，然后减薄的造山带岩石圈会发生张裂造山作用（Zheng and Chen，2017，2021），形成陆内造山带。这时汇聚板块边缘的地热梯度明显升高，出现巴肯型高温低压变质作用和广泛的长英质岩浆作用（Zheng and Zhao，2017；Zheng，2021b；Zheng and Gao，2021）。虽然陆内造山带的形成属于俯冲过去时，但是张裂造山带与增生和碰撞造山带在地貌效应上都以深部物质上升到浅部出现正地形为特征。前人对造山带的研究基本上注重的是俯冲进行时，忽视了俯冲带的结构和过程会随时间而变化。即使对正在进行俯冲的构造带，其结构、过程和产物也随时间而变化（Zheng，2019）。例如，早期低角度挤压体制下的冷俯冲在晚期会变成高角度拉张体制下的热俯冲，古俯冲带在岩石圈减薄后会产生大陆张裂带，其中成功的张裂带就成为大陆裂解和海底扩张带，而夭折的张裂带则成为陆内再活化带、热造山带或超热造山带等（Zheng and Chen，2017，2021）。

就增生和碰撞造山带而言，虽然这两类陆缘造山带都形成于板块俯冲，但是它们在形成之后常常受到造山带去根作用和大陆张裂作用的改造而变成陆内造山带（Zheng，2021b）。传统的威尔逊旋回由海底扩张、大洋俯冲、大陆碰撞再到大陆裂解组成，在大陆裂解之前的大陆张裂也是威尔逊旋回的组成部分，但是两者在形成和演化的产物上存在明显差异。由于造山带形成机制和叠加改造的多样性，恢复陆缘和陆内造山带所经历的构造演化历史需要多学科的交叉融合（Yin and Harrison，2000；Song et al.，2006；Zheng and Chen，2017，2021；Zheng and Zhao，2017；Zhao L et al.，2017，2020）。汇聚板块边缘在俯冲进行时与俯冲过去时之间如何发生时间和空间上的演化？如何从大陆碰撞带演化到海底扩张带？这些是俯冲带时空演化研究的焦点和前沿。

2. 面临的挑战和重点研究方向

（1）处于俯冲进行时的陆缘造山带结构、成分及演化

在板块俯冲过程中，其自身的密度、挥发分含量等物理化学性质不断发生变化，导致俯冲角度、温压结构、释放的流体组成也在改变。不同的俯冲带起始物质、板块边界特征、上覆岩石圈地幔性质等因素都会导致俯冲带演化过程的差异。确定这些因素各自对俯冲带几何结构、地热梯度、流体活动及元素迁移富集、壳幔相互作用方式和过程等方面所起的作用具有极大的挑

战（Zhao et al.，2015，2017；Zheng，2019）。就大洋俯冲带而言，增生造山带的形成在早期是俯冲大洋板块表层的火山岩和沉积物受到刮削并堆砌到活动大陆前缘形成增生楔，在晚期则是通过大陆弧岩浆作用将幔源物质增生到地壳层位，在大洋俯冲过程中还可能出现多块体拼贴类构造增生。东太平洋俯冲带之上的南美安第斯造山带属于典型的增生造山带，在亚洲大陆内部的增生造山带有中亚造山带和秦岭造山带。就大陆俯冲带而言，碰撞造山带是两个大陆块体之间俯冲碰撞的产物，典型代表如喜马拉雅造山带、西阿尔卑斯造山带、大别－苏鲁造山带和柴北缘造山带。在两个大陆碰撞之前有时存在古大洋板块俯冲形成的增生造山带，如喜马拉雅造山带以北的冈底斯造山带。但是，在华南－华北两个大陆碰撞形成大别－苏鲁造山带之前，虽然也存在古大洋板片的俯冲，但是并未形成增生造山带。综合研究这些不同陆缘造山带的地质结构、形成过程及其产物，能够对俯冲进行时的俯冲带结构、物质组成、壳幔相互作用和构造演化提供重要制约。

（2）处于俯冲过去时的陆内造山带的继承和改造

大陆碰撞过程有早期阶段挤压和晚期阶段拉张之分，碰撞造山之后由于造山带去根作用和大陆张裂作用的改造而变成陆内造山带。对碰撞造山带而言，在其形成之前有的出现过增生造山，在其形成之后则普遍出现张裂造山。这三种类型的造山作用也是威尔逊旋回的主要组成部分（Zheng et al.，2019a），这些复合的地质过程会造成陆内造山带在地质结构、物质组成和构造特征等的多期叠加（Rubatto et al.，2011；Chen et al.，2016；Zheng，2021b）。如何准确限定陆内造山带的结构和组成？如何恢复复合造山带中不同块体经历的俯冲、碰撞和折返过程？如何建立造山带演化与镁铁质和长英质岩浆作用的联系？造山带产出矿床的成矿时限和物质来源与造山带时空演化之间的联系是什么？这些都是造山带研究的热点和前沿方向。鉴于造山带的复杂性，需要结合地球物理、地质学和地球化学等多学科综合研究，才能更好地解析造山带的前世今生。

（3）大陆碰撞带演化到海底扩张带的过程和机制

碰撞造山带形成后由于重力不稳定性和地幔对流侵蚀的双重影响，会发生造山带去根作用（Zheng et al.，2019a）。目前已经认识到，造山带在碰撞后伸展阶段可能会形成大陆张裂带（Zheng and Chen，2017，2021），如果进一步成功裂解则形成洋盆和海底扩张带（Zheng et al.，2019a）。认识这个地球动力学演化过程对发展板块构造理论非常关键。大陆碰撞带构造垮塌的过程和机制是什么？大陆碰撞造山带如何转化为大陆张裂带？大陆碰撞带最终转化为海底扩张带的过程和机制是什么？这是需要重点关注的科学问题。对全球典型大陆碰撞造山带、大陆张裂带的构造演化特征进行研究，结合地球动力学模拟，能更好地解释大陆碰撞造山带演化到海底扩张带的过程和机制。

（4）板内镁铁质岩浆的起源

板内镁铁质岩浆岩包括洋岛玄武岩和大陆玄武岩，通常都具有地壳组分的地球化学特征。具有洋岛玄武岩微量元素分布特征的大陆玄武岩，其富集组分的形成可能与俯冲密切相关（Zheng，2019；Zheng et al.，2020）。在现代大洋板块中的洋岛玄武岩中发现 ^{182}W 和 ^{142}Nd 异常以及硫同位素非质量分馏信号，其起源被认为与地球物理探测揭示的核幔边界大型低剪切波速异常带有关（Cabral et al.，2013；Mundl et al.，2017；Peters et al.，2018），这对地幔对流方式和效率、俯冲物质加入源区的时限和壳幔相互作用方式等问题都提出了新的挑战。再循环地壳组分如何进入板内岩浆的地幔源区？它们受地幔对流的影响有多大？它们何时、以何种方式进入板内岩浆的地幔源区？如何通过地球物理探测和地球动力学模拟制约板内镁铁质岩浆岩的起源深度、构造环境和岩浆演化过程？这些是目前岩浆岩研究的前沿领域。未来的研究一方面要结合数值模拟制约板内镁铁质岩浆的源区组分和构造演化特征，另一方面要充分结合地球物理探测、地幔动力学模拟、高温高压实验等方面的结果，共同揭示俯冲物质对镁铁质岩浆的贡献方式和过程。

3. 应对策略与预期目标

俯冲带的时空演化涉及俯冲带的过去和现在的多阶段演化。这需要首先厘清俯冲带和造山带现今的结构和产物，继而查明从俯冲带到造山带所经历的时空演化。推进俯冲带的时空演化研究，需要从以下几个方面着手。

1）剖析增生造山带的结构、过程和产物及其控制因素，这是认识古大洋俯冲带的重要基础。选取增生造山带（如中亚造山带），通过地质学、地球化学、地球物理学、地球动力学等手段进行深入制约。

2）制约从大洋俯冲到大陆碰撞这个构造转换过程的机制和效应，限定大陆碰撞造山的过程及控制因素、俯冲地壳的化学分异特征及其对地幔楔和俯冲带岩浆岩的影响。

3）针对不同类型的复合造山带，进行多学科交叉研究。造山带形成和演化的多阶段特征要求地质学、地球化学、地球物理学和地球动力学等学科共同参与、综合制约。选择全球范围内典型的复合造山带（如特提斯造山带、中亚造山带、中国中央造山带），对岩石圈结构和成分、地壳碰撞和隆升历史、俯冲带流体活动特征和交代行为、碰撞后变质和岩浆作用等方面进行综合剖析，为俯冲带时空演化研究提供典型范例。

4）加强对汇聚板块边缘构造单元形成和演化的研究。汇聚板块边缘岩石构造单元的形成和演化记录了大洋 / 大陆板片俯冲、地体拼贴和碰撞的历史，记录了俯冲带在结构和过程两个方面随时间变化所产生的构造和化学变化。对这些造山带中产出的变质岩、交代岩和岩浆岩进行研究，将揭示造山带在结构和成分上的继承和发展关系，限定其中变质岩和岩浆岩与板块俯冲的联系，特别是从陆缘造山带演化为陆内造山带的过程和机制，发展板块构造理论。

第二节　学科发展的总体思路和发展目标

一、学科发展的总体思路

以汇聚板块边缘为主题，以俯冲带结构、过程、产物和俯冲带动力学为核心研究内容，重点关注以下研究领域：①俯冲带几何结构、温压结构和地质结构；②俯冲地壳变质脱水和部分熔融、俯冲带双向交代作用、地幔楔部分熔融、俯冲带流体活动与元素迁移；③洋壳俯冲带变质岩、陆壳俯冲带变质岩、俯冲带镁铁质岩浆岩、俯冲带长英质岩浆岩和俯冲带热液矿床；④俯冲带地壳再循环形式、板片再循环机制、俯冲带地热梯度演化、俯冲带构造体制、俯冲带动力来源以及俯冲带时空演化。

板块俯冲带是典型的交叉学科，要充分利用地质学、地球化学、地球物理学等学科交叉的优势，扩大学科研究领域，增长学科生命力，保持学科活力，不断培养新的学科生长点。

二、学科新生长点

根据学科现状和发展态势，板块俯冲带在未来 5～10 年主要的生长点包括以下几点。

1）早期地球构造体制。地球是太阳系中目前唯一具有板块构造的动态行星。板块构造的出现使地球从停滞层盖转变成活动层盖，是地球从非宜居行星转变为宜居行星的关键一步。地球上存在古代和现代两种类型的板块构造体制，它们出现的时间、机制和构造体制演化是固体地球科学的前沿问题，也是理解为什么地球能够演化为宜居行星的关键。

2）俯冲带地球动力学。板块俯冲进入地幔会在弧下形成小地幔楔，在地幔过渡带堆叠形成大地幔楔。地球深部动力学与俯冲进入深部地幔的物质，特别是挥发分，具有密切关系。俯冲板片及其中的挥发分在地幔中的命运如何？它对周围地幔变形、物质和能量交换的作用如何？对地球深部物质组成和地球动力学有什么影响？这是理解地幔演化的核心内容之一。进行岩石圈深部地球探测，如大地电磁测深、高精度地震层析成像等，结合地球动力学模拟、地质学和地球化学分析，将极大地推动对俯冲带地球动力学的认识。

3）俯冲带资源能源。板块俯冲带与资源能源的形成和保存联系密切。未来，将着重探讨不同阶段俯冲过程中关键金属成矿的物质来源和成矿过程，以及俯冲带结构和过程与油气形成和保存之间的联系，深入揭示矿产资源成矿机理、油气成藏机理，为矿产资源和油气资源勘查提供指导。

4）俯冲带与宜居地球。俯冲带使碳、氢、氧、硫等关键挥发分在地球内部和表生圈层得以循环，不仅使地表维持了丰富液态水的存在，还通过调节表层大气和海洋成分，逐渐使地球演化为适合生物生存和生命演化的活跃行星。俯冲带海沟处拥有地球上最深的生物圈，为理解极端生命演化等生命交叉前沿领域提供了重要场所。俯冲带也是火山、地震、山体滑坡和泥石流等重大地质灾害多发区域，是识灾、防灾、减灾的重要场所。通过对俯冲带灾害的研究，可以使人类拥有更美好的生存家园。

5）分析测试平台。高精尖的分析测试设备和技术对俯冲带乃至整个地球科学的研究都异常重要。针对俯冲带涉及的复杂多相多组分体系，需要在矿物微区结构、主微量元素和同位素组成的高精度多功能分析、微量样品多种地球化学组成的同时高精度分析，接近俯冲带实际的数值模拟和高温高压实验模拟等方面建立和完善分析测试技术，成立重大共享科研平台，推动俯冲带乃至整个地球科学领域科研的原始创新。

6）高温高压实验。高温高压实验可以对俯冲板片在地球深部的物质状态、俯冲过程中地球化学行为进行重要制约，也可以揭示一些新奇的现象，如超高压岩石中无机成因碳水化合物的发现。在地球深部也发现了富氧的氧化物和含氢的过氧化物，开启了探索地球深部氢氧循环新化学的序幕。这些是俯冲带研究的新前沿。同时，高温高压实验也需要向接近地质实际的复杂多组分体系发展，为理解俯冲带物质状态、过程和产物提供重要依据。

7）数值模拟。数值模拟在俯冲带研究中起着重要作用。一方面，基于动力学的数值模拟可以定量化约束俯冲带的一级结构和构造演化；另一方面，动力学模型可以与热力学模型耦合，模拟板块俯冲过程中岩石经历的温度压力轨迹、部分熔融过程和熔体组成，进而制约俯冲带变质脱水和元素迁移行为，限定流体的物理化学性质，制约壳幔演化过程。

8）大数据与机器学习。地质大数据可以充分利用现有发表数据，基于统计理论和机器学习，更换传统的研究思维和研究范式，有望对解决俯冲带启动和板块构造演化中的一些难点问题产生重要推动作用。

三、学科发展目标

本学科的发展目标是：继续保持俯冲带科学在地球科学中的核心地位，抢占国际基础科学前沿，服务国家在矿产资源、能源、地质减灾等方面的战略需求，紧跟现代化进程和相关学科的发展趋势，加速和深化学科交叉，在前沿基础理论方面取得新突破，在国民经济发展上创造新成就，在保障国家战略发展上做出新贡献。

四、平台建设

为了推动板块俯冲带研究的稳步快速发展，以及在前沿科学及应用领域等特定方向进行强化，赶超和引领国际研究前沿，建议围绕汇聚板块边缘科学研究，设立类似国家自然科学基金委员会基础科学中心，进一步加强稳定支持，建设高水平研究队伍，稳定产出高水平成果，并通过各种方式加强国际合作交流，持续提升国际影响力。

第三节　未来学科发展的重要研究方向

板块构造是20世纪十大自然科学进展之一，是地球科学的一次伟大革命。板块构造由大陆漂移、海底扩张、板块俯冲“三位一体”组成，其中板块俯冲带是板块构造的核心，成为近50年来地球科学研究的前沿和热点。中国科学家虽然没有能够参与建立板块构造理论这场伟大革命，但是通过参加大陆深俯冲研究对发展板块构造理论做出积极贡献。有鉴于此，中国科学院院士郑永飞教授于2017年牵头组织了中国科学院学部“板块俯冲带”科学与技术前沿论坛，在此基础上于2018～2019年组织实施了国家自然科学基金委员会－中国科学院学部联合支持的“板块俯冲带”学科发展战略研究。

本报告根据研究现状和发展态势，充分总结学科发展规律，提出18个科学挑战。板块俯冲带学科未来的重要研究方向，可以概括为以下4大部分。

一、俯冲带结构

俯冲带的结构信息主要包括俯冲带几何结构、温压结构和地质结构三大方面。通过提高俯冲带地震的定位和获得足够高分辨率的俯冲带图像，利用新的实验观测从而获得更清晰的俯冲带几何结构信息。在此过程中，新的地球物理成像技术、新的算法以及新的观测都至关重要；俯冲带温压结构需要发展更好的岩石学和地球化学温度计，限定精确的剪切生热效率、动态演化过程等，最终达到实验与观测的吻合，为俯冲带研究的其他领域提供准确的参考温度场。通过地质填图和精细的构造地质学、岩石学、地球化学和地质年代学研究，有望建立不同俯冲环境下俯冲带地质结构的三维模型，并通过地球动力学模拟验证。

二、俯冲带过程

俯冲板片携带含碳、氢、氮和硫等挥发分向地球深部俯冲。随着俯冲过程中温度和压力的升高，俯冲板片发生显著的变质脱挥发分反应，释放出富水溶液、含水熔体以及超临界流体等多种类型的流体，具体过程包括：①俯冲地壳变质脱水和部分熔融；②俯冲带双向交代作用；③地幔楔部分熔融；④俯冲带流体活动与元素迁移。未来，需要通过精细的岩石学观测、实验岩石学模拟、高精度元素－同位素组成分析和地球动力学模拟等手段，准确刻画俯冲带内以及俯冲板片－地幔楔界面发生的物理－化学过程。

三、俯冲带产物

随着俯冲过程的持续，俯冲板片中的岩石会在较高温度压力下发生变质重结晶，形成俯冲带变质岩，同时板片释放的流体交代地幔楔形成俯冲带岩浆岩。流体交代和部分熔融又会引起成矿元素迁移富集，形成俯冲带热液矿床。因此，俯冲带产物主要包括：①洋壳俯冲带变质岩；②陆壳俯冲带变质岩；③俯冲带镁铁质岩浆岩；④俯冲带长英质岩浆岩；⑤俯冲带热液矿床，当然也包括俯冲带火山和地震灾害。通过利用最新手段对俯冲带不同产物进行综合研究，有利于人们重建俯冲带结构和详细的俯冲过程，有利于预测俯冲带资源能源形成、减少或避免俯冲带灾害给人类带来的损失。

四、俯冲带动力学

俯冲带是地壳物质再循环和地球各圈层发生物质交换的核心场所，也是一个持续动态的动力学过程，主要内容包括：①俯冲带地壳再循环形式；②板片再循环机制；③俯冲带地热梯度演化；④俯冲带构造体制；⑤俯冲带动力来源；⑥俯冲带时空演化。未来利用多学科交叉手段，定量限定俯冲带内各种动力学过程，从而使俯冲带的观测和模拟相吻合，进一步推进对地球各圈层间物质循环、资源能源形成及地球宜居性演化影响的新认识。

基于俯冲带在板块构造中的核心地位，板块俯冲带是固体地球科学最重要的研究领域之一。只有不断地破解上述科学挑战，人类才会更好地认识地球的演化，适应地球的变化，预测地球的未来。因此，利用新的方法和手段加强对俯冲带的综合研究，势必使得固体地球科学研究更上一层楼。

第五章 资助机制与政策建议

第一节 加强学科布局，明确学科战略定位

加强板块俯冲带的研究，需要进一步优化学科布局，明确学科战略定位。在学科布局上，进一步挖掘俯冲带科学的内涵，不断探索学科发展的新生长点，促进学科交叉，完善相关制度建设。

学科方向上，板块俯冲带是穿越地球圈层构造的“利器”，涉及不同的深度、温度和压力条件，因此需要不同学科相互补充、共同研究。在继续强化俯冲带岩石地球化学研究的同时，加强俯冲带的地球物理探测研究，着力提升俯冲带相关的高温高压实验和动力学数值模拟的研究水平。同时，联合不同学科进行交叉研究有利于更好地揭示板块俯冲带的结构、过程及产物。

能力建设上，针对板块俯冲带的核心内容继续加大资助，特别是稳定持续资助，支持建设国际一流创新研究平台，加强人才团队建设，主动发起板块俯冲带国际大科学计划，推动国际合作交流，促进具有重大国际影响力的战略科学家的涌现。建立遴选 1～2 个俯冲带研究的顶级团队，类似国家自然科学基金委员会基础科学中心的模式给予长期稳定支持，促进重大原创成果的出现。

科研环境上，引导弘扬科学家精神，完善科研评价体制，尊重科研发展规律，鼓励原始创新和集成创新研究。鼓励围绕俯冲带关键地区和关键问题的多学科交叉研究，利用国家自然科学基金委员会成立交叉科学部的机会，设立专门研究基金，推动俯冲带相关学科交叉研究。

组织保障上，科学技术部、中国科学院和国家自然科学基金委员会资助

的重大项目、重大研究计划，与国家发展战略相关的重大设施和重大项目方面，可以适当给予板块俯冲带领域倾斜和引导，着力俯冲过程的全链条设计和交叉学科研究。

深远目标上，积极服务国家重大发展战略（倡议），如“一带一路”倡议和“深海深地”战略；加强对矿产资源，特别是国家亟须的稀有稀散稀土金属等关键矿产资源的基础研究和应用研究，为国家经济发展和国防安全提供战略和科学保障；强化俯冲带火山和地震等自然灾害的研究，提升防灾减灾及灾害预警能力，保障人类的生命和财产安全。

第二节　推动学科交叉，加速地球系统科学建设

板块俯冲带涉及地球各个圈层之间的相互作用，是地球圈层之间相互联系最重要的纽带，是开展学科交叉的理想对象。因此，应从各个方面推动促进学科交叉的措施，推动以俯冲带科学为内涵的地球系统科学建设。

学科内涵上，俯冲带科学不仅是地球科学内部的交叉学科，而且是地球科学与海洋科学、环境科学、生命科学等相互交叉的学科，它们都依托于物理、化学、数学、信息科学和工程科学的发展。因此，加强多学科交叉融合，是发展俯冲带科学的关键。

人才培养和队伍建设上，针对俯冲带研究的复杂性，要强化多学科交叉融合的思路，一方面开展俯冲带科学相关的专题教学培养体系，另一方面注重在科研实践中促进交叉型人才的产出；着力构建多学科强交叉的俯冲带研究队伍。

外部环境建设上，围绕俯冲带不同阶段涉及的多物理、化学和地质过程的耦合，鼓励和加强多学科交叉研究。特别是鼓励申请板块俯冲带相关的交叉研究基金，推动俯冲带相关学科交叉研究。

组织保障上，在科学技术部、中国科学院和国家自然科学基金委员会资助的重大项目、重大计划，与国家发展战略相关的重大设施和重大项目方面，引导设计汇聚板块边缘结构和过程及其效应的各个交叉研究方向，在重大项目执行过程中推动学科交叉，促进重大成果的产出。

综上，板块俯冲带研究是一个系统工程，涉及多学科的研究内容及其强耦合作用，必然需要多学科交叉融合才能促进俯冲带科学的内涵式发展。

第三节　加快人才队伍建设，服务国家发展战略

板块俯冲带学科发展的核心在人才培养和队伍建设。经过近几十年的发展，板块俯冲带学科已经培养了一批骨干人才，形成了一定规模的人才梯队。但诸如高温高压实验、数值模拟、物理模拟等方面的人才，尤其是复合型人才仍然比较欠缺，创新团队建设仍需加强。因此，加快紧缺人才的引进和培养，加强创新团队建设，并完善有利于创新人才成长的评价体系，将是今后板块俯冲带学科发展的一项重要任务。

一、紧缺人才的引进和培养

回顾国内板块俯冲带学科过去半个多世纪的发展历史，人们在地质学、地球化学、地球物理学等传统领域培养和造就了一大批人才，取得了一系列成果，国际影响力显著提升，为发展板块构造理论提供了重要支撑。但在地球动力学数值模拟方面，中国的人才储备和人才培养仍然匮乏，国际上以瑞士苏黎世联邦理工学院的Gerya教授课题组为例，对于板块俯冲带的起始及演化等一系列科学问题进行了多项引领研究。在未来一段时间内，国内相关院校及科研单位应加强与国际著名研究机构进行联合培养，尤其是大力挖掘和培养新一代数值模拟后备人才。在高温高压实验方面，尽管国内近些年有了快速发展，组建了众多高温高压实验室，但与板块俯冲带相关的科研人才队伍仍然欠缺。

二、创新团队建设

创新团队建设是推动板块俯冲带学科发展的重要驱动力，国内相关院校及科研单位应积极申报和参与科学技术部重点研发项目、国家自然科学基金委员会创新研究群体及教育部支持的长江学者奖励计划创新团队建设。通过加大资助，持续稳定地支持板块俯冲带领域科研人才发展。通过支持建设国际一流创新研究平台，加强人才团队建设。

三、国际化人才

国际化和国际合作是科学研究的未来发展趋势。板块俯冲带研究需要一

支具有全球视野的人才队伍，从而产出具有重大国际影响力的成果，进而走向世界。一方面，通过参与和发起国际大科学计划，推动国际合作交流，培养具有国际视野的战略科学家，经常活跃在国际学术舞台。另一方面，是现有研究队伍的国际化，吸引国外相关领域优秀学者来华展开长期合作研究。此外，需要将研究区域扩展到国外，不仅要了解中国的板块俯冲带研究，还要同国外不同地区的典型板块俯冲带进行系统对比研究。

四、制度保障

在外部环境上，通过加强制度建设，引导科研人员弘扬科学家精神，完善科研评价体制，形成好的学风和学术生态，尊重科研发展规律，鼓励原始创新和集成创新研究。

五、复合型人才

鼓励多学科交叉复合型人才的培养。通过科学技术部、中国科学院和国家自然科学基金委员会资助的重大项目、重大计划，与国家发展战略相关的重大设施和重大项目的执行，在项目中培养交叉复合型人才。

青年人才是祖国人才的生力军。鼓励以学术优先来资助项目，谁有能力谁揭榜，促进青年复合型人才的快速成长。

第四节　鼓励开展多方位国际合作与基础设施建设，促进学科引领

板块俯冲带学科的发展离不开广泛的国际合作，离不开重大基础设施的建设。中国科学院、自然资源部和高校的国家重点实验室、省部级重点实验室拥有国内最先进的仪器设备和最好的分析技术，是板块俯冲带研究的重要基地。今后，仍要继续支持俯冲带相关的重大科学仪器（如高精度二次离子探针）、综合性大科学平台的建设，支持依托这些平台开发新的技术，拓展板块俯冲带学科新前沿，引领国际板块俯冲带学科发展。

板块俯冲带学科的发展也离不开广泛的国际合作。与国际同行相比，中国在现代俯冲带研究领域比较薄弱，但在古俯冲带研究领域成就突出。通过国际合作，能够促进中国科学家对现代俯冲带的研究，并加强与古俯冲带进

行对比研究。由于板块俯冲带的复杂性，需要对世界典型地区的不同类型俯冲带进行全面深入研究，这需要国际合作。同样，科学思想的碰撞、研究方法和思维的交流也会极大地促进板块俯冲带学科的发展。目前，在一些领域已经开始融入国际学术舞台，少数领域开始发起一些以中国科学家主导的国际合作计划，引领国际相关领域发展。今后，需要瞄准板块俯冲带相关的关键科学问题，发起更多国际大科学计划，同时组织更多的国际会议，深度融入国际学术组织，并在其中发挥重要作用。

总之，通过加强板块俯冲带相关的基础设施建设、开展多方位的国际合作，培养更多具有国际影响力的高水平研究队伍，推动板块俯冲带领域的人才培养。

参考文献

常承法，郑锡澜 . 1973. 中国西藏南部珠穆朗玛峰地区构造特征 . 地质科学，1：1-12.

陈华勇，吴超 . 2020. 俯冲带斑岩铜矿系统成矿机理与主要挑战 . 中国科学：地球科学，50（7）：865-886.

陈凌，王旭，梁晓峰，等 . 2020. 俯冲构造 vs. 地幔柱构造——板块运动驱动力探讨 . 中国科学：地球科学，50（4）：501-514.

陈衍景，李诺 . 2009. 大陆内部浆控高温热液矿床成矿流体性质及其与岛弧区同类矿床的差异 . 岩石学报，25（10）：2477-2508.

陈衍景 . 2013. 大陆碰撞成矿理论的创建及应用 . 岩石学报，29（1）：1-17.

丁志峰，曾融生 . 1996. 青藏高原横波分裂的观测研究 . 地球物理学报，39（2）：211-220.

郭令智，施央申，马瑞士 . 1980. 华南大地构造格架和地壳演化 . 国际交流地质学术论文集（Ⅰ）. 北京：地质出版社 .

侯增谦，高永丰，孟祥金，等 . 2004. 西藏冈底斯中新世斑岩铜矿带：埃达克质斑岩成因与构造控制 . 岩石学报，20（2）：239-248.

侯增谦，赵志丹，高永丰，等 . 2006. 印度大陆板片前缘撕裂与分段俯冲：来自冈底斯新生代火山 - 岩浆作用证据 . 岩石学报，22（4）：761-774.

侯增谦 . 2010. 大陆碰撞成矿论 . 地质学报，84（1）：30-58.

侯增谦，郑远川，杨志明，等 . 2012. 大陆碰撞成矿作用：Ⅰ. 冈底斯新生代斑岩成矿系统 . 矿床地质，31（4）：647-670.

冷伟，毛伟 . 2015. 俯冲带热结构的动力学模型研究 . 中国科学：地球科学，45（6）：736-751.

李继磊 . 2020. 蓝片岩——俯冲带高压低温变质作用和地球动力学过程的记录 . 中国科学：地球科学，50（12）：1692-1708.

李三忠，索艳慧，朱俊江，等 . 2020. 海沟系统研究的进展与前沿 . 中国科学：地球科学，50（12）：1874-1892.

李忠海 . 2014. 大陆俯冲 - 碰撞 - 折返的动力学数值模拟研究综述 . 中国科学：地球科学，

44（5）：817-841.
李忠海，刘明启，Gerya T. 2015. 俯冲隧道中物质运移和流体－熔体活动的动力学数值模拟. 中国科学：地球科学，45（7）：881-899.
刘贻灿，张成伟. 2020. 深俯冲地壳的折返：研究现状与展望. 中国科学：地球科学，50（12）：1748-1769.
王强，唐功建，郝露露，等. 2020. 洋中脊或海岭俯冲与岩浆作用及金属成矿. 中国科学：地球科学，50（10）：1401-1423.
王汝成，谢磊，诸泽颖，等. 2019. 云母：花岗岩－伟晶岩稀有金属成矿作用的重要标志矿物. 岩石学报，35（1）：69-75.
王瑞，朱弟成，王青，等. 2020. 特提斯造山带斑岩成矿作用. 中国科学：地球科学，50（12）：1919-1946.
王泽九，吴功建，肖序常. 1995. 格尔木—额济纳旗地学断面多学科综合调查研究概况. 地球物理学报，38（A2）：1-2.
魏春景，郑永飞. 2020. 大洋俯冲带变质作用、流体行为与岩浆作用. 中国科学：地球科学，50（1）：1-27.
温汉捷，周正兵，朱传威，等. 2019. 稀散金属超常富集的主要科学问题. 岩石学报，35（11）：3271-3291.
吴福元，徐义刚，高山，等. 2008. 华北岩石圈减薄与克拉通破坏研究的主要学术争论. 岩石学报，24（6）：1145-1174.
吴福元，王建刚，刘传周，等. 2019. 大洋岛弧的前世今生. 岩石学报，35（1）：1-15.
吴功建，高锐，余钦范，等. 1991. 青藏高原"亚东—格尔木地学断面综合地球物理调查与研究". 地球物理学报，34（5）：552-562.
肖序常，李廷栋，李光岑，等. 1988. 喜马拉雅岩石圈构造演化. 北京：地质出版社.
肖益林，陈仁旭，陈伊翔，等. 2020. 自然界岩石样品中的超临界流体记录. 矿物岩石地球化学通报，39（3）：448-462.
徐义刚，李洪颜，洪路兵，等. 2018. 东亚大地幔楔与中国东部新生代板内玄武岩成因. 中国科学：地球科学，48（7）：825-843.
徐义刚，王强，唐功建，等. 2020. 弧玄武岩的成因：进展与问题. 中国科学：地球科学，50（12）：1818-1844.
许文良，赵子福，戴立群. 2020. 碰撞后镁铁质岩浆作用：大陆造山带岩石圈地幔演化的物质记录. 中国科学：地球科学，50（12）：1906-1918.
许志琴，杨经绥，张建新，等. 1999. 阿尔金断裂两侧构造单元的对比及岩石圈剪切机制. 地质学报，73（3）：193-205.
杨经绥，许志琴，耿全如，等. 2006. 中国境内可能存在一条新的高压/超高压（?）变质带——青藏高原拉萨地体中发现榴辉岩带. 地质学报，80（12）：1787-1792.

杨经绥，徐向珍，戎和，等 . 2013. 蛇绿岩地幔橄榄岩中的深部矿物：发现与研究进展 . 矿物岩石地球化学通报，32（2）：159-170.

袁学诚，李善芳，华九如 . 2008. 秦岭陆内造山带岩石圈结构 . 中国地质，35（1）：1-17.

袁忠信，张敏，万德芳 . 2003. 低 ^{18}O 碱性花岗岩成因讨论——以内蒙巴尔哲碱性花岗岩为例 . 岩石矿物学杂志，22（2）：119-124.

翟明国，吴福元，胡瑞忠，等 . 2019. 战略性关键金属矿产资源：现状与问题 . 中国科学基金，2：106-111.

张洪瑞，侯增谦 . 2018. 大陆碰撞带成矿作用：年轻碰撞造山带对比研究 . 中国科学：地球科学，48（12）：1629-1654.

张建新 . 2020. 俯冲隧道研究：进展、问题及其挑战 . 中国科学：地球科学，50（12）：1671-1691.

张立飞，姜文波，魏春景，等 . 1998. 新疆阿克苏前寒武纪蓝片岩地体中迪尔闪石的发现及其地质意义 . 中国科学 D 辑：地球科学，28（12）：539-545.

张立飞 . 2007. 极端条件下的变质作用——变质地质学研究的前沿 . 地学前缘，14（1）：33-42.

张立飞，王杨 . 2020. 俯冲带高压 – 超高压变质地体的抬升折返机制：问题和探讨 . 中国科学：地球科学，50（12）：1727-1747.

赵文津，吴珍汉，史大年，等 . 2008. 国际合作 INDEPTH 项目横穿青藏高原的深部探测与综合研究 . 地球学报，29（3）：328-342.

郑建平，熊庆，赵伊，等 . 2019. 俯冲带橄榄岩及其记录的壳幔相互作用 . 中国科学：地球科学，49（7）：1037-1058.

郑永飞，叶凯，张立飞 . 2009. 发展板块构造：从洋壳俯冲到大陆碰撞 . 科学通报，54(13)：1799-1803.

郑永飞，赵子福，陈伊翔 . 2013. 大陆俯冲隧道过程:大陆碰撞过程中的板块界面相互作用 . 科学通报，58（23）：2233-2239.

郑永飞，陈伊翔，戴立群，等 . 2015. 发展板块构造理论：从洋壳俯冲带到碰撞造山带 . 中国科学：地球科学，45（6）：711-735.

郑永飞，陈仁旭，徐峥，等 . 2016. 俯冲带中的水迁移 . 中国科学：地球科学，46（3）：253-286.

郑永飞，徐峥，赵子福，等 . 2018. 华北中生代镁铁质岩浆作用与克拉通减薄和破坏 . 中国科学：地球科学，48（4）：379-414.

郑永飞，陈伊翔 . 2019. 大陆俯冲带壳幔相互作用 . 地球科学，44（12）：3961-3983.

朱弟成，王青，赵志丹 . 2017. 岩浆岩定量限定陆 – 陆碰撞时间和过程的方法和实例 . 中国科学：地球科学，47（6）：657-673.

Abers G A，van Keken P E，Hacker B R. 2017. The cold and relatively dry nature of mantle forearcs in subduction zones. Nature Geoscience，10（5）：333-337.

Abratis M，Wörner G. 2001. Ridge collision，slab-window formation，and the flux of Pacific asthenosphere into the Caribbean realm. Geology，29（2）：127-130.

Agard P，Yamato P，Soret M，et al. 2016. Plate interface rheological switches during subduction infancy：control on slab penetration and metamorphic sole formation. Earth and Planetary Science Letters，451：208-220.

Agard P，Plunder A，Angiboust S，et al. 2018. The subduction plate interface：rock record and mechanical coupling（from long to short timescales）. Lithos，320-321：537-566.

Agius M R，Rychert C A，Harmon N，et al. 2017. Mapping the mantle transition zone beneath Hawaii from Ps receiver functions：evidence for a hot plume and cold mantle downwellings. Earth and Planetary Science Letters，474：226-236.

Aguillon-Robles A，Calmus T，Benoit M，et al. 2001. Late miocene adakites and Nb-enriched basalts from Vizcaino Peninsula，Mexico：indicators of East Pacific Rise subduction below Southern Baja California?. Geology，29（6）：531-534.

Ahrens T J，Schubert G. 1975. Rapid formation of eclogite in a slightly wet mantle. Earth and Planetary Science Letters，27（1）：90-94.

Allègre C J，Turcotte D L. 1986. Implications of a two component marble-cake mantle. Nature，323（6084）：123-127.

Amstutz A. 1951. Sur l'évolution des structures alpines. Archives des Sciences，4（5）：323-329.

Anderson D L. 1982. Hotspots，polar wander，Mesozoic convection and the geoid. Nature，297（5865）：391-393.

Anderson D L. 1994. The sublithospheric mantle as the source of continental flood basalts；the case against the continental lithosphere and plume head reservoirs. Earth and Planetary Science Letters，123（1-3）：269-280.

Anderson D L. 2001. Top-down tectonics?. Science，293（5537）：2016-2018.

Anderson D L. 2007. The eclogite engine：chemical geodynamics as a Galileo thermometer. Geological Society of America Special Papers，430（3）：47-64.

Andrews E R，Billen M I. 2009. Rheologic controls on the dynamics of slab detachment. Tectonophysics，464（1-4）：60-69.

Angiboust S，Agard P，Yamato P，et al. 2012. Eclogite breccias in a subducted ophiolite：a record of intermediate depth earthquakes? . Geology，40（8）：707-710.

Annen C，Blundy J D，Sparks R S J. 2006. The genesis of intermediate and silicic magmas in deep crustal hot zones. Journal of Petrology，47（3）：505-539.

Arculus R J. 1985. Oxidation status of the mantle：past and present. Annual Review of Earth and

Planetary Sciences, 13（2）: 75-95.

Arculus R J. 2003. Use and abuse of the terms calcalkaline and calcalkalic. Journal of Petrology, 44（5）: 929-935.

Arculus R J, Ishizuka O, Bogus K A, et al. 2015. A record of spontaneous subduction initiation in the Izu-Bonin-Mariana arc. Nature Geoscience, 8（9）: 728-733.

Arculus R J, Gurnis M, Ishizuka O, et al. 2019. How to create new subduction zones : a global perspective. Oceanography, 32（1）: 160-175.

Atwater T, Stock J. 1998. Pacific-North America plate tectonics of the Neogene southwestern United States : an update. International Geology Review, 40（5）: 375-402.

Audetat A, Pettke T, Heinrich C A, et al. 2008. The composition of magmatic-hydrothermal fluids in barren and mineralized intrusions. Economic Geology, 103（5）: 877-908.

Bachmann O, Bergantz G. 2008. The magma reservoirs that feed supereruptions. Elements, 4 : 17-21.

Bai W J, Zhou M F, Robinson P J. 1993. Possibly diamond-bearing mantle peridotites and podiform chromitites in the Luobusa and Dongqiao ophiolites, Tibet. Canadian Journal of Earth Sciences, 30（8）: 1650-1659.

Baker E T, Embley R W, Walker S L, et al. 2008. Hydrothermal activity and volcano distribution along the Mariana arc. Journal of Geophysical Research-Solid Earth, 113（B8）: B08S09.

Baldwin S L, Monteleone B D, Webb L E, et al. 2004. Pliocene eclogite exhumation at plate tectonic rates in eastern Papua New Guinea. Nature, 431（7006）: 263-267.

Ballhaus C. 1993. Redox states of lithospheric and asthenospheric upper mantle. Contributions to Mineralogy and Petrology, 114（3）: 331-348.

Barbarin B. 1999. A review of the relationships between granitoid types, their origins and their geodynamic environments. Lithos, 46（3）: 605-626.

Barruol G, Bonnin M, Pedersen H, et al. 2011. Belt-parallel mantle flow beneath a halted continental collision : the Western Alps. Earth and Planetary Science Letters, 302（3-4）: 429-438.

Beaumont C, Jamieson R A, Butler J P, et al. 2009. Crustal structure : a key constraint on the mechanism of ultra-high-pressure rock exhumation. Earth and Planetary Science Letters, 287（1-2）: 116-129.

Bebout G E, Scholl D W, Kirby S H, et al. 1996. Subduction: Top to Bottom. Geophysical Monograph, 96 : 1-384.

Bebout G E. 2007. Metamorphic chemical geodynamics of subduction zones. Earth and Planetary Science Letters, 260（3）: 373-393.

Bebout G E. 2014. Chemical and isotopic cycling in subduction zones. Treatise on Geochemistry, 4：703-747.

Bebout G E，Penniston-Dorland S C. 2016. Fluid and mass transfer at subduction interfaces-the field metamorphic record. Lithos，240-243：228-258.

Bebout G E，Scholl D W，Stern R J，et al. 2018. Twenty years of subduction zone science：subduction top to bottom 2（ST2B-2）. GSA Today，28（2）：4-10.

Bédard J H. 2018. Stagnant lids and mantle overturns：implications for Archaean tectonics, magma genesis，crustal growth，mantle evolution，and the start of plate tectonics. Geoscience Frontiers，9（1）：19-49.

Behn M D，Kelemen P B，Hirth G，et al. 2011. Diapirs as the source of the sediment signature in arc lavas. Nature Geosciences，4（9）：641-646.

Beinlich A，John T，Vrijmoed J C，et al. 2020. Instantaneous rock transformations in the deep crust driven by reactive fluid flow. Nature Geosciences，13（4）：307-311.

Benioff H. 1954. Orogenesis and deep crustal structure：additional evidence from seismology. Geological Society of America Bulletin，65（5）：385-400.

Betts P G，Mason W G，Moresi L. 2012. The influence of a mantle plume head on the dynamics of a retreating subduction zone. Geology，40（8）：739-742.

Bézos A，Escrig S，Langmuir C H，et al. 2009. Origins of chemical diversity of back-arc basin basalts：a segment-scale study of the Eastern Lau Spreading Center. Journal of Geophysical Research-Solid Earth，114（B6）：B06212.

Bird P. 1979. Continental delamination and the Colorado Plateau. Journal of Geophysical Research-Solid Earth，84（B13）：7561-7571.

Bird P. 1988. Formation of the Rocky Mountains，western United States：a continuum computer model. Science，239（4847）：1501-1507.

Blanco-Quintero I F，Garcia-Casco A，Gerya T V. 2011. Tectonic blocks in serpentinite mélange（eastern Cuba）reveal large-scale convective flow of the subduction channel. Geology，39（1）：79-82.

Blundy J，Cashman K V，Rust A，et al. 2010. A case for CO_2-rich arc magmas. Earth and Planetary Science Letters，290（3-4）：289-301.

Bodinier J L，Godard M. 2014. Orogenic，ophiolitic，and abyssal peridotites. Treatise on Geochemistry，3：103-167.

Bonin B. 2004. Do coeval mafic and felsic magmas in post-collisional to within-plate regimes necessarily imply two contrasting，mantle and crustal，sources?A review. Lithos，78（1-2）：1-24.

Bonin B，Janousek V，Moyen J F. 2020. Chemical variation，modal composition and classification of granitoids. Geological Society Special Publications，491：9-51.

Bott M H P. 1993. Modelling the plate-driving mechanism. Journal of Geological Society，150（5）：941-951.

Brenan J M，Shaw H F，Ryerson F J，et al. 1995. Mineral-aqueous fluid partitioning of trace elements at 900 ℃ and 2.0 GPa：constraints on the trace element chemistry of mantle and deep crustal fluids. Geochimica et Cosmochimica Acta，59（16）：3331-3350.

Brisbourne A M，Clinton J，Hetenyi G，et al. 2012. AlpArray-technical strategies for large-scale European co-operation in broadband seismology. EGU General Assembly Conference Abstracts，14：11345.

Brown M. 2006. Duality of thermal regimes is the distinctive characteristic of plate tectonics since the Neoarchean. Geology，34（11）：961-964.

Brown D，Ryan P D，Afonso J C，et al. 2011. Arc-continent collision：the making of an orogeny. In：Brown D，Ryan P D.（Editors）. Arc-Continent Collision. Berlin Heidelberg：Springer-Verlag：477-493.

Brudzinski M R，Chen W P. 2003. A petrologic anomaly accompanying outboard earthquakes beneath Fiji-Tonga：corresponding evidence from broadband P and S waveforms. Journal of Geophysical Research-Solid Earth，108（B6）：2299.

Butler J P，Beaumont C，Jamieson R A. 2013. The Alps 1：a working geodynamic model for burial and exhumation of（ultra）high-pressure rocks in Alpine-type orogens. Earth and Planetary Science Letters，377-378：114-131.

Cabral R A，Jackson M G，Rose-Koga E F，et al. 2013. Anomalous sulphur isotopes in plume lavas reveal deep mantle storage of Archaean crust. Nature，496（7446）：490-493.

Caby R. 1994. Precambrian coesite from northern Mali：first record and implications for plate tectonics in the trans-Saharan segment of the Pan-African belt. European Journal of Mineralogy，6：235-244.

Camacho A，Baadsgaard H，Davis D W，et al. 2012. Radiogenic isotope systematics of the Tanco and Silverleaf granitic pegmatites，Winnipeg River pegmatite district，Manitoba. Can Mineral，50（6）：1775-1792.

Capitanio F A，Faccenna C，Zlotnik S，et al. 2011. Subduction dynamics and the origin of Andean orogeny and the Bolivian orocline. Nature，480（7375）：83-86.

Carswell D A，O'Brien P J，Wilson R N，et al. 1997. Thermobarometry of phengite-bearing eclogites in the Dabie Mountains of central China. Journal of Metamorphic Geology，15（2）：239-252.

Cashman K，Sparks R，Blundy J. 2017. Vertically extensive and unstable magmatic systems：a unified view of igneous processes. Science，355（6331）：eaag3055.

Castro A，Gerya T，García-Casco A，et al. 2010. Melting relations of MORB-sediment mélanges in underplated mantle wedge plumes：implications for the origin of Cordilleran-type batholiths. Journal of Petrology，51（6）：1267-1295.

Castro A. 2020. The dual origin of I-type granites：the contribution from experiments. Geological Society Special Publications，491（1）：101-145.

Cawood P A，Hawkesworth C J，Dhuime B. 2013. The continental record and the generation of continental crust. Geological Society of America Bulletin，125（1-2）：14-32.

Cawood P A，Hawkesworth C J，Pisarevsky S A，et al. 2018. Geological archive of the onset of plate tectonics. Philosophical Transactions of the Royal Society A: Mathematical Physical and Engineering Sciences，376（2132）：20170405.

Cawood P A. 2020. Metamorphic rocks and plate tectonics. Science Bulletin，65（12）：968-969.

Chapman J B，Ducea M N. 2019. The role of arc migration in Cordilleran orogenic cyclicity. Geology，47（7）：627-631.

Chase C G. 1978. Extension behind island arcs and motions relative to hot spots. Journal of Geophysical Research-Solid Earth，83（B11）：5385-5386.

Chemenda A I，Burg J P，Mattauer M. 2000. Evolutionary model of the Himalaya-Tibet system-geopoem：based on new modelling，geological and geophysical data. Earth and Planetary Science Letters，174（3）：397-409.

Chen Y，Ye K，Wu T F，et al. 2013. Exhumation of oceanic eclogites：thermodynamic constraints on pressure，temperature，bulk composition and density. Journal of Metamorphic Geology，31(5)：549-570.

Chen Y，Li W，Yuan X，et al. 2015. Tearing of the Indian lithospheric slab beneath southern Tibet revealed by SKS-wave splitting measurements. Earth and Planetary Science Letters，413：13-24.

Chen Y X，Schertl H P，Zheng Y F，et al. 2016. Mg-O isotopes trace the origin of Mg-rich fluids in the deeply subducted continental crust of Western Alps. Earth and Planetary Science Letters，456：157-167.

Chen R X，Li H Y，Zheng Y F，et al. 2017. The crust-mantle interaction in a continental subduction channel：evidence from orogenic peridotites in North Qaidam，northern Tibet. Journal of Petrology，58(2)：191-226.

Chen Y X，Zhou K，Gao X Y. 2017. Partial melting of ultrahigh-pressure metamorphic rocks during continental collision：evidence，time，mechanism，and effect. Journal of Asian Earth Sciences，145：177-191.

Chen Y X，Lu W，He Y，et al. 2019. Tracking Fe mobility and Fe speciation in subduction zone fluids at the slab-mantle interface in a subduction channel：a tale of whiteschist from the Western Alps. Geochimica et Cosmochimica Acta，267：1-16.

Chen K，Tang M，Lee C T A，et al. 2020. Sulfide-bearing cumulates in deep continental arcs：the missing copper reservoir. Earth and Planetary Science Letters，531：115971.

Cheng H，King R L，Nakamura E，et al. 2008. Coupled Lu-Hf and Sm-Nd geochronology constrains garnet growth in ultra-high-pressure eclogites from the Dabie orogen. Journal of Metamorphic Geology，26（7）：741-758.

Chevrot S，Sylvnder M，Diaz J，et al. 2015. The Pyrenean architecture as revealed by teleseismic P-to-Sconverted waves recorded along two dense transects. Geophysical Journal International，200（2）：1096-1107.

Chopin C. 1984. Coesite and pure pyrope in high-grade blueschists of the Western Alps：a first record and some consequences. Contributions to Mineralogy and Petrology，86（2）：107-118.

Chopin C. 2003. Ultrahigh-pressure metamorphism：tracing continental crust into the mantle. Earth and Planetary Science Letters，212（1-2）：1-14.

Christie D M，Fisher C R，Lee S M，et al. 2006. Back-arc spreading systems：geological，biological，chemical，and physical interactions. Geophysical Monograph Series，166：1-303. Published by the American Geophysical Union in Washington DC.

Chung S L，Chu M F，Zhang Y Q，et al. 2005. Tibetan tectonic evolution inferred from spatial and temporal variations in post-collisional magmatism. Earth-Science Reviews，68（3-4），173-196.

Chung S L，Liu D，Ji J，et al. 2003. Adakites from continental collision zones：melting of thickened lower crust beneath southern Tibet. Geology，31（11）：1021-1024.

Clemens J D，Stevens G. 2012. What controls chemical variation in granitic magmas?. Lithos，134-135：317-329.

Clemens J D，Stevens G，Bryan S E. 2019. Conditions during the formation of granitic magmas by crustal melting-hot or cold；drenched，damp or dry?. Earth-Science Reviews 200（1）：102982.

Cloos M，Shreve R L. 1988a. Subduction-channel model of prism accretion，melange formation，sediment subduction，and subduction erosion at convergent plate margins：1. Background and description. Pure and Applied Geophysics，128（3）：455-500.

Cloos M，Shreve R L. 1988b. Subduction-channel model of prism accretion，melange formation，sediment subduction，and subduction erosion at convergent plate margins：2. Implications and discussion. Pure and Applied Geophysics，128（3）：501-545.

Coats R R. 1962. Magma type and crustal structure in the Aleutian arc. Geophysical Monograph，

6：92-109.

Codillo E A，Le Roux V，Marschall H R. 2018. Arc-like magmas generated by mélange-peridotite interaction in the mantle wedge. Nature Communications，9（1）：2864.

Coleman R G，Wang X M. 1995. Ultrahigh Pressure Metamorphism. Cambridge：Cambridge University Press.

Coleman D S，Gray W，Glazner A F. 2004. Rethinking the emplacement and evolution of zoned plutons：geochronologic evidence for incremental assembly of the Tuolumne Intrusive Suite，California. Geology，32（5）：433-436.

Collins W J，Richards S W. 2008. Geodynamic significance of S-type granites in circum-Pacific orogens. Geology，36（7）：559-562.

Collins W J，Murphy J B，Johnson T E，et al. 2020. Critical role of water in the formation of continental crust. Nature Geoscience，13（5）：331-338.

Condie K C. 1997. Plate Tectonics and Crustal Evolution. 4th Edition. Butterworth-Heinemann.

Condie K C，Kröner A. 2008. When did plate tectonics begin? Evidence from the geologic record. Geological Society of America Special Paper，440：249-263.

Coney P J，Reynolds S J. 1977. Cordilleran Benioff zones. Nature，270（5636）：403-406.

Cong B L，Zhai M G，Carswell D A，et al. 1995. Petrogenesis of ultrahigh-pressure rocks and their country rocks at Shuanghe in Dabieshan，Central China. European Journal of Mineralogy，7（7）：119-138.

Conrad C P，Lithgow-Bertelloni C. 2002. How mantle slabs drive plate tectonics. Science，298（5591）：207-209.

Conrad C P，Lithgow-Bertelloni C. 2004. The temporal evolution of plate driving forces：importance of "slab suction" versus "slab pull" during the Cenozoic. Journal of Geophysical Research-Solid Earth，109：B10407.

Cooke D R，Hollings P，Walshe J L. 2005. Giant porphyry deposits：characteristics，distribution，and tectonic controls. Economic Geology，100（5）：801-818.

Cooke D R，Hollings P，Wilkinson J J，et al. 2014. Geochemistry of Porphyry Deposits. 2nd Edition. Treatise on Geochemistry，13，357-381.

Cooper K M，Kent A J. 2014. Rapid remobilization of magmatic crystals kept in cold storage. Nature，506（7489）：480-483.

Cox A，Hart R B. 1986. Plate Tectonics：How It Works. Oxford：Blackwell Scientific Publications.

Cruz-Uribe A M，Marschall H R，Gaetani G A，et al. 2018. Generation of alkaline magmas in subduction zones by partial melting of melange diapirs-an experimental study. Geology，46（4）：343-346.

Cui W, Feng L, Hu Z, et al. 2013. On 7000 m sea trials of the manned submersible Jiaolong. Marine Technology Society, 47 (1): 67-82.

Dai L Q, Zhao Z F, Zheng Y F, et al. 2011. Zircon Hf-O isotope evidence for crust-mantle interaction during continental deep subduction. Earth and Planetary Science Letters, 308 (1-2): 229-244.

Dai L Q, Zhao Z F, Zheng Y F. 2014. Geochemical insights into the role of metasomatic hornblendite in generating alkali basalts. Geochemistry Geophysics Geosystems, 15 (10): 3762-3779.

Dai L Q, Zhao Z F, Zheng Y F, et al. 2017. Geochemical distinction between carbonate and silicate metasomatism in generating the mantle sources of alkali basalts. Journal of Petrology, 58 (5): 863-884.

Dai L M, Wang L L, Lou D, et al., 2020. Slab rollback versus delamination: contrasting fates of flat-slab subduction and implications for South China evolution in the Mesozoic. Journal of Geophysical Research-Solid Earth, 125 (4): e2019JB019164.

Danyushevsky L V. 2001. The effect of small amounts of H_2O on crystallisation of mid-ocean ridge and backarc basin magmas. Journal of Volcanology and Geothermal Research, 110 (3-4): 265-280.

Danyushevsky L V, Perfit M R, Eggins S M, et al. 2003. Crustal origin for coupled 'ultra-depleted' and 'plagioclase' signatures in MORB olivine-hosted melt inclusions: evidence from the Siqueiros Transform Fault, East Pacific Rise. Contribution to Mineralogy and Petrology, 144 (5): 619-637.

Dasgupta R, Hirschmann M M. 2006. Melting in the Earth's deep upper mantle caused by carbon dioxide. Nature, 440 (7084): 659-662.

Dasgupta R, Hirschmann M M. 2007. Effect of variable carbonate concentration on the solidus of mantle peridotite. American Mineralogist, 92 (2-3): 370-379.

Dasgupta R, Hirschmann M M. 2010. The deep carbon cycle and melting in Earth's interior. Earth and Planetary Science Letters, 298 (1-2): 1-13.

Dasgupta R. 2013. Ingassing, storage, and outgassing of terrestrial carbon through geologic time. Reviews in Mineralogy & Geochemistry, 75(1): 183-229.

Dasgupta R. 2018. Volatile-bearing partial melts beneath oceans and continents—where, how much, and of what compositions?. American Journal of Science, 318: 141-165.

Davies J H, von Blanckenburg F. 1995. Slab breakoff: a model of lithosphere detachment and its test in the magmatism and deformation of collisional orogens. Earth and Planetary Science Letters, 129: 85-102.

Davies J H. 1999. Simple analytic model for the thermal structure of subduction zones.

Geophysical Journal of the Royal Astronomical Society, 139（3）: 823-828.

Debret B, Bolfan-Casanova N, Padrón-Navarta J A, et al. 2015. Redox state of iron during high-pressure serpentinite dehydration. Contributions to Mineralogy and Petrology, 169(4): 36.

DeCelles P G, Ducea M N, Kapp P, et al. 2009. Cyclicity in cordilleran orogenic systems. Nature Geoscience, 2（4）: 251-257.

Defant M J, Drummond M S. 1990. Derivation of some modern arc magmas by melting of young subducted lithosphere. Nature, 347 : 662-665.

Defant M J, Kepezhinskas P. 2001. Evidence suggests slab melting in arc magmas. EOS, 82(6): 65-69.

Dewey J, Spall H. 1975. Pre-Mesozoic plate tectonics : how far back in Earth history can the Wilson Cycle be extended?. Geology, 3（8）: 422-424.

Dewey J F. 1980. Episodicity, sequence and style at convergent plate margins. Geological Association of Canada Special Paper, 20 : 553-574.

Dickinson W R. 1995. Forearc basins. In : Busby C J, Ingersoll R V.(Editors). Tectonics of Sedimentary Basins. Cambridge : Massachusetts, Blackwell Science : 211-261.

Dilek Y, Yang J S. 2018. Ophiolites, diamonds, and ultrahigh-pressure minerals : new discoveries and concepts on upper mantle petrogenesis. Lithosphere, 10 : 3-13.

Dill H G. 2010. The "chessboard" classification scheme of mineral deposits : mineralogy and geology from aluminum to zirconium. Earth Science Review, 100（1-4）: 1-420.

Dobrzhinetskaya L, Green II H W, Wang S. 1996. Alpe Arami : a peridotite massif from depths of more than 300 km. Science, 271（5257）: 1841-1845.

Du J G, Audétat A. 2020. Early sulfide saturation is not detrimental to porphyry Cu-Au formation. Geology, 48（5）: 519-524.

Ducea M N, Saleeby J B, Bergantz G. 2015a. The architecture, chemistry, and evolution of continental magmatic arcs. The Annual Review of Earth and Planetary Sciences, 43 : 299-331.

Ducea M N, Paterson S R, DeCelles P G. 2015b. High-volume magmatic events in subduction systems. Elements, 11（2）: 99-104.

Duesterhoeft E, Quinteros J, Oberhänsli R, et al. 2014. Relative impact of mantle densification and eclogitization of slabs on subduction dynamics : a numerical thermodynamic/thermokinematic investigation of metamorphic density evolution. Tectonophysics, 637 : 20-29.

Duretz T, Schmalholz S M, Gerya T V. 2012. Dynamics of slab detachment. Geochemistry, Geophysics, Geosystems, 13（3）: 2453.

Durkin K，Castillo P R，Straub S M，et al. 2020. An origin of the along-arc compositional variation in the Izu-Bonin arc system. Geoscience Frontiers，11（5）：1621-1634.

Dziak R P，Bohnenstiehl D R，Baker E T，et al. 2015. Long-term explosive degassing and debris flow activity at West Mata submarine volcano. Geophysical Research Letters，42：1480-1487.

Eiler J. 2003. Inside the subduction factory. Geophysical Monograph Series，138：1-311. Published by American Geophysical Union in Washington DC.

Elburg M A，van Leeuwen T，Foden J，et al. 2002. Origin of geochemical variability by arc-continent collision in the Biru area，southern Sulawesi（Indonesia）. Journal of Petrology，43（4）：581-606.

Elburg M A，van Bergen M J，Foden J D. 2004. Subducted upper and lower continental crust contributes to magmatism in the collision sector of the Sunda-Banda arc，Indonesia. Geology，32（1）：41-44.

Elliott T，Plank T，Zindler A，et al. 1997. Element transport from slab to volcanic front at the Mariana arc. Journal of Geophysical Research-Solid Earth，102（B7）：14991-15019.

England P C，Thompson A B. 1984. Pressure-temperature-time paths of regional metamorphism：part I. Heat transfer during the evolution of regions of thickened continental crust. Journal of Petrology，25（4）：894-928.

England P，Wilkins C. 2004. A simple analytical approximation to the temperature structure in subduction zones. Geophysical Journal International，159（3）：1138-1154.

English J M，Johnston S T，Wang K. 2003. Thermal modeling of the Laramide orogeny testing the flat-slab subduction hypothesis. Earth and Planetary Science Letters，214（3-4）：619-632.

Ernst W G. 1988. Tectonic history of subduction zones inferred from retrograde blueschist P-T paths. Geology，16（12）：1081-1084.

Ero K A，Ekwueme B N. 2009. Mineralization of pegmatites in parts of the Oban Massif, Southeastern Nigeria：a preliminary analysis. Chinese Journal of Geochemistry，28（2）：146-153.

Escrig S，Bézos A，Goldstein S L，et al. 2009. Mantle source variations beneath the Eastern Lau Spreading Center and the nature of subduction components in the Lau basin-Tonga arc system. Geochemistry Geophysics Geosystems，10（4）：Q04014.

Evans K A，Tomkins A G. 2011. The relationship between subduction zone redox budget and arc magma fertility. Earth and Planetary Science Letters，308（3-4）：401-409.

Evans K A，Reddy S M，Tomkins A G，et al. 2017. Effects of geodynamic setting on the redox state of fluids released by subducted mantle lithosphere. Lithos，278-281：26-42.

Falloon T J，Green D H. 1989. The solidus of carbonated，fertile peridotite. Earth and Planetary Science Letters，94（3-4）：364-370.

Fang H J，van der Hilst R D. 2019. Earthquake depth phase extraction with P wave autocorrelation provides insight into mechanisms of intermediate-depth earthquakes. Geophysical Research Letters，46（24）：14440-14449.

Fang W，Dai L Q，Zheng Y F，et al. 2020. Tectonic transition from oceanic subduction to continental collision：new geochemical evidence from Early-Middle Triassic mafic igneous rocks in southern Liaodong Peninsula，east-central China. GSA Bulletin 132（7-8）：1469-1488.

Faryad S W，Cuthbert S J. 2020. High-temperature overprint in（U）HPM rocks exhumed from subduction zones; a product of isothermal decompression or a consequence of slab break-off（slab rollback）?. Earth-Science Reviews，202：103108.

Ferrando S，Frezzotti M L，Dallai L，et al. 2005. Multiphase solid inclusions in UHP rocks（Su-Lu，China）：remnants of supercritical silicate-rich aqueous fluids released during continental subduction. Chemical Geology，223（1-3）：68-81.

Ferrari L. 2004. Slab detachment control on mafic volcanic pulse and mantle heterogeneity in central Mexico. Geology，32（1）：77-80.

Flament N，Gurnis M，Mueller R D，et al. 2015. Influence of subduction history on South American topography. Earth and Planetary Science Letters，430：9-18.

Foley S F，Barth M G，Jenner G A. 2000. Rutile/melt partition coefficients for trace elements and an assessment of the influence of rutile on the trace element characteristics of subduction zone magmas. Geochimica et Cosmochimica Acta，64（5）：933-938.

Foley S F，Tiepolo M，Vannucci R. 2002. Growth of early continental crust controlled by melting of amphibolite in subduction zones. Nature，417（6891）：837-840.

Forsyth D，Uyeda S. 1975. On the relative importance of the driving forces of plate motions. Geophysical Journal International，43（1）：163-200.

Foulger G R，Natland J H. 2003. Is "hotspot" volcanism a consequence of plate tectonics? . Science，300（5621）：921-922.

Fretzdorff S，Livermore R A，Devey C W，et al. 2002. Petrogenesis of the back-arc East Scotia ridge，South Atlantic Ocean. Journal of Petrology，43（8）：1435-1467.

Frezzotti M L，Selverstone J，Sharp Z D，et al. 2011. Carbonate dissolution during subduction revealed by diamond-bearing rocks from the Alps. Nature Geoscience，4（10）：703-706.

Frezzotti M L，Ferrando S. 2015. The chemical behavior of fluids released during deep subduction based on fluid inclusions. American Mineralogist，100（2-3）：352-377.

Frisch W, Meschede M, Blakey R C. 2011. Plate Tectonics. Springer-Verlag Berlin, Heidelberg.

Frohlich C. 2006. A simple analytical method to calculate the thermal parameter and temperature within subducted lithosphere. Physics of the Earth and Planetary Interiors 155（3-4）: 281-285.

Frost D J, McCammon C A. 2008. The redox state of Earth's mantle. Annual Review of Earth and Planetary Sciences, 36 : 389-420.

Fukao Y, Obayashi M, Inoue H, et al. 1992. Subducting slabs stagnant in the mantle transition zone. Journal of Geophysical Research-Solid Earth, 97（B2）: 4809-4822.

Fukao Y, Obayashi M, Nakakuki T, et al. 2009. Stagnant slab : a review. Annual Review of Earth and Planetary Sciences, 37 : 19-46.

Fukao Y, Obayashi M. 2013. Subducted slabs stagnant above, penetrating through, and trapped below the 660 km discontinuity. Journal of Geophysical Research-Solid Earth, 118（11）: 5920-5938.

Fumagalli P, Poli S. 2005. Experimentally determined phase relations in hydrous peridotites to 6.5 GPa and their consequences on the dynamics of subduction zones. Journal of Petrology, 46（3）: 555-578.

Gaetani G A, Grove T L. 2003. Experimental constrains on melt generation in the mantle wedge. Geophysical Monograph Series, 138 : 107-134. Published by American Geophysical Union in Washington DC.

Gale A, Dalton C A, Langmuir C H, et al. 2013. The mean composition of ocean ridge basalts. Geochemistry Geophysics Geosystems, 14（3）: 489-518.

Gamble J, Woodhead J, Wright I, et al. 1996. Basalt and sediment geochemistry and magma petrogenesis in a transect from oceanic island arc to rifted continental margin arc : the Kermadec-Hikurangi margin, SW Pacific. Journal of Petrology, 37（6）: 1523-1546.

Ganade de Araujo C E, Rubatto R, Hermann J, et al. 2014. Ediacaran 2,500-km-long synchronous deep continental subduction in the West Gondwana Orogen. Nature Communications, 5 : 5198.

Ganguly J, Freed A M, Saxena S K. 2009. Density profiles of oceanic slabs and surrounding mantle : integrated thermodynamic and thermal modeling, and implications for the fate of slabs at the 660 km discontinuity. Physics of the Earth and Planetary Interiors, 172（2-4）: 257-267.

Gao S, Rudnick R L, Yuan H L, et al. 2004. Recycling lower continental crust in the North China craton. Nature, 432（7019）: 892-897.

Gao J, Long L L, Klemd R, et al. 2009. Tectonic evolution of the South Tianshan orogen and adjacent regions, NW China : geochemical and age constraints of granitoid rocks. International Journal of Earth Sciences, 98（6）: 1221-1238.

Gao X Y, Zheng Y F, Chen Y X. 2011. U-Pb ages and trace elements in metamorphic zircon and titanite from UHP eclogite in the Dabie orogen：constraints on P-T-t path. Journal of Metamorphic Geology, 29（7）：721-740.

Gao R, Lu Z W, Klemperer S L, et al. 2016. Crustal-scale duplexing beneath the Yarlung Zangbo suture in the western Himalaya. Nature Geoscience, 9（7）：555-560.

Gao X, Wang K L. 2017. Rheological separation of the megathrust seismogenic zone and episodic tremor and slip. Nature, 543：416-419.

Garzanti E, Radeff G, Malusà M G. 2018. Slab breakoff：a critical appraisal of a geological theory as applied in space and time. Earth-Science Reviews, 177：303-319.

Gazel E, Hayes J L, Hoernle K, et al. 2015. Continental crust generated in oceanic arcs. Nature Geoscience, 8（4）：321-327.

Geersen J. 2019. Sediment-starved trenches and rough subducting plates are conducive to tsunami earthquakes. Tectonophysics, 762：28-44.

Gerya T V, Stöckhert B, Perchuk A L. 2002. Exhumation of high-pressure metamorphic rocks in a subduction channel：a numerical simulation. Tectonics, 21（6）：1-19.

Gerya T V, Yuen D A. 2003. Rayleigh-Taylor instabilities from hydration and melting propel 'cold plumes' at subduction zones. Earth and Planetary Science Letters, 212（1-2）：47-62.

Gerya T, Stöckhert B. 2006. Two-dimensional numerical modeling of tectonic and metamorphic histories at active continental margins. International Journal of Earth Sciences, 95（2）：250-274.

Gerya T V, Perchuk L L, Burg J P. 2008. Transient hot channels：perpetrating and regurgitating ultrahigh-pressure, high temperature crust–mantle associations in collision belts. Lithos, 103：236-256.

Gerya T V, Stern R J, Baes M, et al. 2015. Plate tectonics on the Earth triggered by plume-induced subduction initiation. Nature, 527：221-225.

Gill J B. 1974. Role of underthrust oceanic crust in the genesis of a Fijian calc-alkaline suite. Contributions to Mineralogy and Petrology, 43（1）：29-45.

Gill J B. 1981. Orogenic Andesites and Plate Tectonics. Heidelberg Berlin：Springer-Verlag.

Gill R. 2010. Igneous Rocks and Processes：A Practical Guide. London：John Wiley & Sons.

Goes S, Capitanio F A, Morra G, et al. 2011. Signatures of downgoing plate-buoyancy driven subduction in Cenozoic plate motions. Physics of the Earth and Planetary Interiors, 184（1-2）：1-13.

Goes S, Agrusta R, van Hunen J, et al. 2017. Subduction-transition zone interaction：a review. Geosphere, 13（3）：644-664.

Golden J, McMillan M, Downs R T, et al. 2013. Rhenium variations in molybdenite（MoS2）：

Evidence for progressive subsurface oxidation. Earth and Planetary Science Letters, 366：1-5.

Gomberg J，the Cascadia and Beyond Working Group. 2010. Slow-slip phenomena in Cascadia from 2007 and beyond：a review. Geological Society of America Bulletin，122(7-8)：963-978.

Gómez-Tuena A，Straub S M，Zellmer G F. 2013. An introduction to orogenic andesites and crustal growth. Geological Society Special Publications，385（1）：1-13.

Gordon S M，Little T A，Hacker B R，et al. 2012. Multi-stage exhumation of young UHP-HP rocks：timescales of melt crystallization in the D'Entrecasteaux Islands，southeastern Papua New Guinea. Earth and Planetary Science Letters，351-352：237-246.

Green T H，Ringwood A E. 1968. Genesis of the calc-alkaline igneous rock suite. Contributions to Mineralogy and Petrology，18：105-162.

Green D H. 1972. Magmatic activity as the major process in the chemical evolution of the earth's crust and mantle. Tectonophysics，13（1-4）：47-71.

Green D H，Rosenthal A，Kovács I. 2012. Comment on "The beginnings of hydrous mantle wedge melting" by CB Till，TL Grove，AC Withers，Contributions to Mineralogy and Petrology, DOI 10.1007/s00410-011-0692-6. Contributions to Mineralogy and Petrology，164：1077-1081.

Green D H，Hibberson W O，Rosenthal A，et al. 2014. Experimental study of the influence of water on melting and phase assemblages in the upper mantle. Journal of Petrology, 55（10）: 2067-2096.

Groome W G，Thorkelson D J. 2009. The three-dimensional thermo-mechanical signature of ridge subduction and slab window migration. Tectonophys，464（1-4）：70-83.

Grove T，Parman S，Bowring S，et al. 2002. The role of an H_2O-rich fluid component in the generation of primitive basaltic andesites and andesites from the Mt. Shasta region，N California. Contributions to Mineralogy and Petrology，142（4）：375-396.

Grove T L，Till C B，Lev E，et al. 2009. Kinematic variables and water transport control the formation and location of arc volcanoes. Nature，459：694-697.

Grove T L，Till C B，Krawczynski M J. 2012. The role of H_2O in subduction zone magmatism. Annual Review of Earth and Planetary Sciences，40：413-439.

Guest A，Schubert G，Gable C W. 2003. Stress field in the subducting lithosphere and comparison with deep earthquakes in Tonga. Journal of Geophysical Research-Solid Earth，108（B6）：2288.

Guillot S，Hattori K，Agard P，et al. 2009. Exhumation processes in oceanic and continental subduction contexts：a review. In：Lallemand S，Funiciello F.(Editors). Subduction Zone Geodynamics. Berlin Heidelberg：Springer-Verlag：175-205.

Guilmette C, Smit M A, van Hinsbergen D J J, et al. 2018. Forced subduction initiation recorded in the sole and crust of the Semail Ophiolite of Oman. Nature Geoscience, 11（9）: 688-695.

Guo S, Ye K, Yang Y H, et al. 2014. In situ Sr isotopic analyses of epidote : tracing the sources of multi-stage fluids in ultrahigh-pressure eclogite（Ganghe, Dabie terrane）. Contributions to Mineralogy and Petrology, 167（2）: 975.

Guo Z F, Wilson M, Zhang L H, et al. 2014. The role of subduction channel mélanges and convergent subduction systems in the petrogenesis of post-collisional K-rich mafic magmatism in NW Tibet. Lithos, 198-199 : 184-201.

Guo Z F, Wilson M, Zhang M L, et al. 2015. Post-collisional ultrapotassic mafic magmatism in South Tibet : products of partial melting of pyroxenite in the mantle wedge induced by roll-back and delamination of the subducted Indian continental lithosphere slab. Journal of Petrology, 56（7）: 1365-1405.

Guo S, Zhao K D, John T, et al. 2019. Metasomatic flow of metacarbonate-derived fluids carrying isotopically heavy boron in continental subduction zones : insights from tourmaline-bearing ultra-high pressure eclogites and veins（Dabie terrane, eastern China）. Geochimica et Cosmochimica Acta, 253 : 159-200.

Gurnis M. 1989. A reassessment of the heat-transport by variable viscosity convection with plates and lids. Geophysical Research Letters, 16（2）: 179-182.

Gutscher M A, Spakman W, Bijwaard H. 2000. Geodynamics of flat subduction : seismicity and tomographic constraints from the Andean margin. Tectonics, 19（5）: 814-833.

Hacker B R, Peacock S M, Abers G A, et al. 2003. Subduction factory 2. Are intermediate-depth earthquakes in subducting slabs linked to metamorphic dehydration reactions?. Journal of Geophysical Research-Solid Earth, 108 : 2030.

Hacker B R. 2008. H_2O subduction beyond arcs. Geochemistry Geophysics Geosystems, 9（3）: Q03001.

Hacker B R, Kelemen P B, Behn M D. 2011. Differentiation of the continental crust by relamination. Earth and Planetary Science Letters, 307（3-4）: 501-516.

Hacker B R, Gerya T V. 2013. Paradigms, new and old, for ultrahigh-pressure tectonism. Tectonophysics, 603 : 79-88.

Hall P S, Kincaid C. 2001. Diapiric flow at subduction zones : a recipe for rapid transport. Science, 292（5526）: 2472-2475.

Hall P S. 2012. On the thermal evolution of the mantle wedge at subduction zones. Physics of the Earth and Planetary Interiors, 198-199 : 9-27.

Halpaap F, Rondenay S, Perrin A, et al. 2019. Earthquakes track subduction fluids from slab source to mantle wedge sink. Science Advances，5（4）：eaav7369.

Harris N B W，Pearce J A，Tindle A G. 1986. Geochemical characteristics of collision-zone magmatism. Geological Society Special Publication，19：67-81.

Hasegawa A，Umino N，Takagi A. 1978. Double-planed structure of the deep seismic zone in the northeastern Japan arc. Tectonophysics，47（2）：43-58.

Hasenclever J，Morgan J P，Hort M，et al. 2011. 2D and 3D numerical models on compositionally buoyant diapirs in the mantle wedge. Earth and Planetary Science Letters，311（1-2）：53-68.

Hastie A R，Fitton J G，Mitchell S F，et al. 2015. Can fractional crystallization，mixing and assimilation processes be responsible for Jamaican-type adakites? Implications for generating Eoarchaean continental crust. Journal of Petrology，56（7）：1251-1284.

Hawkesworth C J，Gallagher K，Hergt J M，et al. 1993. Mantle and slab contributions in arc magmas. Annual Review of Earth and Planetary Sciences，21（1）：175-204.

Hawkesworth C J，Brown M. 2018. Earth dynamics and the development of plate tectonics. Philosophical Transactions Mathematical Physical and Engineering Sciences，376（2132）：1-5.

Hayes G P，Wald D J，Johnson R L. 2012. Slab1.0：a three-dimensional model of global subduction zone geometries. Journal of Geophysical Research-Solid Earth，117（B1）：B01302.

Hayes G P，Moore G L，Portner D E，et al. 2018. Slab2，a comprehensive subduction zone geometry model. Science，362（6410）：58-61.

He C S，Zheng Y F. 2018. Seismic evidence for the absence of deeply subducted continental slabs in the lower lithosphere beneath the Central Orogenic Belt of China. Tectonophysics，723（4）：178-189.

He Y S，Li S G，Hoefs J，et al. 2011. Post-collisional granitoids from the Dabie orogen：new evidence for partial melting of a thickened continental crust. Geochimica et Cosmochimica Acta，75（13）：3815-3838.

Hermann J，Spandler C J，Hack A，et al. 2006. Aqueous fluids and hydrous melts in high-pressure and ultra-high pressure rocks：implications for element transfer in subduction zones. Lithos，92（3-4）：399-417.

Hermann J，Spandler C J. 2008. Sediment melts at sub-arc depths：an experimental study. Journal of Petrology，49：717-740.

Hermann J，Rubatto D. 2009. Accessory phase control on the trace element signature of sediment melts in subduction zones. Chemical Geology，265（3-4）：512-526.

Hermann J，Zheng Y F，Rubatto D. 2013. Deep fluids in subducted continental crust.

Elements, 9 (4): 281-287.

Hermann J, Rubatto D. 2014. Subduction of continental crust to mantle depth. Treatise on Geochemistry, 4: 309-340.

Hernández-Uribe D, Hernández-Montenegro J D, Cone K A, et al. 2020. Oceanic slab-top melting during subduction: implications for trace-element recycling and adakite petrogenesis. Geology, 48 (3): 216-220.

Herzberg C. 2004. Partial crystallization of mid-ocean ridge basalts in the crust and mantle. Journal of Petrology, 45 (12): 2389-2405.

Herzberg C, Condie K, Korenaga J. 2010. Thermal history of the earth and its petrological expression. Earth and Planetary Science Letters, 292 (1-2): 79-88.

Hess H H. 1962. History of ocean basins. In: Engel A E J, James H L, Leonard B F.(Editors). Petrologic Studies: A Volume in Honor of A. F. Buddington. Boulder, CO: Geological Society of America: 599-620.

Hildreth W, Moorbath S. 1988. Crustal contributions to arc magmatism in the Andes of Central Chile. Contributions to Mineralogy and Petrology, 98: 455-489.

Hirahara Y, Kimura J I, Senda R, et al. 2015. Geochemical variations in Japan Sea back-arc basin basalts formed by high-temperature adiabatic melting of mantle metasomatized by sediment subduction components. Geochemistry Geophysics Geosystems, 16 (5): 1324-1347.

Hirschmann M M, Kogiso T, Baker M B, et al. 2003. Alkalic magmas generated by partial melting of garnet pyroxenite. Geology, 31 (6): 481-484.

Hirschmann M M. 2006. Water, melting, and the deep earth H_2O cycle. Annual Review of Earth and Planetary Science, 2334 (34): 54629-54653.

Hoang N, Uto K. 2006. Upper mantle isotopic components beneath the Ryukyu arc system: evidence for 'back-arc' entrapment of Pacific MORB mantle. Earth and Planetary Science Letters, 249 (3-4): 229-240.

Hochstaedter A G, Gill J B, Peters R, et al. 2001. Across-arc geochemical trends in the Izu-Bonin arc: contributions from the subducting slab. Geochemistry Geophysics Geosystems, 2: 2000GC000105.

Hoernle K, Abt D L, Fischer K M, et al. 2008. Arc-parallel flow in the mantle wedge beneath Costa Rica and Nicaragua. Nature, 451: 1094-1097.

Hofmann A W, White W M. 1982. Mantle plumes from ancient oceanic crust. Earth and Planetary Science Letters, 57 (2): 421-436.

Hofmann A W. 1997. Mantle geochemistry: the message from oceanic volcanism. Nature, 385: 219-229.

Holder R M, Viete D R, Brown M, et al. 2019. Metamorphism and the evolution of plate tectonics. Nature, 572 (7769): 378-381.

Holmes A. 1931. Radioactivity and earth movements. Nature, 128 (3229): 496.

Hou Z Q, Gao Y F, Qu X M, et al. 2004. Origin of adakitic intrusives generated during mid-Miocene east-west extension in southern Tibet. Earth and Planetary Science Letters, 220 (1-2): 139-155.

Hou Z Q, Yang Z M, Lu Y J, et al. 2015. A genetic linkage between subduction-and collision-related porphyry Cu deposits in continental collision zones. Geology, 43 (3): 247-250.

Hou Z Q, Zhou Y, Wang R, et al. 2017. Recycling of metal-fertilized lower continental crust: origin of non-arc Au-rich porphyry deposits at cratonic edges. Geology, 45 (6): 563-566.

Houseman G A, McKenzie D P, Molnar P. 1981. Convective instability of a thickened boundary layer and its relevance for the thermal evolution of continental convergent belts. Journal of Geophysical Research-Solid Earth, 86 (B7): 6115-6132.

Huang J L, Zhao D P. 2006. High-resolution mantle tomography of China and surrounding regions. Journal of Geophysical Research-Solid Earth, 111 (B9): B09305.

Humphreys E. 2009. Relation of flat subduction to magmatism and deformation in the western United States. Geological Society of America Memoir, 204: 85-98.

Hwang S L, Shen P, Chu H T, et al. 2000. Nanometersize K-PbO_2-type TiO_2 in garnet: a thermobarometer for ultrahigh-pressure metamorphism. Science, 288: 321-324.

Irifune T. 1993. Phase transformations in the earth's mantle and subducting slabs: implications for their compositions, seismic velocity and density structures and dynamics. Island Arc, 2 (2): 55-71.

Isacks B, Molnar P. 1969. Mantle earthquake mechanisms and the sinking of the lithosphere. Nature, 223: 1121-1124.

Ishikawa T, Nakamura E. 1994. Origin of the slab component in arc lavas from across-arc variation of B and Pb isotopes. Nature, 370: 205-208.

Ishizuka O, Taylor R N, Milton J A, et al. 2003. Fluid-mantle interaction in an intra-oceanic arc: constraints from high-precision Pb isotopes. Earth and Planetary Science Letters, 211: 221-236.

Ishizuka O, Tani K, Reagan M K. 2014. Izu-Bonin-Mariana forearc crust as a modern ophiolite analogue. Elements, 10 (2): 115-120.

Jackson M, Blundy J, Sparks R. 2018. Chemical differentiation, cold storage and remobilization of magma in the Earth's crust. Nature, 564: 405-409.

Jagoutz O, Kelemen P B. 2015. Role of arc processes in the formation of continental crust.

Annual Review of Earth and Planetary Sciences, 43 : 363-404.

Jagoutz O, Klein B. 2018. On the importance of crystallization-differentiation for the generation of SiO_2-rich melts and the compositional build-up of arc (and continental) crust. American Journal of Science, 318 : 29-63.

Jahn B, Cornichet J, Cong B L, et al. 1996. Ultrahigh- ε Nd eclogites from an ultrahigh-pressure metamorphic terrane of China. Chemical Geology, 127 (1-3): 61-79.

Jahn B, Caby R, Monie P. 2001. The oldest UHP eclogites of the world : age of UHP metamorphism, nature of protoliths and tectonic implications. Chemical Geology, 178(1-4): 143-158.

James D E, Sacks I S. 1999. Cenozoic formation of the Central Andes : a geophysical perspective. In : Skinner B J.(Editor). Geology and ore deposits of the Central Andes. Society of Economic Geologists Special Publication, 7 : 1-25.

Jarrard R D. 1986. Relations among subduction parameters. Reviews of Geophysics, 24 (2): 184-217.

John T, Medvedev S, Rupke L H, et al. 2009. Generation of intermediate-depth earthquakes by self-localizing thermal runaway. Nature Geoscience, 2 : 137-140.

John T, Gussone N, Podladchikov Y Y, et al. 2012. Volcanic arcs fed by rapid pulsed fluid flow through subducting slabs. Nature Geoscience, 5 (7): 489-492.

Jolivet L, Faccenna C, Goffé B, et al. 2003. Subduction tectonics and exhumation of high-pressure metamorphic rocks in the Mediterranean orogens. American Journal of Science, 303 (5): 353-409.

Kamenetsky V S, Eggins S M, Crawford A J, et al. 1998. Calcic melt inclusions in primitive olivine at 43°N MAR : evidence for melt-rock reaction/melting involving clinopyroxene-rich lithologies during MORB generation. Earth and Planetary Science Letters, 160 : 115-132.

Kaneko Y, Katayama I, Yamamoto H, et al. 2003. Timing of Himalayan ultrahigh-pressure metamorphism : sinking rate and subduction angle of the Indian continental crust beneath Asia. Journal of Metamorphic Geology, 21 : 589-599.

Kapp P, Yin A, Manning C E, et al. 2003. Tectonic evolution of the early Mesozoic blueschist-bearing Qiangtang metamorphic belt, central Tibet. Tectonics, 22 (4): 1043.

Karato S. 2012. On the origin of the asthenosphere. Earth and Planetary Science Letters, 321-322 : 95-103.

Kattenhorn S A, Prockter L M. 2014. Evidence for subduction in the ice shell of Europa. Nature Geoscience, 7 : 762-767.

Kawamoto T, Kanzaki M, Mibe K, et al. 2012. Separation of supercritical slab-fluids to form aqueous fluid and melt components in subduction zone magmatism. Proceedings of the

National Academy of Sciences，109：18695-18700.

Kay R W. 1978. Aleutian magnesian andesites：melts from subducted Pacific Ocean crust. Journal of Volcanology and Geothermal Research，4（1-2）：117-132.

Kay R W，Kay S M. 1991. Creation and destruction of lower continental crust. Geologische Rundschau，80（2）：259-278.

Kay S M，Jicha B R，Citron G L，et al. 2019. The calc-alkaline hidden Bay and Kagalaska plutons and the construction of the Central Aleutian Oceanic arc crust. Journal of Petrology，60（2）：393-439.

Kearey P，Klepeis K A，Vine F J. 2009. Globle Tectonics. Chiechester：Wiley-Blackwell .

Kelemen P B，Hart S R，Bernstein S. 1998. Silica enrichment in the continental upper mantle via melt/rock reaction. Earth and Planetary Science Letters，164（1）：387-406.

Kelemen P B，Rilling J L，Parmentier E M，et al. 2003. Thermal structure due to solid-state flow in the mantle wedge beneath arcs. Geophysical Monograph，138：293-311.

Kelemen P B，Hanghoj K，Greene A R. 2014. One view of the geochemistry of subduction-related magmatic arcs，with an emphasis on primitive andesite and lower crust. Treatise on Geochemistry，4：749-805.

Kelemen P B，Behn M D. 2016. Formation of lower continental crust by relamination of buoyant arc lavas and plutons. Nature Geoscience，9（3）：197-205.

Keller C B，Schoene B，Barboni M，et al. 2015. Volcanic-plutonic parity and the differentiation of the continental crust. Nature，523（7560）：301-307.

Kelley K A，Plank T，Grove T L，et al. 2006. Mantle melting as a function of water content beneath back-arc basins. Journal of Geophysical Research-Solid Earth，111（B9）：B09208.

Kelley K A，Cottrell E. 2009. Water and the oxidation state of subduction zone magmas. Science，325：605-607.

Kemp A I S，Hawkesworth C J，Foster G L，et al. 2007. Magmatic and crustal differentiation history of granitic rocks from Hf-O isotopes in zircon. Science，315（5814）：980-983.

Kepezhinskas P，Defant M J. 1996. Contrasting styles of mantle metasomatism above subduction zones：constraints from ultramafic xenoliths in Kamchatka. Geophys. Monogr. AGU，96：307-313.

Kepezhinskas P，McDermott F，Defant M J，et al. 1997. Trace element and Sr-Nd-Pb isotopic constraints on a three-component model of Kamchatka arc petrogenesis. Geochimica et Cosmochimica Acta，61（3）：577-600.

Kerrich R，Goldfarb R，Groves D，et al. 2000. The geodynamics of world-class gold deposits：characteristics，space-time distributions，and origins. Reviews in Economic Geology，13：501-551.

Kessel R, Schmidt M W, Ulmer P, et al. 2005. Trace element signature of subduction-zone fluids, melts and supercritical liquids at 120-180 km depth. Nature, 437：724-727.

Kincaid C, Sacks I S. 1997. Thermal and dynamical evolution of the upper mantle in subduction zones. Journal of Geophysical Research-Solid Earth, 1021（B6）: 12295-12315.

Kirby S H, Durham W B, Stern L A. 1991. Mantle phase changes and deep-earthquake faulting in subducting lithosphere. Science, 252（5003）: 216-225.

Kirby S H, Engdahl R E, Denlinger R. 1996. Intermediatedepth Intraslab earthquakes and arc volcanism as physical expressions of crustal and uppermost mantle metamorphism in subducting slabs. Geophysical Monograph, 96：195-214.

Klein E M, Langmuir C H. 1987. Global correlations of ocean ridge basalt chemistry with axial depth and crustal thickness. Journal of Geophysical Research-Solid Earth, 92（B8）: 8089-8115.

Kogiso T, Hirschmann M M, Reiners P W. 2004a. Length scales of mantle heterogeneities and their relationship to ocean island basalt geochemistry. Geochimica et Cosmochimica Acta, 68（2）: 345-360.

Kogiso T, Hirschmann M M, Pertermann M. 2004b. High-pressure partial melting of mafic lithologies in the mantle. Journal of Petrology, 45（12）: 2407-2422.

Kohn M J, Castro A E, Kerswell B C, et al. 2018. Shear heating reconciles thermal models with the metamorphic rock record of subduction. Proceedings of the National Academy of Sciences, 115（46）: 11706-11711.

Konzett J, Ulmer P. 1999. The Stability of hydrous potassic phases in Lherzolitic Mantle—an experimental study to 9.5 GPa in simplified and natural bulk compositions. Journal of Petrology, 40（4）: 629-652.

Korenaga J. 2012. Plate tectonics and planetary habitability：current status and future challenges. Annals of the New York Academy of Sciences, 1260（1）: 87-94.

Korenaga J. 2013. Initiation and evolution of plate tectonics on earth：theories and observations. Annual Review of Earth and Planetary Sciences, 41（1）: 117-151.

Korsakov A V, Hermann J. 2006. Silicate and carbonate melt inclusions associated with diamond in deeply subducted carbonated rocks. Earth and Planetary Science Letters, 241（1-2）: 104-118.

Koulakov I, Sobolev S V. 2006. A tomographic image of Indian lithosphere break-off beneath the Pamir-Hindukush region. Geophysical Journal International, 164（2）: 425-440.

Krawczynski M J, Grove T L, Behrens H. 2012. Amphibole stability in primitive arc magmas：effects of temperature, H_2O content, and oxygen fugacity. Contributions to Mineralogy and Petrology, 164：317-339.

Krogh Ravna E J，Terry M P. 2004. Geothermobarometry of UHP and HP eclogites and schists–an evaluation of equilibria among garnet-clinopyroxene-kyanite-phengite–coesite/quartz. Journal of Metamorphic Geology，22：579-592.

Kuiper Y D，Wakabayashi J. 2018. A comparison between mid-Paleozoic New England，USA，and the modern western USA：subduction of an oceanic ridge-transform fault system. Tectonophysics，745：278-292.

Kuno H. 1959. Origin of Cenozoic petrographic provinces of Japan and surrounding areas. Bulletin Volcanologique，20：37-76.

Kushiro I. 1968. Compositions of magmas formed by partial zone melting of the Earth's upper mantle. Journal of Geophysical Research，73(2): 619-634.

Kushiro I，Sato H. 1978. Origin of some calc-alkaline andesites in Japanese islands. Bulletin Volcanologique，41（4）: 576-585.

Kylander-Clark A R C，Hacker B R，Mattinson C G. 2012. Size and exhumation rate of ultrahigh-pressure terranes linked to orogenic stage. Earth and Planetary Science Letters，321-322：115-120.

Labrousse L，Prouteau G，Ganzhorn A C. 2011. Continental exhumation triggered by partial melting at ultrahigh pressure. Geology，39（12）: 1171-1174.

Lai X D，Yang X Y，Sun W D. 2012. Geochemical constraints on genesis of dolomite marble in the Bayan Obo REE-Nb-Fe deposit，Inner Mongolia：implications for REE mineralization. Journal of Asian Earth Sciences，57：90-102.

Lambart S，Koornneef J M，Millet M A，et al. 2019. Highly heterogeneous depleted mantle recorded in the lower oceanic crust. Nature Geoscience，12：482-486.

Langmuir C H，Bezos A，Escrig S，et al. 2006. Chemical systematics and hydrous melting of the mantle in bac-arc basins. In: Christie D M, Fisher C R, Lee S M, Givens S. (Eds.), Back-arc Spreading Systems: Geological, Biological, Chemical, and Physical Interactions. Geophysical Monograph Series，166：87-146. Published by American Geophysical Union in Washington DC.

Lay T，Kanamori H，Ammon C J，et al. 2012. Depth-varying rupture properties of subduction zone megathrust faults. Journal of Geophysical Research-Solid Earth，117（B4）: B04311.

Lay T. 2015. The surge of great earthquakes from 2004 to 2014. Earth and Planetary Science Letters，409（1）: 133-146.

Le Pape F，Jones A G，Vozar J，et al. 2012. Penetration of crustal melt beyond the Kunlun Fault into northern Tibet. Nature Geoscience，5（5）: 330-335.

Le Pichon X. 1968. Sea-floor spreading and continental drift. Journal of Geophysical Research，73：3661-3705.

Le Pichon X, Francheteau J, Bonnin, J. 1973. Plate Tectonics. Amsterdam : Elsevier.

Leat P T, Pearce J A, Barker P F, et al. 2004. Magma genesis and mantle flow at a subducting slab edge : the South Sandwich arc-basin system. Earth and Planetary Science Letters, 227 (1-2): 17-35.

Lee C T A, Cheng X, Horodyskyj U. 2006. The development and refinement of continental arcs by primary basaltic magmatism, garnet pyroxenite accumulation, basaltic recharge and delamination : insights from the Sierra Nevada, California. Contributions to Mineralogy and Petrology, 151 : 222-242.

Lee C T A, Morton D M, Kistler R W, et al. 2007. Petrology and tectonics of Phanerozoic continent formation : from island arcs to accretion and continental arc magmatism. Earth and Planetary Science Letters, 263 (3-4): 370-387.

Lee C T A, Luffi P, Plank, T, et al. 2009. Constraints on the depths and temperatures of basaltic magma generation on Earth and other terrestrial planets using new thermobarometers for mafic magmas. Earth and Planetary Science Letters, 279 (1-2): 20-33.

Lee C T A, Luffi P, Chin E J. 2011. Building and destroying continental mantle. Annual Review of Earth and Planetary Sciences, 39 : 59-90.

Lee C T A, Bachmann O. 2014. How important is the role of crystal fractionation in making intermediate magmas? Insights from Zr and P systematics. Earth and Planetary Science Letters, 393 : 266-274.

Lenze A, Stöckhert B. 2007. Microfabrics of UHP metamorphic granites in the Dora Maira Massif, western Alps-no evidence of deformation at great depth. Journal of Metamorphic Geology, 25 (4): 461-475.

Li S G, Xiao Y, Liou D, et al. 1993. Collision of the North China and Yangtse Blocks and formation of coesite-bearing eclogites : timing and processes. Chemical Geology, 109 (1): 89-111.

Li S G, Wang S S, Chen Y Z, et al. 1994. Excess argon in phengite from eclogite : evidence from dating of eclogite minerals by Sm-Nd, Rb-Sr and $^{40}Ar/^{39}Ar$ methods. Chemical Geology, 112 (3): 343-350.

Li S G, Jagoutz E, Chen Y, et al. 2000. Sm-Nd and Rb-Sr isotopic chronology and cooling history of ultrahigh pressure metamorphic rocks and their country rocks at Shuanghe in the Dabie Mountains, Central China. Geochimica et Cosmochimica Acta, 64 (6): 1077-1093.

Li X, Bock G, Vafidis A, et al. 2003. Receiver function study of the Hellenic subduction zone : imaging crustal thickness variations and the oceanic Moho of the descending African lithosphere. Geophysical Journal of the Royal Astronomical Society, 155 (2): 733-748.

Li Z X，Evans D A D，Zhang S. 2004. A 90° Spin on Rodinia：possible causal links between the Neoproterozoic supercontinent，superplume，true polar wander and low-latitude glaciation. Earth and Planetary Science Letters，220（3）：409-421.

Li C，Zhai Q G，Dong Y S，et al. 2006. Discovery of eclogite and its geological significance in Qiangtang area，central Tibet. Chinese Science Bulletin，51（9）：1095-1100.

Li Z X，Li X H. 2007. Formation of the 1300-km-wide intracontinental orogen and postorogenic magmatic province in Mesozoic South China：a flat-slab subduction model. Geology，35（2）：179-182.

Li Z X，Bogdanova S V，Collins A S，et al. 2008. Assembly，configuration，and break-up history of Rodinia：a synthesis. Precambrian Research，160（1-2）：179-210.

Li C，van Der Hilst R D. 2010. Structure of the upper mantle and transition zone beneath Southeast Asia from traveltime tomography. Journal of Geophysical Research-Solid Earth，115：B07308.

Li Z H，Di Leo J F，Ribe N M. 2014. Subduction-induced mantle flow，finite strain，and seismic anisotropy：numerical modeling. Journal of Geophysical Research-Solid Earth，119（6）：5052-5076.

Li H Y，Xu Y G，Ryan J G，et al. 2016a. Olivine and melt inclusion chemical constraints on the source of intracontinental basalts from the eastern North China Craton：discrimination of contributions from the subducted Pacific slab. Geochimica et Cosmochimica Acta，178：1-19.

Li H Y，Zhou Z，Ryan J G，et al. 2016b. Boron isotopes reveal multiple metasomatic events in the mantle beneath the eastern North China Craton. Geochimica et Cosmochimica Acta，194：77-90.

Li J L，Klemd R，Gao J，et al. 2016. Poly-cyclic metamorphic evolution of eclogite：evidence for multistage burial-exhumation cycling in a subduction channel. Journal of Petrology，57（1）：119-146.

Li S G，Yang W，Ke S，et al. 2017. Deep carbon cycles constrained by a large-scale mantle Mg isotope anomaly in eastern China. National Science Review, 4（1）：111-120.

Li H Y，Taylor R N，Prytulak J，et al. 2019a. Radiogenic isotopes document the start of subduction in the Western Pacific. Earth and Planetary Science Letters，518：197-210.

Li H Y，Li J，Ryan J G，et al. 2019b. Molybdenum and boron isotope evidence for fluid-fluxed melting of intraplate upper mantle beneath the eastern North China Craton. Earth and Planetary Science Letters，520：105-114.

Li Z H，Gerya T，Connolly J A D. 2019. Variability of subducting slab morphologies in the mantle transition zone：insight from petrological-thermomechanical modeling. Earth-Science Reviews，196：102874.

Li J L，Schwarzenbach E M，John T，et al. 2020. Uncovering and quantifying the subduction zone sulfur cycle from the slab perspective. Nature Communications，11（1）：514.

Liang X F，Chen Y，Tian X B，et al. 2016. 3D imaging of subducting and fragmenting Indian continental lithosphere beneath southern and central Tibet using body-wave finite-frequency tomography. Earth and Planetary Science Letters，443：162-175.

Liegeois J P. 1998. Preface：some words on post-collisional magmatism. Lithos，45：15-17.

Linnen R L，Lichtervelde M V，Černý P. 2012. Granitic pegmatites as sources of strategic metals. Elements，8（4）：275-280.

Liou J G，Tsujimori T，Zhang R Y，et al. 2004. Global UHP metamorphism and continental subduction/collision：the Himalayan model. International Geology Review，46（1）：1-27.

Liou J G，Ernst W G，Zhang R Y，et al. 2009. Ultrahigh-pressure minerals and metamorphic terranes—the view from China. Journal of Asian Earth Sciences，35：199-231.

Liou J G，Tsujimori T，Yang J，et al. 2014. Recycling of crustal materials through study of ultrahigh-pressure minerals in collisional orogens，ophiolites，and mantle xenoliths：a review. Journal of Asian Earth Sciences，96：386-420.

Lissenberg C J，Dick H J B. 2008. Melt-rock reaction in the lower oceanic crust and its implications for the genesis of mid-ocean ridge basalt. Earth and Planetary Science Letters，271（1-4）：311-325.

Lithgow-Bertelloni C，Richards M A. 1998. The dynamics of Cenozoic and Mesozoic plate motions. Reviews of Geophysics，36（1）：27-78.

Liu L，Sun Y，Xiao P X，et al. 2002. Discovery of ultrahigh-pressure magnesite-bearing garnet lherzolite（>3.8 GPa）in the Altyn Tagh，Northwest China. Chinese Science Bulletin，47（11）：881-886.

Liu L，Chen D L，Zhang A D，et al. 2005. Ultrahigh pressure gneissic K-feldspar garnet clinopyroxenite in the Altyn Tagh，NW China evidence from clinopyroxene exsolution in garnet. Science China Earth Science，48（7）：1000-1010.

Liu F L，Xu Z Q，Liou J G，et al. 2007. Ultrahigh-pressure mineral assemblages in zircons from the surface to 5158 m depth in cores of the main drill hole，Chinese Continental Scientific Drilling Project，southwestern Sulu belt，China. International Geological Review，49（5）：454-478.

Liu L，Zhang J，Green Ii H W，et al. 2007. Evidence of former stishovite in metamorphosed sediments，implying subduction to ＞350 km. Earth and Planetary Science Letters，263（3-4）：180-191.

Liu X W，Jin Z M，Green H W. 2007. Clinoenstatite exsolution in diopsidic augite of Dabieshan：garnet peridotite from depth of 300 km. American Mineralogist，92（4）：546-

552.

Liu L J, Spasojevic S, Gurnis M. 2008. Reconstructing Farallon plate subduction beneath North America back to the late Cretaceous. Science, 322 : 934-938.

Liu F L, Liou J G. 2011. Zircon as the best mineral for P-T-time history of UHP metamorphism : a review on mineral inclusions and U-Pb SHRIMP ages of zircons from the Dabie-Sulu UHP rocks. Journal of Asian Earth Sciences, 40 (1): 1-39.

Liu L J, Stegman D R. 2012. Origin of Columbia River flood basalt controlled by propagating rupture of the Farallon slab. Nature, 482 : 386-389.

Liu L, Liao X Y, Wang Y W, et al. 2016. Early Paleozoic tectonic evolution of the North Qinling Orogenic Belt in Central China : insights on continental deep subduction and multiphase exhumation. Earth-Science Reviews, 159 : 58-81.

Liu S, Currie C A. 2016. Farallon plate dynamics prior to the Laramide orogeny : numerical models of flat subduction. Tectonophys, 666 : 33-47.

Liu X, Zhao D P. 2016. P and S wave tomography of Japan subduction zone from joint inversions of local and teleseismic travel times and surface-wave data. Physics of the Earth and Planetary Interiors, 252 : 1-22.

Liu F L, Zhang L F, Li X L, et al. 2017. The metamorphic evolution of Paleoproterozoic eclogites in Kuru-Vaara, northern Belomorian Province, Russia : constraints from P-T pseudosections and zircon dating. Precambrian Research, 289 : 31-47.

Liu M Q, Li Z H, Yang S H. 2017. Diapir versus along-channel ascent of crustal material during plate convergence : constrained by the thermal structure of subduction zones. Journal of Asian Earth sciences, 145 : 16-36.

Liu L, Zhang J F, Cao Y T, et al. 2018. Evidence of former stishovite in UHP eclogite from the South Altyn Tagh, western China. Earth and Planetary Science Letters, 484 : 353-362.

Liu H, Zartman R E, Ireland T R, et al. 2019. Global atmospheric oxygen variations recorded by Th/U systematics of igneous rocks. Proceedings of the National Academy of Sciences, 116 (38): 18854-18859.

Livermore R. 2018. The Tectonic Plates Are Moving. Oxford : Oxford University Press.

London D. 2009. The origin of primary textures in granitic pegmatites. Canadian Mineralogist, 47 : 697-724.

Lupton J E, Arculus R J, Evans L J, et al. 2012. Mantle hotspot neon in basalts from the Northwest Lau Back-arc Basin. Geophysical Research Letter, 39 (8): L08308.

Lyons T W, Reinhard C T, Planavsky N J. 2014. The rise of oxygen in Earth's early ocean and atmosphere. Nature, 506 : 307-315.

Lytle M L, Kelley K A, Hauri E H, et al. 2012. Tracing mantle sources and Samoan influence in the northwestern Lau back-arc basin. Geochemistry Geophysics Geosystems，13（10）：Q10019.

Lyu C，Pedersen H A，Paul A，et al. 2017. Shear wave velocities in the upper mantle of the Western Alps：new constraints using array analysis of seismic surface waves. Geophysical Journal International，210（1）：321-331.

Ma L，Wang Q，Li Z X，et al. 2017. Subduction of Indian continent beneath southern Tibet in the latest Eocene（~35 Ma）：insights from the Quguosha gabbros in southern Lhasa block. Gondwana Research，41：77-92.

Macera P M，Gasperini D，Ranalli G，et al. 2008. Slab detachment and mantle plume upwelling in subduction zones：an example from the Italian South-Eastern Alps. Journal of Geodynamics，45（1）：32-48.

Macpherson C G，Hilton D R，Mattey D P，et al. 2000. Evidence for an ^{18}O-depleted mantle plume from contrsting $^{18}O/^{16}O$ ratios of back-arc lavas from the Manus basin and Mariana trough. Earth and Planetary Science Letters，176（2）：171-183.

Magee C，Stevenson C T E，Ebmeier S K，et al. 2018. Magma plumbing systems：a geophysical perspective. Journal of Petrology，59（6）：1217-1251.

Magni V，van Hunen J，Funiciello F，et al. 2012. Numerical models of slab migration in continental collision zones. Solid Earth，3（2）：293-306.

Magni V，Faccenna C，van Hunen J，et al. 2013. Delamination vs. break-off：the fate of continental collision. Geophysical Research Letters，40（2）：285-289.

Maierova P，Schulmann K，Gerya T. 2018. Relamination styles in collisional orogens. Tectonics，37（1-2）：224-250.

Malaspina N，Hermann J，Scambelluri M，et al. 2006. Polyphase inclusions in garnet-orthopyroxenite（Dabie Shan，China）as monitors for metasomatism and fluid-related trace element transfer in subduction zone peridotite. Earth and Planetary Science Letters，249（3-4）：173-187.

Mallik A，Dasgupta R，Tsuno K，et al. 2016. Effects of water，depth and temperature on partial melting of mantle-wedge fluxed by hydrous sediment-melt in subduction zones. Geochimica et Cosmochimica Acta，195：226-243.

Malusà M G，Faccenna C，Baldwin S L，et al. 2015. Contrasting styles of（U）HP rock exhumation along the Cenozoic Adria-Europe plate boundary（Western Alps，Calabria，Corsica）. Geochemistry Geophysics Geosystems，16（6）：1786-1824.

Mandler B，Grove T L. 2016. Controls on the stability and composition of amphibole in the Earth's mantle. Contributions to Mineralogy and Petrology，171（8-9）：68.

Mann P，Taira A. 2004. Global tectonic significance of the Solomon Islands and Ontong Java

Plateau convergent zone. Tectonophysics，389（3-4）：137-190.

Manning C E. 2004. The chemistry of subduction-zone fluids. Earth and Planetary Science Letters，223（1-2）：1-16.

Margheriti L，Lucente F P，Pondrelli S. 2003. SKS splitting measurements in the Apenninic–Tyrrhenian domain（Italy）and their relationship with lithospheric subduction and mantle convection. Journal of Geophysical Research-Solid Earth，108（B4）：2218.

Marschall H R，Schumacher J C. 2012. Arc magmas sourced from melange diapirs in subduction zones. Nature Geosciences，5：862-867.

Martin H，Smithies R H，Rapp R，et al. 2005. An overview of adakite tonalite-trondhjemite-granodiorite，TTG and sanukitoid：relationships and some implications for crustal evolution. Lithos，79（1-2）：1-24.

Martin H，Moyen J F，Guitreau M，et al. 2014. Why Archaean TTG cannot be generated by MORB melting in subduction zones. Lithos，198-199：1-13.

Martinez F，Taylor B. 2002. Mantle wedge control on back-arc crustal accretion. Nature，416（6879）：417-420.

Martinez F，Taylor B. 2006. Modes of crustal accretion in back-arc basins：inferences from the Lau Basin. Geophysical Monograph，166：5-30.

Maruyama S. 1994. Plume tectonics. Journal of Geological Society of Japan，100（1）：24-49.

Maruyama S，Liou J G，Terabayashi M. 1996. Blueschists and eclogites of the world and their exhumation. International Geology Review，38（6）：485-594.

Maruyama S，Santosh M，Zhao D. 2007. Superplume，supercontinent，and postperovskite：mantle dynamics and anti-plate tectonics on the core–mantle boundary. Gondwana Research，11（1-2）：7-37.

Maruyama S，Hasegawa A，Santosh M，et al. 2009. The dynamics of big mantle wedge，magma factory，and metamorphic-metasomatic factory in subduction zones. Gondwana Research，16：414-430.

McBirney A R. 1969. Compositional variations in Cenozoic calc-alkaline suites of Central America. Oregon Department of Geology and Mineral Industries Bulletins，65：1-185.

McCulloch M T，Gamble J A. 1991. Geochemical and geodynamical constraints on subduction zone magmatism. Earth and Planetary Science Letters，102（3-4）：358-374.

McGuire J，Plank T，Barrientos S，et al. 2017. The SZ4D Initiative：Understanding the Processes that Underlie Subduction Zone Hazards in 4D. The IRIS Consortium.

McKenzie D P，Parker R L. 1967. North pacific—an example of tectonics on a sphere. Nature，216：1267-1280.

McKenzie D P. 1969. Speculations on the consequences and causes of plate motions. Geophysical

Journal International，18（1）：1-32.

McKenzie D，Bickle M J. 1988. The volume and composition of melt generated by extension of the lithosphere. Journal Petrology，29（3）：625-679.

Menant A，Angiboust S，Monié P，et al. 2018. Brittle deformation during Alpine basal accretion and the origin of seismicity nests above the subduction interface. Earth and Planetary Science Letters，487：84-93.

Métrich N，Schiano P，Clocchiatti R，et al. 1999. Transfer of sulfur in subduction settings：an example from Batan Island（Luzon volcanic arc，Philippines）. Earth and Planetary Science Letters，167（1-2）：1-14.

Millensifer T A，Sinclair D，Jonasson I，et al. 2014. Rhenium. In：Gunn G.(Editor). Critical Metals Handbook. Oxford：John Wiley & Sons：340-360.

Miyashiro A. 1961. Evolution of metamorphic belts. Journal Petrology，2（3）：277-311.

Molnar P，Freedman D，Shih J S. 1979. Lengths of intermediate and deep seismic zones and temperatures in downgoing slabs of lithosphere. Geophysical Journal International，56（1）：41-54.

Molnar P，England P. 1990. Temperatures，heat flux，and frictional stress near major thrust faults. Journal of Geophysical Research-Solid Earth，95（B4）：4833-4856.

Molnar P，England P. 1995. Temperatures in zones of steady-state underthrusting of young oceanic lithosphere. Earth and Planetary Science Letters，131（1-2）：57-70.

Moore W B，Webb A A. 2013. Heat-pipe Earth. Nature，501：501-505.

Moores E M，Yıkılmaz M B，Kellogg L H. 2013. Tectonics：50 years after the revolution. Special Paper of the Geological Society of America，500：321-369.

Morgan W J. 1968. Rises，trenches，great faults and crustal blocks. Journal of Geophysical Research，73（6）：1959-1982.

Morgan W J. 1971. Convection plumes in the lower mantle. Nature，230（5288）：42-43.

Morgan W J. 1972. Deep mantle convection plumes and plate motions. AAPG Bulletin，56（2）：203-213.

Mori L，Gomez-Tuena A，Schaaf P，et al. 2009. Lithospheric removal as a trigger for flood basalt magmatism in the Trans-Mexican Volcanic Belt. Journal of Petrology，50（11）：2157-2186.

Morishige M，van Keken P E. 2014. Along-arc variation in the 3-D thermal structure around the junction between the Japan and Kurile arcs. Geochemistry Geophysics Geosystems，15：2225-2240.

Morris J D，Leeman W P，Tera F. 1990. The subducted component in island arc lavas：constraints from Be isotopes and B-Be systematics. Nature，344（6261）：31-36.

Moyen J F，Martin H. 2012. Forty years of TTG research. Lithos，148：312-336.

Moyen J F，Laurent O，Chelle-Michou C，et al. 2017. Collision vs. subduction-related magmatism：two contrasting ways of granite formation and implications for crustal growth. Lithos，277：154-177.

Mposkos E D，Kostopoulos D K. 2001. Diamond，former coesite and supersilicic garnet in metasedimentary rocks from the Greek Rhodope：a new ultrahigh-pressure metamorphic province established. Earth and Planetary Science Letters，192（4）：497-506.

Müller R D，Sdrolias M，Gaina C，et al. 2008. Long-term sea-level fluctuations driven by ocean basin dynamics. Science，319（5868）：1357-1362.

Müller R D，Seton M，Zahirovic S，et al. 2016. Ocean basin evolution and global-scale plate reorganization events since Pangea breakup. Annual Review of Earth and Planetary Sciences，44：107-138.

Mundl A，Touboul M，Jackson M G，et al. 2017. Tungsten-182 heterogeneity in modern ocean island basalts. Science，356：66-69.

Nandedkar R H，Ulmer P，Müntener O. 2014. Fractional crystallization of primitive hydrous arc magmas：an experimental study at 0.7 GPa. Contributions to Mineralogy and Petrology，167（6）：1015.

Nebel O，Arculus R J. 2015. Selective ingress of a Samoan plume component into the northern Lau backarc basin. Nature Communications，6：6554.

Negredo A M，Replumaz A，Villasenor A，et al. 2007. Modeling the evolution of continental subduction processes in the Pamir-Hindu Kush region. Earth and Planetary Science Letters，259（1-2）：212-225.

Németh K. 2012. W. Frisch，M. Meschede，R.C. Blakey：plate tectonics—continental drift and mountain building. Bulletin of Volcanology，74（1）：305-307.

Newman S，Stolper E，Stern R. 2000. H_2O and CO_2 in magmas from the Mariana arc and back arc systems. Geochemistry Geophysics Geosystems，1（5）：1013.

Ni H W，Zhang L，Xiong X L，et al. 2017. Supercritical fluids at subduction zones：evidence，formation condition，and physicochemical properties. Earth-Science Reviews，167：62-71.

Nielsen S G，Marschall H R. 2017. Geochemical evidence for melange melting in global arcs. Science Advances，3（4）：e1602402.

Nishikawa T，Matsuzawa T，Ohta K，et al. 2019. The slow earthquake spectrum in the Japan Trench illuminated by the S-net seafloor observatories. Science，365：808-813.

Nishio Y，Sasaki S，Gamo T，et al. 1998. Carbon and helium isotope systematics of North Fiji Basin basalt glasses：carbon geochemical cycle in the subduction zone. Earth and Planetary

Science Letters, 154 (1-4): 127-138.

Niu Y L, O'Hara M J, Pearce J A. 2001. Initiation of subduction zones as a consequence of lateral compositional buoyancy contrast within the lithosphere : a petrological perspective. Journal of Petrology, 44 : 764-778.

Niu Y L, O'Hara M J. 2008. Global correlations of ocean ridge basalt chemistry with axial depth : a new perspective. Journal of Petrology, 49 : 633-664.

Niu Y L, Zhao Z, Zhu D C, et al. 2013. Continental collision zones are primary sites for net continental crust growth—a testable hypothesis. Earth-Science Reviews, 127 : 96-110.

Niu Y L. 2017. Slab breakoff : a causal mechanism or pure convenience?. Science Bulletin, 62 (7): 456-461.

Noda A. 2016. Forearc basins : types, geometries, and relationships to subduction zone dynamics. Geological Society of America Bulletin, 128 (5-6): 879-895.

O'Neill C, Marchi S, Bottke W, et al. 2019. The role of impacts on Archaean tectonics. Geology, 48 (2): 174-178.

O'Reilly S Y, Griffin W L. 2013. Mantle metsomatism. In : Harlov D E, Austrheim H.(Editors). Metasomatism and the Chemical Transformation of Rock. Berlin Heidelberg : Springer-Verlag Berlin Heidelberg : 471-533.

Ogasawara Y, Fukasawa K, Maruyama S. 2002. Coesite exsolution from supersilicic titanite in UHP marble from the Kokchetav Massif, northern Kazakhstan. American Mineralogist, 87 (4): 454-461.

Okay A I, Xu S T, Sengor A M C. 1989. Coesite from the Dabie Shan eclgoites, central China. European Journal of Mineralogy, 1 : 595-598.

Okumura T, Watanabe H, Chen C, et al. 2016. The world's deepest brucite-carbonate chimneys at a serpentinite-hosted system, the Shinkai Seep Field, Southern Mariana Forearc. Geochemistry Geophysics Geosystems, 17 : 3775-3796.

Oreskes N. 2003. Plate Tectonics : An Insider's History of the Modern Theory of the Earth. Boulder : Westview Press.

Parkinson I J, Arculus R J. 1999. The redox state of subduction zones : insights from arc-peridotites. Chemical Geology, 160 (4): 409-423.

Pawley A R, Wood B J. 1995. The high-pressure stability of talc and 10 Å phase; potential storage sites for H_2O in subduction zones. American Mineralogist, 80 (9): 998-1003.

Peacock S M. 1991. Numerical simulation of subduction zone pressure-temperature-time paths : constraints on fluid production and arc magmatism. Philosophical Transactions of the Royal Society, 335(1638): 341-353.

Peacock S M, Wang K. 1999. Seismic consequences of warm versus cool subduction metamorphism：examples from southwest and northeast Japan. Science，286(5441)：937-939.

Peacock S M. 2001. Are the lower planes of double seismic zones caused by serpentine dehydration in subducting oceanic mantle? .Geology，29（4)：299-302.

Peacock S M. 2003. Thermal structure and metamorphic evolution of subducting slabs. Geophysical Monography，138：7-22.

Pearce J A. 1983. Role of the sub-continental lithosphere in magma genesis at active continental margins. In：Hawkesworth C J，Norry M J.(Editors). Continental Basalts & Mantle Xenoliths. Nantwich：Shiva：230-249.

Pearce J A，Harris N B W，Tindle A G. 1984. Trace element discrimination diagrams for the tectonic interpretation of granitic rocks. Journal of Petrology，25：956-983.

Pearce J A，Peate D W. 1995. Tectonic implications of the composition of volcanic arc magmas. Annual Review of Earth and Planetary Sciences，23：251-285.

Pearce J A，Stern R J. 2006. Origin of back-arc basin magmas：trace element and isotope perspectives. Geophysical Monograph，166：63-86.

Pekov I V，Kononkova N N. 2010. Rubidium mineralization in rare-element granitic pegmatites of the Voron'i Tundras，Kola Peninsula，Russia. Geochemistry International，48（7)：695-713.

Peng Z，Gomberg J. 2010. An integrated perspective of the continuum between earthquakes and slow-slip phenomena. Nature Geoscience，3：599-607.

Penniston-Dorland S C，Kohn M J，Manning C E. 2015. The global range of subduction zone thermal structures from exhumed blueschists and eclogites：rocks are hotter than models. Earth and Planetary Science Letters，428：243-254.

Penniston-Dorland S C，Kohn M J，Piccoli P M. 2018. A mélange of subduction temperatures：evidence from Zr-in-rutile thermometry for strengthening of the subduction interface. Earth and Planetary Science Letters，482：525-535.

Pesicek J D，Zhang H J，Thurber C H. 2014. Multiscale seismic tomography and earthquake relocation incorporating differential time data：application to the Maule subduction zone，Chile. Bulletin of the Seismological Society of America，104（2)：1037-1044.

Peters B J，Carlson R W，Day J M D，et al. 2018. Hadean silicate differentiation preserved by anomalous $^{142}Nd/^{144}Nd$ ratios in the Reunion hotspot source. Nature，555：89-93.

Petford N，Atherton M. 1996. Na-rich partial melts from newly underplated basaltic crust：the Cordillera Blanca Batholith，Peru. Journal of Petrology，37（6)：1491-1521.

Pilet S，Baker M B，Stopler E M. 2008. Metasomatized lithosphere and the origin of alkaline lavas. Science，320（5878)：916-919.

Pirard C, Hermann J. 2015. Focused fluid transfer through the mantle above subduction zones. Geology, 43(10): 915-918.

Piromallo C, Morelli A. 2003. P-wave tomography of the mantle under the Alpine-Mediterranean area. Journal of Geophysical Research-Solid Earth, 108(B2): 2065.

Plank T, Langmuir C H. 1993. Tracing trace-elements from sediment input to volcanic output at subduction zones. Nature, 362: 739-743.

Plank T. 2005. Constraints from thorium/lanthanum on sediment cycling at subduction zones and evolution of the continents. Journal of Petrology, 46(5): 921-944.

Plank T, Kelley K A, Zimmer M M, et al. 2013. Why do mafic arc magmas contain ~4 wt% water on average?. Earth and Planetary Science Letters, 364: 168-179.

Plank T, Manning C E. 2019. Subducting carbon. Nature, 574(7778): 343-352.

Poli S, Schmidt M W. 2002. Petrology of subducted slabs. Annual Review of Earth and Planetary Science, 30: 207-235.

Poli S. 2015. Carbon mobilized at shallow depths in subduction zones by carbonatitic liquids. Nature Geoscience, 8: 633-636.

Raimbourg H, Famin V, Palazzin G, et al. 2018. Fluid properties and dynamics along the seismogenic plate interface. Geosphere, 14(2): 469-491.

Rampone E, Hofmann A W. 2012. A global overview of isotopic heterogeneities in the oceanic mantle. Lithos, 148: 247-261.

Ranero C R, Morgan J P, McIntosh K, et al. 2003. Bending-related faulting and mantle serpentinization at the Middle America trench. Nature, 425: 367-373.

Rapp R P, Shimizu N, Norman M D. 2003. Growth of early continental crust by partial melting of eclogite. Nature, 425: 605-609.

Ratajeski K, Sisson T W, Glazner A F. 2005. Experimental and geochemical evidence for derivation of the El Capitan Granite, California, by partial melting of hydrous gabbroic lower crust. Contributions to Mineralogy and Petrology, 149: 713-734.

Rea D K, Ruff L J. 1996. Composition and mass flux of sediment entering the world's subduction zones: implications for global sediment budgets, great earthquakes, and volcanism. Earth and Planetary Science Letters, 140(1-4): 1-12.

Reagan M K, Heaton D E, Schmitz M D, et al. 2019. Forearc ages reveal extensive short-lived and rapid seafloor spreading following subduction initiation. Earth and Planetary Science Letters, 506: 520-529.

Replumaz A, Negredo A M, Guillot S, et al. 2010. Multiple episodes of continental subduction during India/Asia convergence: insight from seismic tomography and tectonic reconstruction. Tectonophysics, 483(1): 125-134.

Ribe N M, Stutzmann E, Ren Y, et al. 2007. Buckling instabilities of subducted lithosphere beneath the transition zone. Earth and Planetary Science Letters, 254（1-2）: 173-179.

Ribe N M. 2010. Bending mechanics and mode selection in free subduction : a thin-sheet analysis. Geophysical Journal International, 180 : 559-576.

Richards J P. 2003. Tectono-magmatic precursors for porphyry Cu-（Mo-Au）deposit formation. Economic Geology, 98（8）: 1515-1533.

Richards J P, Kerrich R. 2007. Adakite-like rocks : their diverse origins and questionable role in metallogenesis. Economic Geology, 102（4）: 537-576.

Richards J P, Spell T, Rameh E, et al. 2012. High Sr/Y magmas reflect arc maturity, high magmatic water content, and porphyry Cu ± Mo ± Au potential : examples from the Tethyan arcs of central and eastern Iran and western Pakistan. Economic Geology, 107（2）: 295-332.

Richards J P. 2015. Tectonic, magmatic, and metallogenic evolution of the Tethyan orogen : from subduction to collision. Ore Geology Reviews, 70 : 323-345.

Ring U, Pantazides H, Glodny J, et al. 2020. Forced return flow deep in the subduction channel, Syros, Greece. Tectonics, 39（1）: e2019TC005768.

Ringwood A E. 1975. Composition and Petrology of the Earth's Mantle. McGraw-Hill.

Ringwood A E. 1990. Slab-mantle interactions : 3. Petrogenesis of intraplate magmas and structure of the upper mantle. Chemical Geology, 82 : 187-207.

Rogers G, Dragert H. 2003. Episodic tremor and slip on the Cascadia subduction zone : the chatter of silent slip. Science, 300（5627）: 1942-1943.

Rondenay S, Abers G A, Keken P E. 2008. Seismic imaging of subduction zone metamorphism. Geology, 36（4）: 275-278.

Rubatto D, Hermann J. 2003. Zircon formation during fluid circulation in eclogites（Monviso, Western Alps）: implications for Zr and Hf budget in subduction zones. Geochimica et Cosmochimica Acta, 67（12）: 2173-2187.

Rubatto D, Regis D, Hermann J, et al. 2011. Yo-yo subduction recorded by accessory minerals in the Italian Western Alps. Nature Geoscience, 4 : 338-342.

Rudnick R L. 1995. Making continental crust. Nature, 378 : 573-578.

Rudnick R L, Gao S. 2003. Composition of the continental crust. Treatise Geochemistry, 3 : 1-64.

Rumble D, Liou J G, Jahn B M. 2003. Continental crust subduction and ultrahigh pressure metamorphism. Treatise on Geochemistry, 3 : 293-319.

Rusk B G, Reed M H, Dilles J H, et al. 2004. Compositions of magmatic hydrothermal fluids determined by LA-ICP-MS of fluid inclusions from the porphyry copper-molybdenum deposit at Butte, MT. Chemical Geology, 210 : 173-199.

Ryan J G, Morris J, Tera F, et al. 1995. Cross-arc geochemical variations in the Kurile arc as a function of slab depth. Science, 270 : 625-627.

Saffer D M, Tobin H J. 2011. Hydrogeology and mechanics of subduction zone forearcs : fluid flow and pore pressures. Annual Review of Earth and Planetary Sciences, 39 (1): 157-186.

Sage F, Collot J Y, Ranero C R. 2006. Interplate patchiness and subduction-erosion mechanisms : evidence from depth-migrated seismic images at the central Ecuador convergent margin. Geology, 34 (4): 997-1000.

Saito S, Tani K. 2017. Transformation of juvenile Izu-Bonin-Mariana oceanic arc into mature continental crust : an example from the Neogene Izu collision zone granitoid plutons, Central Japan. Lithos, 277 : 228-240.

Sajona F G, Maury R C, Bellon H, et al. 1996. High field strength element enrichment of Pliocene-Pleistocene Island arc basalts, Zamboanga Peninsula, western Mindanao (Philippines) . Journal of Petrology, 37 (3): 693-726.

Saleeby J, Ducea M, Clemens-Knott D. 2003. Production and loss of high-density batholithic root, southern Sierra Nevada, California. Tectonics, 22 (6) .doi:10:1029/2002TC001374.

Salimbeni S, Pondrelli S, Margheriti L. 2013. Hints on the deformation penetration induced by subductions and collision processes : seismic anisotropy beneath the Adria region (Central Mediterranean) . Journal of Geophysical Research-Solid Earth, 118 (11): 5814-5826.

Sanfilippo A, Dick H J B, O'Hara Y. 2013. Melt-rock reaction in the mantle : mantle troctolites from the Parece Vela ancient back-arc spreading center. Journal of Petrology, 54 (5): 861-885.

Sano Y, Williams S N. 1996. Fluxes of mantle and subducted carbon along convergent plate boundaries. Geophysical Research Letters, 23 (20): 2749-2752.

Savage M K. 1999. Seismic anisotropy and mantle deformation : what have we learned from shear wave splitting?. Reviews of Geophysics, 37 (1): 65-106.

Savov I P, Ryan J G, D'Antonio M, et al. 2007. Shallow slab fluid release across and along the Mariana arc-basin system : insights from geochemistry of serpentinized peridotites from the Mariana fore arc. Journal of Geophysical Research-Solid Earth, 112 : B09205.

Scambelluri M, Pettke T, van Roermund H L M. 2008. Majoritic garnets monitor deep subduction fluid flow and mantle dynamics. Geology, 36 (1): 59-62.

Schellart W P, Freeman J, Stegman D R, et al. 2007. Evolution and diversity of subduction zones controlled by slab width. Nature, 446 : 308-311.

Schellart W P, Moresi L. 2013. A new driving mechanism for backarc extension and backarc shortening through slab sinking induced toroidal and poloidal mantle flow : results from dynamic subduction models with an overriding plate. Journal of Geophysical Research : Solid Earth, 118 (6): 3221-3248.

Schmidt M W, Poli S. 1998. Experimentally based water budgets for dehydrating slabs and consequences for arc magma generation. Earth and Planetary Science Letters, 163 (1-4): 361-379.

Schmidt M W, Poli S. 2003. Generation of mobile components during subduction of oceanic crust. Treatise on Geochemistry, 3 : 567-591.

Schmidt M W, Poli S. 2014. Devolatilization during subduction. Treatise on Geochemistry, 4 : 669-701.

Schmidt M W, Jagoutz O. 2017. The global systematics of primitive arc melts. Geochemistry Geophysics Geosystems, 18 (8): 2817-2854.

Schneider F M, Yuan X, Schurr B, et al., 2013. Seismic imaging of subducting continental lower crust beneath the Pamir. Earth and Planetary Science Letters, 375 : 101-112.

Schott B, Schmeling H. 1998. Delamination and detachment of a lithospheric root. Tectonophysics, 296 (3-4): 225-247.

Scholl D W, von Huene R. 2007. Crustal recycling at modern subduction zones applied to the past—issues of growth and preservation of continental basement crust, mantle geochemistry, and supercontinent reconstruction. Geological Society of America Memoir, 200 : 9-32.

Schwarz-Schampera U, Herzig P M. 2002. Indium : Geology, Mineralogy, and Economics. Berlin : Springer.

Seghedi I, Downes H, Szakacs A, et al. 2004. Neogene-quaternary magmatism and geodynamics in the Carpathian-Pannonian region : a synthesis. Lithos, 72 (3-4): 117-146.

Sheard E R, Williams-Jones A E, Heiligmann M, et al. 2012. Controls on the concentration of zirconium, niobium, and the rare earth elements in the Thor Lake rare metal deposit, northwest Territories, Canada. Economic Geology, 107 (1): 81-104.

Sheen A I, Kendall B, Reinhard C T, et al. 2018. A model for the oceanic mass balance of rhenium and implications for the extent of Proterozoic ocean anoxia. Geochimica et Cosmochimica Acta, 227 : 75-95.

Shervais J W, Reagan M, Haugen E, et al. 2019. Magmatic response to subduction initiation : Part 1. Fore-arc basalts of the Izu-Bonin arc from IODP Expedition 352. Geochemistry Geophysics Geosystems, 20 : 314-338.

Shillington D J, Bécel A, Nedimović M R, et al. 2015. Link between plate fabric, hydration and subduction zone seismicity in Alaska. Nature Geoscience, 8 : 961-964.

Shirey S B, Richardson S H. 2011. Start of the Wilson cycle at 3 Ga shown by diamonds from subcontinental mantle. Science，333：434-436.

Shreve R L，Cloos M. 1986. Dynamics of sediment subduction，melange formation，and prism accretion. Journal of Geophysical Research-Solid Earth，91（B10）：10229-10245.

Sillitoe R H. 1997. Characteristics and controls of the largest porphyry copper-gold and epithermal gold deposits in the circum-Pacific region. Australian Journal of Earth Sciences，44：373-388.

Sillitoe R H. 2010. Porphyry copper system. Economic Geology，105（1）：3-41.

Sinclair W D，Jonasson I R，Kirkham R V，et al. 2009. Rhenium and other platinum-group metals in porphyry deposits. Geological Survey of Canada Valley，California，and a mixing origin for the Sierra Nevada batholith. Contributions to Mineralogy and Petrology，126：81-108.

Sisson T W，Grove T L，Coleman D S. 1996. Hornblende gabbro sill complex at Onion Valley，California，and a mixing origin for the Sierra Nevada batholith. Contributions to Mineralogy and Petrology，26：81-108.

Sizova E，Gerya T，Brown M. 2012. Exhumation mechanisms of melt-bearing ultrahigh pressure crustal rocks during collision of spontaneously moving plates. Journal of Metamorphic Geology，30（9）：927-955.

Smith D C. 1984. Coesite in clinopyroxene in the Caledonides and its implications for geodynamics. Nature，310（5979）：641-644.

Smithies R H，Champion D C，Sun S S. 2004. Early evidence for LILE-enriched mantle source regions：diverse magmas from the c. 3.0 Ga Mallina Basin，Pilbara Craton，NW Australia. Journal of Petrology，45（8）：1515-1537.

Sobolev N V，Shatsky V S. 1990. Diamond inclusions in garnets from metamorphic rocks：a new environment for diamond formation. Nature，343：742-746.

Sobolev A V，Chaussidon M. 1996. H_2O concentrations in primary melts from supra-subduction zones and mid-ocean ridges：implications for H_2O storage and recycling in the mantle. Earth and Planetary Science Letters，137（1-4）：45-55.

Sobolev A V，Hofmann A W，Sobolev S V，et al. 2005. An olivine-free mantle source of Hawaiian shield basalts. Nature，434：590-597.

Sobolev A V，Hofmann A W，Kuzmin D V，et al. 2007. The amount of recycled crust in sources of mantle-derived melts. Science，316（5823）：412-417.

Soesoo A，Bons P D，Gray D R，et al. 1997. Divergent double subduction：tectonic and petrologic consequences. Geology，25（28）：755-758.

Solomon M. 1990. Subduction，arc reversal，and the origin of porphyry copper-gold deposits in island arcs. Geology，18（7）：630-633.

Song S G, Yang J S, Xu Z Q, et al. 2003. Metamorphic evolution of the coesite-bearing ultrahigh-pressure terrane in the North Qaidam, Northern Tibet, NW China. Journal of Metamorphic Geology, 21 (6): 631-644.

Song S G, Zhang L F, Chen J, et al. 2005. Sodic amphibole exsolutions in garnet from garnet-peridotite, North Qaidam UHPM belt, NW China: implications for ultradeep-origin and hydroxyl defects in mantle garnets. American Mineralogist, 90 (5-6): 814-820.

Song S G, Zhang L F, Niu Y L, et al. 2006. Evolution from oceanic subduction to continental collision: a case study from the Northern Tibetan Plateau based on geochemical and geochronological data. Journal of Petrology, 47 (6): 435-455.

Song S G, Yang L M, Zhang Y Q, et al. 2017. Qi-Qin Accretionary Belt in Central China Orogen: accretion by trench jam of oceanic plateau and formation of intra-oceanic arc in the Early Paleozoic Qin-Qi-Kun Ocean. Science Bulletin, 62 (15): 1035-1038.

Spandler C, Pirard C. 2013. Element recycling from subducting slabs to arc crust: a review. Lithos, 170-171: 208-223.

Spengler D, van Roermund H L M, Drury M R, et al. 2006. Deep origin and hot melting of an Archaean orogenic peridotite massif in Norway. Nature, 440: 913-917.

Stadler G, Gurnis M, Burstedde C, et al. 2010. The dynamics of plate tectonics and mantle flow: from local to global scales. Science, 329: 1033-1038.

Stampfli G M, Borel G D. 2002. A plate tectonic model for the Paleozoic and Mesozoic constrained by dynamic plate boundaries and restored synthetic oceanic isochrons. Earth and Planetary Science Letters, 196: 17-33.

Stern R J. 2002. Subduction zones. Reviews of Geophysics, 40: 1012.

Stern R J, Fouch M J, Klemperer S. 2003. An overview of the Izu-Bonin-Marian subduction factory. In: Eiler J. (Ed.), Inside the Subduction Factory. AGU Geophysical Monograph, Washington DC, 138: 175-222.

Stern R J. 2004. Subduction initiation: spontaneous and induced. Earth and Planetary Science Letters, 226 (3-4): 275-292.

Stern R J. 2005. Evidence from ophiolites, blueschists, and ultrahigh-pressure metamorphic terranes that the modern episode of subduction tectonics began in Neoproterozoic time. Geology, 33 (7): 557-560.

Stern R J. 2007. When and how did plate tectonics begin? Theoretical and empirical considerations. Chinese Science Bulletin, 52 (5): 578-591.

Stern R J, Scholl D W. 2010. Yin and yang of continental crust creation and destruction by plate tectonic processes. International Geology Review, 52 (1): 1-31.

Stern R J, Gerya T. 2018. Subduction initiation in nature and models: a review. Tectonophiscs,

746：173-198.

Stixrude L，Lithgow-Bertelloni C. 2012. Geophysics of chemical heterogeneity in the mantle. Annual Review of Earth and Planetary Sciences，40：569-595.

Stolper D A，Keller C. 2018. A record of deep-ocean dissolved O_2 from the oxidation state of iron in submarine basalts. Nature，553：323-327.

Stolper D A，Bucholz C E. 2019. Neoproterozoic to early Phanerozoic rise in island arc redox state due to deep ocean oxygenation and increased marine sulfate levels. Proceedings of the National Academy of Sciences，116（18）：8746-8755.

Stracke A，Hofmann A W，Hart S R. 2005. FOZO，HIMU，and the rest of the mantle zoo. Geochemistry，Geophysics，Geosystems，6（5）：Q05007.

Straub S M，Gomez-Tuena A，Stuart F M，et al. 2011. Formation of hybrid arc andesites beneath thick continental crust. Earth and Planetary Science Letters，303（3-4）：337-347.

Stüeken E E，Kipp M A，Koehler M C，et al. 2016. The evolution of Earth's biogeochemical nitrogen cycle. Earth-Science Reviews，160：220-239.

Su B，Chen Y，Guo S，et al. 2019. Garnetite and pyroxenite in the mantle wedge formed by slab-mantle interactions at different melt/rock ratios. Journal of Geophysical Research-Solid Earth，124（7）：6504-6522.

Sun W D，Ling M X，Chung S L，et al. 2012. Geochemical constraints on adakites of different origins and copper mineralization. Journal of Geology，12（1）：105-120.

Sun W J，Li S Z，Liu X，et al. 2015. Deep structures and surface boundaries among Proto-Tethyan micro-blocks：constraints from seismic tomography and aeromagnetic anomalies in the Central China Orogen. Tectonophysics，659：109-121.

Suzuki K，Kitajima K，Sawaki Y，et al. 2015. Ancient oceanic crust in island arc lower crust：evidence from oxygen isotopes in zircons from the Tanzawa Tonalitic Pluton. Lithos，228-229：43-54.

Syracuse E M，van Keken P E，Abers G A. 2010. The global range of subduction zone thermal models. Physics of the Earth And Planetary Interiors，183（1-2）：73-90.

Syracuse E M，Maceira M，Prieto G A，et al. 2016. Multiple plates subducting beneath Colombia，as illuminated by seismicity and velocity from the joint inversion of seismic and gravity data. Earth and Planetary Science Letters，444: 139-149.

Tackley P J. 1998. Self-consistent generation of tectonic plates in three-dimensional mantle convection. Earth and Planetary Science Letters，157（1-2）：9-22.

Tamura A，Arai S. 2006. Harzburgite-dunite-orthopyroxenite suite as a record of supra-subduction zone setting for the Oman ophiolite mantle. Lithos，90（1-2）：43-56.

Tang G J，Chung S L，Hawkesworth C J，et al. 2017. Short episodes of crust generation during protracted accretionary processes: evidence from Central Asian Orogenic Belt, NW China. Earth and Planetary Science Letters, 464: 142-154.

Tang C A, Webb A A G, Moore W B, et al. 2020. Breaking Earth's shell into a global plate network. Nature Communications, 11: 3621.

Tang Y W, Chen L, Zhao Z F, et al. 2020. Geochemical evidence for the production of granitoids through reworking of the juvenile mafic arc crust in the Gangdese orogen, southern Tibet. Gelogical Society of America Bulletin, 132 (7-8): 1347-1364.

Tassara A, Götze H J, Schmidt S, et al. 2006. Three-dimensional density model of the Nazca plate and the Andean continental margin. Journal of Geophysical Research-Solid Earth, 111 (B9): B09404.

Tatsumi Y. 1981. Melting experiments on a high-Magnesian andesite. Earth and Planetary Science Letters, 54: 357-365.

Tatsumi Y, Sakuyama M, Fukuyama H, et al. 1983. Generation of arc basalt magmas and thermal structure of the mantle wedge in subduction zones. Journal of Geophysical Research-Solid Earth, 88 (B7): 5815-5825.

Tatsumi Y. 1986. Formation of the volcanic front in subduction zones. Geophysical Research Letters, 13 (8): 717-720.

Tatsumi Y. 2005. The subduction factory: how it operates in the evolving Earth. GSA Today, 15 (7): 4-10.

Tatsumi Y. 2006. High-Mg andesites in the Setouchi volcanic belt, Southwest Japan: analogy to Archean magmatism and continental crust formation?. Annual Review of Earth and Planetary Sciences, 34: 467-499.

Taylor S R, White A J R. 1965. Geochemistry of andesites and the growth of continents. Nature, 208 (5007): 271-273.

Taylor B. 1995. Backarc Basins Tectonics and Magmatism. New York: Springer.

Taylor S R, McLennan S M. 1995. The geochemical evolution of the continental crust. Reviews of Geophysics, 33 (2): 241-265.

Taylor B, Martinez F. 2003. Back-arc basin basalt systematics. Earth and Planetary Science Letters, 210 (3-4): 481-497.

Thomas R, Davidson P, Hahn A. 2008. Ramanite- (Cs) and ramanite- (Rb): new cesium and rubidium pentaborate tetrahydrate minerals identified with Raman spectroscopy. American Mineralogist, 93 (7): 1034-1042.

Thomas R, Davidson P, Schmidt C. 2011. Extreme alkali bicarbonate- and carbonate-rich fluid inclusions in granite pegmatite from the Precambrian Rønne granite, Bornholm Island, Denmark. Contributions to Mineralogy and Petrology, 161 (2): 315-329.

Thomson A R，Walter M J，Kohn S C，et al. 2016. Slab melting as a barrier to deep carbon subduction. Nature，529：76-79.

Thorkelson D J. 1996. Subduction of diverging plates and the principles of slab window formation. Tectonophys，255（1-2）：47-63.

Tian L，Castillo P R，Hilton D R，et al. 2011. Major and trace element and Sr-Nd isotope signatures of the northern Lau Basin lavas：implications for the composition and dynamics of the back-arc basin mantle. Journal of Geophysical Research-Solid Earth，116：B11201.

Till C B，Grove T L，Withers A C. 2012. The beginnings of hydrous mantle wedge melting. Contributions to Mineralogy and Petrolog，163：669-688.

Todd E，Gill J B，Wysoczanski R J，et al. 2010. Sources of constructional cross-chain volcanism in the southern Havre Trough：new insights from HFSE and REE concentration and isotope systematics. Geochemistry Geophysics Geosystems，11：Q04009.

Tonarini S，Leeman W P，Leat P T，2011. Subduction erosion of forearc mantle wedge implicated in the genesis of the South Sandwich Island（SSI）arc：evidence from boron isotope systematics. Earth and Planetary Science Letters，301（1-2）：275-284.

Turcotte D L，Schubert G. 2002. Geodynamics. 2nd Edition. New York：Cambridge University Press.

Turner F J，Verhoogen J. 1960. Igneous and Metamorphic Petrology. New York：McGraw-Hill.

Turner S J，Langmuir C H，Dungan M A，et al. 2017. The importance of mantle wedge heterogeneity to subduction zone magmatism and the origin of EM1. Earth and Planetary Science Letters，472：216-228.

Ulmer P，Trommsdorff V. 1995. Serpentine stability to mantle depths and subduction-related magmatism. Science，268（5212）：858-861.

Umino S，Kushiro I. 1989. Experimental studies on boninitepetrogenesis. In：Crawford A J.(Editor). Boninites. New York：Chapman and Hall：89-111.

Uyeda S，Kanamori H. 1979. Back-arc opening and the mode of subduction. Journal of Geophysical Research-Solid Earth，84（B3）：1049-1061.

van de Zedde D M A，Wortel M J R. 2001. Shallow slab detachment as a transient source of heat at midlithospheric depth. Tectonics，20（6）：868-882.

van der Hilst R D，Widyantoro S，Engdahl E R. 1997. Evidence for deep mantle circulation from global tomography. Nature，386：578-584.

van Hunen J，van den Berg A P，Vlaar N J. 2002. On the role of subducting oceanic plateaus in the development of shallow flat subduction. Tectonophys，352（3-4）：317-333.

van Hunen J，van den Berg A P，Vlaar N J. 2004. Various mechanisms to induce present-day shallow flat subduction and implications for the younger Earth：a numerical parameter study.

Physics of the Earth and Planetary Interiors, 146 (1-2) : 179-194.

van Hunen J, Allen M B. 2011. Continental collision and slab break-off : a comparison of 3-D numerical models with observations. Earth and Planetary Science Letters, 302 (1-2) : 27-37.

van Keken P E, Hacker B R, Syracuse E M, et al. 2011. Subduction factory : 4. Depth-dependent flux of H_2O from subducting slabs worldwide. Journal of Geophysical Research-Solid Earth, 116 (B1) : B01401.

van Keken P E, Wada I, Abers G A, et al. 2018. Mafic high-pressure rocks are preferentially exhumed from warm subduction settings. Geochemistry Geophysics Geosystems, 19 (9) : 2934-2961.

van Keken P E, Wada I, Sime N, et al. 2019. Thermal structure of the forearc in subduction zones : a comparison of methodologies. Geochemistry Geophysics Geosystems, 20 (7) : 3268-3288.

van Roermund H L M, Drury M R. 1998. Ultra-high pressure (P> 6 GPa) garnet peridotites in Western Norway : exhumation of mantle rocks from >185 km depth. Terra Nova, 10 : 295-301.

van Summeren J, Conrad C P, Lithgow-Bertelloni C. 2012. The importance of slab pull and a global asthenosphere to plate motions. Geochemistry Geophysics Geosystems, 13 (2) : Q0AK03.

Vannucchi P, Sage F, Morgan J P, et al. 2012. Toward a dynamic concept of the subduction channel at erosive convergent margins with implications for interplate material transfer. Geochemistry Geophysics Geosystems, 13 (2) : Q02003.

Veith K F. 1977. The nature of the dual zone of seismicity in the Kurils arc. EOS, 58 : 1232.

Vielzeuf D, Schmidt M W. 2001. Melting relations in hydrous systems revisited : application to metapelites, metagreywackes and metabasalts. Contributions to Mineralogy and Petrology, 141 : 251-267.

von Blanckenburg F, Davies J H. 1995. Slab breakoff : a model for syncollisional magmatism and tectonics in the Alps. Tectonics, 14 (1) : 120-131.

von Huene R, Scholl D W. 1991. Observations at convergent margins concerning sediment subduction, subduction erosion, and the growth of continental crust. Reviews of Geophysics, 29 (3) : 279-316.

Wada I W, Wang K, He J, et al. 2008. Weakening of the subduction interface and its effects on surface heat flow, slab dehydration, and mantle wedge serpentinization. Journal of Geophysical Research-Solid Earth, 113 (B4) : B04402.

Wada I W, Wang K. 2009. Common depth of slab-mantle decoupling : reconciling diversity and uniformity of subduction zones. Geochemistry Geophysics Geosystems，10：Q10009.

Wada I W，King S. 2015. Dynamics of subducting slabs：numerical modeling and constraints from seismology，geoid，topography，geochemistry and petrology. Treatise on Geophysics，7：325-370.

Wadati K. 1935. On the activity of deep-focus earthquakes in the Japan Island and neighbourhood. Geophysical Magazine，8：305-326.

Wallace P J. 2005. Volatiles in subduction zone magmas：concentrations and fluxes based on melt inclusion and volcanic gas data. Journal of Volcanology and Geothermal Research，140（1-3）：217-240.

Wallis S，Tsuboi M，Suzuki K F，et al. 2005. Role of partial melting in the evolution of the Sulu（eastern China）ultrahigh-pressure terrane. Geology，33（2）：129-132.

Wang Q，Wyman D A，Zhao Z H，et al. 2007. Petrogenesis of Carboniferous adakites and Nb-enriched arc basalts in the Alataw area，northern Tianshan Range（western China）：implications for Phanerozoic crustal growth in the Central Asia orogenic belt. Chemical Geology，236（1-2）：42-64.

Wang Y，Fan W，Zhang G，et al. 2013. Phanerozoic tectonics of the South China Block：key observations and controversies. Gondwana Research，23：1273-1305.

Wang R，Richards J P，Zhou L M，et al. 2015. The role of Indian and Tibetan lithosphere in spatial distribution of Cenozoic magmatism and porphyry Cu-Mo ± Au deposits in the Gangdese belt，southern Tibet. Earth-Science Reviews，150：68-94.

Wang R，Tafti R，Hou Z Q，et al. 2017. Across-arc geochemical variation in the Jurassic magmatic zone，southern Tibet：implication for continental arc-related porphyry Cu-Au mineralization. Chemical Geology，451：116-134.

Wang Y H，Zhang L F，Zhang J J，et al. 2017. The youngest eclogite in central Himalaya：P-T path，U-Pb zircon age and its tectonic implication. Gondwana Research，41：188-206.

Wang R，Weinberg R F，Collins W J，et al. 2018. Origin of post-collision magmas and formation of porphyry Cu deposits in southern Tibet. Earth-Science Reviews，181：122-143.

Wang X J，Chen L H，Hofmann A W，et al. 2018. Recycled ancient ghost carbonate in the Pitcairn mantle plume. Proceedings of the National Academy of Sciences，115（35）：8682-8687.

Wang Y，Zhang L F，Li Z H，et al. 2019. The exhumation of subducted oceanic eclogite：insights from phase equilibrium and thermomechanical modeling. Tectonics，38：1764-1797.

Wang C G，Lo Cascio M，Liang Y，et al. 2020. An experimental study of peridotite dissolution in eclogite-derived melts：implications for styles of melt-rock interaction in lithospheric mantle beneath the North China Craton. Geochimica et Cosmochimica Acta，278：157-176.

Wannamaker P E，Caldwell T G，Jiracek G R，et al. 2009. Fluid and deformation regime of an advancing subduction system at Marlborough，New Zealand. Nature，460：733-737.

Warren C J. 2013. Exhumation of（ultra-）high-pressure terranes：concepts and mechanisms. Solid Earth，4：75-92.

Wei W，Unsworth M，Jones A，et al. 2001. Detection of widespread fluids in the Tibetan crust by magnetotelluric studies. Science，292（5517）：716-718.

Wei C J，Wang W，Clarke G L，et al. 2009. Metamorphism of high/ultrahigh-pressure pelitic-felsic schist in the south Tianshan orogen，NW China：Phase equilibria and P-T path. Journal of Petrology，50（10）：1973-1991.

Wei C J，Clarke G L. 2011. Calculated phase equilibria for MORB compositions：a reappraisal of the metamorphic evolution of lawsonite eclogite. Journal of Metamorphic Geology，29（9）：939-952.

White D A，Roeder D H，Nelson T H，et al. 1970. Subduction. Geological Society of America Bulletin，81：3431-3432.

White W M. 2010. Oceanic island basalts and mantle plumes：the geochemical perspective. Annual Review of Earth and Planetary Sciences，38：133-160.

Wiens D A，Kelley K A，Plank T. 2006. Mantle temperature variations beneath back-arc spreading centers inferred from seismology，petrology，and bathymetry. Earth and Planetary Science Letters，248（1-2）：30-42.

Williams S，Flament N，Müller R D，et al. 2015. Absolute plate motions since 130 Ma constrained by subduction zone kinematics. Earth and Planetary Science Letters，418：66-77.

Wilson J T. 1963. Evidence from islands on the spreading of ocean floors. Nature，197（4867）：536-538.

Wilson J T. 1966. Did the Atlantic close and then re-open?. Nature，211：676-681.

Wilson J T. 1968. Static or mobile earth：the current scientific revolution. Proceedings of the American Philosophical Society，112：309-320.

Wilson J T. 1973. Mantle plumes and plate motions. Tectonophysics，19（2）：149-164.

Wilson C R，Spiegelman M，van Keken P E，et al. 2014. Fluid flow in subduction zones：the role of solid rheology and compaction pressure. Earth and Planetary Science Letters，401：261-274.

Wilson R W，Houseman G A，Buiter S J H，et al. 2019. Fifty years of the Wilson cycle concept in plate tectonics：an overview. Geological Society，London，Special

Publications, 470 : 1-17.

Wittig N, Pearson D G, Webb M, et al. 2008. Origin of cratonic lithospheric mantle roots : a geochemical study of peridotites from the North Atlantic Craton, West Greenland. Earth and Planetary Science Letters, 274 (1-2): 24-33.

Woodhead J, Stern R J, Pearce J, et al. 2012. Hf-Nd isotope variation in Mariana Trough basalts : the importance of "ambient mantle" in the interpretation of subduction zone magmas. Geology, 40 (6): 539-542.

Workman R K, Hart S R, Jackson M, et al. 2004. Recycled metasomatized lithosphere as the origin of the enriched mantle II (EM2) end-member : evidence from the Samoan Volcanic Chain. Geochemistry, Geophysics, Geosystems, 5 (4): Q04008.

Wu R X, Zheng Y F, Wu Y B, et al. 2006. Reworking of juvenile crust : element and isotope evidence from Neoproterozoic granodiorite in South China. Precambrian Research, 146 (3-4): 179-212.

Wu Y B, Zheng Y F, Zhang S B, et al. 2007. Zircon U-Pb ages and Hf isotope compositions of migmatite from the North Dabie terrane in China : constraints on partial melting. Journal of Metamorphic Geology, 25 (9): 991-1009.

Wu Y, Fei Y W, Jin Z M, et al. 2009. The fate of subducted upper continental crust : an experimental study. Earth and Planetary Science Letters, 282 (1-4): 275-284.

Wu F Y, Liu X C, Liu Z C, et al. 2020. Highly fractionated Himalayan leucogranites and associated rare-metal mineralization. Lithos, 352-353 : 105319.

Wyllie P J. 1971. Role of water in magma generation and initiation of diapiric uprise in the mantle. Journal of Geophysical Research, 76 (5): 1328-1338.

Wyllie P J. 1984. Constraints imposed by experimental petrology on possible and impossible magma sources and products. Philosophical Transactions of the Royal Society, 310 (1514): 439-456.

Xia Q X, Zheng Y F, Hu Z. 2010. Trace elements in zircon and coexisting minerals from low-T/UHP metagranite in the Dabie orogen : implications for action of supercritical fluid during continental subduction-zone metamorphism. Lithos, 114 : 385-412.

Xia B, Zhang L F, Du Z X, et al. 2019. Petrology and age of Precambrian Aksu blueschist, NW China. Precambrian Research, 326 : 295-311.

Xiao Y L, Hoefs J, van den Kerkhof, A M, et al. 2000. Fluid history of UHP metamorphism in Dabie Shan, China : a fluid inclusion and oxygen isotope study on the coesite-bearing eclogite from Bixiling. Contributions to Mineralogy and Petrology, 139 (1): 1-16.

Xiong J W, Chen Y X, Ma H Z, et al. 2022. Tourmaline boron isotopes trace metasomatism by serpentinite-derived fluid in continental subduction zone. Geochimica et Cosmochimica

Acta，320：122-142.

Xu Z. 1987. Etude tectonique et microtectonique de la chaine paleozoique et triasique des Quilings（Chine）：unpubl. Diplome de Doctorat，Universite des Sciences et Techniques du Languedoc，Montpellier，France：93-107.

Xu S T，Okay A I，Ji S Y，et al. 1992. Diamond from the Dabie Shan metamorphic rocks and its implication for tectonic setting. Science，256（5053）：80-82.

Xu Y G，Lan J B，Yang Q J，et al. 2008. Eocene break-off of the Neo-Tethyan slab as inferred from intraplate-type mafic dykes in the Gaoligong orogenic belt，eastern Tibet. Chemical Geology，255（3-4）：439-453.

Xu Z Q，Yang W C，Ji S C，et al. 2009. Deep root of a continent-continent collision belt：evidence from the Chinese Continental Scientific Drilling（CCSD）deep borehole in the Sulu ultrahigh-pressure（HP-UHP）metamorphic terrane，China. Tectonophysics，475（2）：204-219.

Xu Z，Zheng Y F，Zhao Z F. 2017. The origin of Cenozoic continental basalts in east-central China：constrained by linking Pb isotopes to other geochemical variables. Lithos，268-271：302-319.

Xu C，Kynický J，Song W L，et al. 2018. Cold deep subduction recorded by remnants of a Paleoproterozoic carbonated slab. Nature Communications，9：2790.

Yang J S，Xu Z Q，Song S G，et al. 2001. Discovery of coesite in the North Qaidam Early Paleozoic ultrahigh pressure（UHP）metamorphic belt，NW China. Comptes Rendus de l'Academie des Sciences Serie II Fascicule A-Sciences de la Terre et des Planetes，333：719-724.

Yang J S，Xu Z Q，Dobrzhnestskaya L，et al. 2003. Discovery of metamorphic diamond in central China：an indication of a >4000 km-long-zone of deep subduction resulting from multiple continental collisions. Terra Nova，15（6）：370-379.

Yang W C，Cheng Z Y，Zhang C H，2003. Geophysical investigation for site-selection of Chinese Continental Scientific Drilling and Dabie-Sulu lithosphere. Acta Geologica Sinica，24（5）：391-404.

Yang J H，Wu F Y，Wilde S A，et al. 2008. Mesozoic decratonization of the North China block. Geology，36（6）：467-470.

Yang L M，Song S G，Su L，et al. 2019. Heterogeneous Oceanic Arc Volcanic Rocks in the South Qilian Accretionary Belt（Qilian Orogen，NW China）. Journal of Petrology，60（1）：85-116.

Ye K，Cong B，Ye D. 2000. The possible subduction of continental material to depths greater than 200 km. Nature，407：734-736.

Yin A，Harrison T M. 2000. Geologic evolution of the Himalayan-Tibetan orogen. Annual Review of Earth and Planetary Sciences，28：211-280.

Yin A. 2012. Structural analysis of the Valles Marineris fault zone：possible evidence for large-scale strike-slip faulting on Mars. Lithosphere，4（4）：286-330.

Yin A，Xie Z M，Meng L S. 2018. A viscoplastic shear-zone model for deep（15-50 km）slow-slip events at plate convergent margins. Earth and Planetary Science Letters，491：81-94.

Yu Y G，Wu Z Q，Wentzcovitch R M. 2008. α-β-γ transformations in Mg_2SiO_4 in Earth's transition zone. Earth and Planetary Science Letters，273（1-2）：115-122.

Yu H L，Zhang L F，Wei C J，et al. 2017. Age and P-T Conditions of the Gridino-type eclogite in the Belomorian Province，Russia. Journal of Metamorphic Geology，35（8）：855-869.

Yu Y，Gao S，Liu K H. 2017. Mantle transition zone discontinuities beneath the Indochina Peninsula：implications for slab subduction and mantle upwelling. Geophysical Research Letters，44（14）：7159-7167.

Yuan X H，Ni J，Kind R，et al. 1997. Lithospheric and upper mantle structure of southern Tibet from a seismological passive source experiment. Journal of Geophysical Research-Solid Earth，102（B12）：27491-27500.

Yui T F，Rumble D，Lo C H. 1995. Unusually low $\delta^{18}O$ ultra-high-pressure metamorphic rocks from the Sulu Terrain，eastern China. Geochimica et Cosmochimica Acta，59（13）：2859-2864.

Zahirovic S，Müller R D，Seton M，et al. 2015. Tectonic speed limits from plate kinematic reconstructions. Earth and Planetary Science Letters，418：40-52.

Zellmer G F，Edmonds M，Straub S M. 2015. The role of volatiles in the genesis，evolution and eruption of arc magmas. Geological Society Special Publication，410：1-17.

Zhang L F, Ellis D J，Jiang W B. 2002. Ultrahigh-pressure metamorphism in western Tianshan, China: Part I. Evidence from inclusions of coesite pseudomorphs in garnet and from quartz exsolution lamellae in omphacite in eclogites. American Mineralogist，87：853-860.

Zhang S Q，Karato S I. 1995. Lattice preferred orientation of olivine aggregates deformed in simple shear. Nature，375：774-777.

Zhang J X，Zhang Z M，Xu Z Q，et al. 2001. Petrology and geochronology of eclogites from the western segment of the Altyn Tagh，Northwestern China. Lithos，56：187-206.

Zhang L F，Ellis D J，Jiang W B. 2002. Ultrahigh-pressure metamorphism in western Tianshan，China：Part I. Evidence from inclusions of coesite pseudomorphs in garnet and from quartz exsolution lamellae in omphacite in eclogites. American Mineralogist，87（7）：853-860.

Zhang R Y，Liou J G，Yang J S，et al. 2003. Ultrahigh-pressure metamorphism in the forbidden zone : the Xugou garnet peridotite, Sulu terrane, eastern China. Journal of Metamorphic Geology, 21 : 539-550.

Zhang H J, Thurber C H, Shelly D, et al. 2004. High-resolution subducting-slab structure beneath northern Honshu, Japan, revealed by double-difference tomography. Geology, 32 (4): 361-364.

Zhang Z M, Shen K, Sun W D, et al. 2008. Fluid in deeply subducted continental crust : petrology, mineral chemistry and fluid inclusion of UHP metamorphic veins from the Sulu orogen, eastern China. Geochimica et Cosmochimica Acta, 72 (13): 3200-3228.

Zhang D, Audétat A. 2017. What caused the formation of the giant Bingham Canyon porphyry Cu-Mo-Au deposit? Insights from melt inclusions and magmatic sulfides. Economic Geology, 112 (2): 221-244.

Zhang L, Chen R X, Zheng Y F, et al. 2017. Whole-rock and zircon geochemical distinction between oceanic- and continental-type eclogites in the North Qaidam orogen, northern Tibet. Gondwana Research, 44 : 67-88.

Zhang H J, Wang F, Myhill R, et al. 2019. Slab morphology and deformation beneath Izu-Bonin. Nature Communications, 10 (1): 1310.

Zhang G L, He Y M, Ai Y S, et al. 2021. Indian continental lithosphere and related volcanism beneath Myanmar : constraints from local earthquake tomography. Earth and Planetary Science Letters, 567 (1-2): 116987.

Zhao W, Nelson K D, Chen J. 1993. Deep seismic reflection evidence for continental underthrusting beneath southern Tibet. Nature, 366 : 557-559.

Zhao D P, Hasegawa A, Kanamori H. 1994. Deep structure of Japan subduction zone as derived from local, regional, and teleseismic events. Journal of Geophysical Research-Solid Earth, 99 (B11): 22313-22329.

Zhao D P, Xu Y, Wiens D, et al. 1997. Depth extent of the Lau back-arc spreading center and its relation to subduction processes. Science, 278 (5336): 254-257.

Zhao D P, Lei J S, Tang R Y. 2004. Origin of the Changbai intraplate volcanism in Northeast China : evidence from seismic tomography. Chinese Science Bulletin, 49 (13): 1401-1408.

Zhao Z F, Zheng Y F, Chen R X, et al. 2007. Element mobility in mafic and felsic ultrahigh-pressure metamorphic rocks during continental collision. Geochimica et Cosmochimica Acta, 71 (21): 5244-5266.

Zhao L, Allen R M, Zheng T, et al. 2012. High-resolution body-wave tomography models of the upper mantle beneath eastern China and the adjacent areas. Geochemistry Geophysics Geosystems, 13 : Q06007.

Zhao Z F, Zheng Y F, Zhang J, et al. 2012. Syn-exhumation magmatism during continental collision : evidence from alkaline intrusives of Triassic age in the Sulu orogen. Chemical Geology, 328 : 70-88.

Zhao Z F, Dai L Q, Zheng Y F. 2013. Postcollisional mafic igneous rocks record crust-mantle interaction during continental deep subduction. Scientific Reports, 3 (1) : 3413.

Zhao L, Paul A, Guillot S, et al. 2015. First seismic evidence for continental subduction beneath the Western Alps. Geology, 43 (9) : 815-818.

Zhao L, Paul A, Malusà M G, et al. 2016. Continuity of the Alpine slab unraveled by high-resolution P-wave tomography. Journal of Geophysical Research-Solid Earth, 121 (12) : 8721-8737.

Zhao L, Xu X, Malusà M G. 2017. Seismic probing of continental subduction zones. Journal of Asian Earth Sciences, 145 : 37-45.

Zhao Z F, Zheng Y F, Chen Y X, et al. 2017a. Partial melting of subducted continental crust : geochemical evidence from synexhumation granite in the Sulu orogen. Geological Society of America Bulletin, 129 (11-12) : 1692-1707.

Zhao Z F, Liu Z B, Chen Q. 2017b. Melting of subducted continental crust : geochemical evidence from Mesozoic granitoids in the Dabie-Sulu orogenic belt, east-central China. Journal of Asian Earth Sciences, 145 : 260-277.

Zhao L, Malusa M G, Yuan H Y, et al. 2020. Evidence for a serpentinized plate interface favouring continental subduction. Nature Communications, 11 : 2171.

Zheng Y F, Fu B, Gong B, et al. 1996. Extreme $\delta^{18}O$ depletion in eclogite from the Su-Lu terrane in East China. European Journal of Mineralogy, 8 (2) : 317-323.

Zheng Y F, Fu B, Gong B, et al. 2003. Stable isotope geochemistry of ultrahigh pressure metamorphic rocks from the Dabie-Sulu orogen in China : implications for geodynamics and fluid regime. Earth Science Reviews, 62 (1-2) : 105-161.

Zheng H, Li T, Gao R, et al. 2007. Teleseismic P-wave tomography evidence for the Indian lithospheric mantle subducting northward beneath the Qiangtang terrane. Chinese Journal of Geophysics, 50 (5) : 1418-1426.

Zheng Y F, Zhang S B, Zhao Z F, et al. 2007. Contrasting zircon Hf and O isotopes in the two episodes of Neoproterozoic granitoids in South China : implications for growth and reworking of continental crust. Lithos, 96 (1-2) : 127-150.

Zheng Y F, Wu R X, Wu Y B, et al. 2008. Rift melting of juvenile arc-derived crust : geochemical evidence from Neoproterozoic volcanic and granitic rocks in the Jiangnan Orogen, South China. Precambrian Research, 163 (3-4) : 351-383.

Zheng Y F. 2009. Fluid regime in continental subduction zones : petrological insights from ultrahigh-pressure metamorphic rocks. Journal of the Geological Society, 166 (4): 763-782.

Zheng Y F, Chen R X, Zhao Z F. 2009. Chemical geodynamics of continental subduction-zone metamorphism: insights from studies of the Chinese Continental Scientific Drilling (CCSD) core samples. Tectonophysics, 475 (3-4): 327-358.

Zheng Y F, Xia Q X, Chen R X, et al. 2011. Partial melting, fluid supercriticality and element mobility in ultrahigh-pressure metamorphic rocks during continental collision. Earth-Science Reviews, 107 (3-4): 342-374.

Zheng Y F. 2012. Metamorphic chemical geodynamics in continental subduction zones. Chemical Geology, 328: 5-48.

Zheng Y F, Hermann J. 2014. Geochemistry of continental subduction-zone fluids. Earth Planets Space, 66 (1): 93.

Zheng Y F, Chen Y X. 2016. Continental versus oceanic subduction zones. National Science Review, 3 (4): 495-519.

Zheng Y F, Chen R X. 2017. Regional metamorphism at extreme conditions: implications for orogeny at convergent plate margins. Journal of Asian Earth Sciences, 145: 46-73.

Zheng Y F, Zhao Z F. 2017. Introduction to the structures and processes of subduction zones. Journal of Asian Earth Sciences, 145: 1-15.

Zheng Y F. 2018. Fifty years of plate tectonics. National Science Review, 5 (3): 119-119.

Zheng Y F. 2019. Subduction zone geochemistry. Geoscience Frontiers, 10 (4): 1223-1254.

Zheng Y F, Zhao Z F, Chen R X. 2019a. Ultrahigh-pressure metamorphic rocks in the Dabie-Sulu orogenic belt: compositional inheritance and metamorphic modification. Geological Society London, Special Publications, 474 (1): 89-132.

Zheng Y F, Mao J W, Chen Y J, et al. 2019b. Hydrothermal ore deposits in collision orogens. Science Bulletin, 64 (3): 205-212.

Zheng Y F, Xu Z, Chen L, et al. 2020. Chemical geodynamics of mafic magmatism above subduction zones. Journal of Asian Earth Sciences, 194: 104185.

Zheng Y F, Zhao G C. 2020. Two styles of plate tectonics in Earth's history. Science Bulletin, 65 (4): 329-334.

Zheng Y F. 2021a. Plate tectonics. In: Alderton D, Elias S A.(Editors). Encyclopedia of Geology. 2nd Edition. United Kingdom: Academic Press.

Zheng Y F. 2021b. Convergent plate boundaries and accretionary wedges. In: Alderton D, Elias S A.(Editors). Encyclopedia of Geology. 2nd Edition. United Kingdom: Academic Press.

Zheng Y F. 2021c. Metamorphism in subduction zones. In: Alderton D, Elias S A.(Editors). Encyclopedia of Geology. 2nd Edition. United Kingdom: Academic Press.

Zheng Y F. 2021d. Exhumation of ultrahigh-pressure metamorphic terranes. In：Alderton D, Elias S A.(Editors). Encyclopedia of Geology. 2nd Edition. United Kingdom：Academic Press.

Zheng Y F，Chen R X. 2021. Extreme metamorphism and metamorphic facies series at convergent plate boundaries：implications for supercontinent dynamics. Geosphere, 17（6）: 1647-1685.

Zheng Y F，Gao P. 2021. The production of granitic magmas through crustal anatexis at convergent plate boundaries. Lithos，402-403：106232.

Zhong S J，Gurnis M. 1993. Dynamic feedback between a continent-like raft and thermal-convection. Journal of Geophysical Research-Solid Earth，98（B7）：12219-12232.

Zhong S J，Zhang N，Li Z X，et al. 2007. Supercontinent cycles，true polar wander，and very long-wavelength mantle convection. Earth and Planetary Science Letters，261（3-4）：551-564.

Zhong X Y，Li Z H. 2019. Forced subduction initiation at passive continental margins：velocity-driven versus stress-driven. Geophysical Research Letters，46（20）：11054-11064.

Zhong X Y，Li Z H. 2020. Subduction initiation during collision-induced subduction transference：numerical modeling and implications for the Tethyan evolution. Journal of Geophysical Research-Solid Earth，125（20）：e2019JB019288.

Zhou L G，Xia Q X，Zheng Y F，et al. 2015. Tectonic evolution from oceanic subduction to continental collision during the closure of Paleotethyan ocean：geochronological and geochemical constraints from metamorphic rocks in the Hong'an orogen. Gondwana Research，28（1）：348-370.

Zhu J C，Li R K，Li F C，et al. 2001. Topaz-albite granites and rare-metal mineralization in the Limu district，Guanxi Province，southeast China. Mineralium Deposita，36：393-405.

Zhu G Z，Gerya T V，Yuen D A，et al. 2009. Three-dimensional dynamics of hydrous thermal-chemical plumes in oceanic subduction zones. Geochemistry Geophysics Geosystems，10（11）：1-20.

Zhu D C，Zhao Z D，Niu Y L，et al. 2011. The Lhasa Terrane：record of a microcontinent and its histories of drift and growth. Earth and Planetary Science Letters，301（1-2）：241-255.

Zhu D C，Zhao Z D，Niu Y L，et al. 2013. The origin and pre-Cenozoic evolution of the Tibetan Plateau. Gondwana Research，23（4）：1429-1454.

Zhu D C，Wang Q，Zhao Z D，et al. 2015. Magmatic record of India-Asia collision. Scientific Reports 5（1）：14289.

Zhu R X，Fan H R，Li J W，et al. 2015. Decratonic gold deposits. Science China-Earth Sciences，58（9）：1523-1537.

Zhu D C，Li S M，Cawood P A，et al. 2016. Assembly of the Lhasa and Qiangtang terranes in central Tibet by divergent double subduction. Lithos，245：7-17.

Zhu D C，Wang Q，Chung S L，et al. 2019. Gangdese magmatism in southern Tibet and India-Asia convergence since 120 Ma. Geological Society Special Publications，483（1）：583-604.

Zindler A，Hart S. 1986. Chemical geodynamics. Annual Review of Earth and Planetary Sciences，14：493-571.

关键词索引

D

F

G

H

J

K

L

M

N

P

Q

R

S

T

W

X

Y

Z